Ergebnisse der Mathematik und ihrer Grenzgebiete

Band 25

Herausgegeben von

P. R. Halmos · P. J. Hilton · R. Remmert · B. Szőkefalvi-Nagy

Unter Mitwirkung von

L. V. Ahlfors · R. Baer · F. L. Bauer · R. Courant · A. Dold
J. L. Doob · S. Eilenberg · M. Kneser · G. H. Müller
M. M. Postnikov · H. Rademacher · B. Segre · E. Sperner

Geschäftsführender Herausgeber: P. J. Hilton

Roman Sikorski

Boolean Algebras

Third Edition

Springer-Verlag Berlin · Heidelberg · New York 1969

ISBN-13: 978-3-642-85822-2   e-ISBN-13: 978-3-642-85820-8
DOI: 10.1007/ 978-3-642-85820-8

To

Professor Kazimierz Kuratowski

Preface

There are two aspects to the theory of Boolean algebras; the algebraic and the set-theoretical. A Boolean algebra can be considered as a special kind of algebraic ring, or as a generalization of the set-theoretical notion of a field of sets. Fundamental theorems in both of these directions are due to M. H. STONE, whose papers have opened a new era in the development of this theory.

This work treats the set-theoretical aspect, with little mention being made of the algebraic one.

The book is composed of two chapters and an appendix.

Chapter I is devoted to the study of Boolean algebras from the point of view of finite Boolean operations only; a greater part of its contents can be found in the books of BIRKHOFF [2] and HERMES [1]. Chapter II seems to be the first systematic study of Boolean algebras with infinite Boolean operations.

To understand Chapters I and II it suffices only to know fundamental notions from general set theory and set-theoretical topology. No knowledge of lattice theory or of abstract algebra is presumed. Less familiar topological theorems are recalled, and only a few examples use more advanced topological means; but these may be omitted. All theorems in both chapters are given with full proofs.

On the other hand, no complete proofs are given in the Appendix, which contains mainly a short exposition of some of the applications of Boolean algebras to other parts of mathematics with references to the literature. An elementary knowledge of the theories discussed is assumed.

I am very much indebted to Professor PAUL R. HALMOS for suggesting that I write this book.

I wish to express my thanks to H. BASS, A. BIAŁYNICKI-BIRULA and R. WHERRITT for the revision of the manuscript, and to J. BROWKIN, R. ENGELKING and T. TRACZYK for help in proofreading.

Warsaw-New Orleans-Princeton ROMAN SIKORSKI
1957–1958

Preface to the second edition

Chapter I and the Appendix are almost unchanged. On the contrary, many new results are included in Chapter II; some sections have been extended while others have been completely rewritten. However the general character of Chapter II has been preserved.

I am very grateful to PH. DWINGER, H. GAIFMAN, A. W. HALES, J. D. HALPERN, C. R. KARP, K. MATTHES, R. S. PIERCE, Z. SEMADENI and F. M. YAQUB for valuable information which helped greatly in bringing the material up to date.

I am also obliged to A. E. FARLEY for the revision of the manuscript and to T. TRACZYK for help in proofreading.

Aarhus, 1962   ROMAN SIKORSKI

Preface to the third edition

The third edition has been reprinted from the second by offset lithography. It is unchanged except for the correction of errors and the removal of misprints.

Warsaw, 1968   ROMAN SIKORSKI

Contents

Terminology and notation

Capital latin letters are used to denote sets of points and their Boolean analogue, elements of Boolean algebras. Capital gothic letters denote classes of sets and their Boolean analogue, sets of elements of Boolean algebras (except for filters and ideals). In particular, $\mathfrak{A}$ and $\mathfrak{B}$ (with indices, if necessary) always denote Boolean algebras or fields of sets. The letter $\mathfrak{F}$ always denotes a field of sets.

The symbol "$\cup$" is used both for the set-theoretical union and for the more general notion of Boolean join. In most cases, if both interpretations of "$\cup$" are possible, they coincide. In the opposite case, either it is explicitly stated, or it is evident from the text how the symbol "$\cup$" should actually be interpreted. The same remarks hold for the dual symbol "$\cap$" used both for the set-theoretical intersection and for the more general notion of Boolean meet. The same is true for the symbols "$\bigcup$" and "$\bigcap$" of the corresponding infinite operations (see also notation on p. 55—56 for infinite Boolean joins and meets) and for the symbol "$-$" of complementation and the symbol "$\subset$" of inclusion.

The empty set is denoted by $\wedge$, and so is its Boolean analogue, the zero element. The dual notion, the unit element in a Boolean algebra, is denoted by the dual symbol $\vee$. The letter $\varDelta$ denotes an ideal. The dual symbol $\varV$ denotes a filter. Thus dual Boolean notions and operations are denoted by dual symbols.

$\mathfrak{m}$ always denotes an infinite cardinal. $\mathfrak{n}$ denotes any (finite or infinite) non-zero cardinal (except when other hypotheses are explicitly stated). The cardinal of the set of all integers will be denoted both by $\aleph_0$ and σ. The last notation will be used chiefly in expressions like "σ-measure", "σ-field", "σ-algebra" etc. according to the generally adopted terminology. Sets of cardinality $\aleph_0$ are called enumerable or countable. Sets of greater power are called non-enumerable or uncountable. The cardinal of a set X is denoted by $\overline{\overline{X}}$.

If we concentrate our investigation on subsets of a fixed set X, then X is often called a "space" (no additional structure of X is distinguished, unless it is explicitly stated).

By a topological space we understand a set with a closure operation satisfying the well-known four axioms of KURATOWSKI (see p. 198). However in all cases (except, perhaps, § 41) only Hausdorff spaces play an essential part. For any subset S of a topological space, $\mathbf{C}S$ and $\mathbf{I}S$ denote the closure and the interior of S, respectively.

By an *indexed set* $\{A_t\}_{t \in T}$ we shall understand a mapping which assigns, to every $t \in T$, an element A_t. This notion should not be identified with the set of all A_t, $t \in T$. This is essential, e.g., in § 13, § 16, § 36 and § 38 where indexed sets of Boolean algebras are examined. This is not essential in many other cases, e.g. when joins and meets of indexed sets of elements of a Boolean algebra are examined (Chapter II).

The following abbreviation will be useful, especially in Chapter II: an indexed set $\{A_t\}_{t \in T}$ will be called an $\mathfrak{m}$-*indexed set* if $\overline{\overline{T}} \leq \mathfrak{m}$. The same terminology will be assumed for doubly indexed sets; $\{A_{t,s}\}_{t \in T, s \in S}$ is called an $\mathfrak{m}$-*indexed set* if $\overline{\overline{T}} \leq \mathfrak{m}$ and $\overline{\overline{S}} \leq \mathfrak{m}$ and it is called an $(\mathfrak{m}, \mathfrak{n})$-*indexed set* if $\overline{\overline{T}} \leq \mathfrak{m}$ and $\overline{\overline{S}} \leq \mathfrak{n}$.

If S and T are non-empty sets, then S^T will denote the set of all mappings of T into S. If $f \in S^T$ and $g \in T^U$, then fg denotes the composite mapping given by $fg(u) = f(g(u))$ for $u \in U$. If $T' \subset T$ and $f \in S^T$, then $f \mid T'$ is the mapping f restricted to T'.

Formulas and examples are quoted by giving only their numbers if they are in the same section. Otherwise the number of the section is added.

Chapter I

Finite joins and meets

§ 1. Definition of Boolean algebras

A Boolean algebra is a non-empty set $\mathfrak{A}$ in which two binary operations $\cup$, $\cap$ and one unary operation $-$ are defined which have, roughly speaking, the same properties as the set-theoretical union, intersection and complementation of subsets of a fixed space. Since the elements of $\mathfrak{A}$ have many of the properties of sets, we shall denote them by capital letters $A, B, \ldots$ used generally to denote sets. For arbitrary elements $A, B \in \mathfrak{A}$, $A \cup B$ and $A \cap B$ are elements in $\mathfrak{A}$, uniquely determined by A and B and called respectively the *join* and the *meet* of A and B. For each element $A \in \mathfrak{A}$, $-A$ is an element in $\mathfrak{A}$, uniquely determined by A and called the *complement* of A. The operations $\cup$, $\cap$, $-$ are characterized by a set of axioms assuring that these operations have properties analogoues to those of union, intersection and complementation of sets respectively. Many equivalent sets of axioms characterizing $\cup$, $\cap$, $-$ are known[1]. We assume here the following one[2]:

(A$_1$) $A \cup B = B \cup A$, $A \cap B = B \cap A$,

(A$_2$) $A \cup (B \cup C) = (A \cup B) \cup C$, $A \cap (B \cap C) = (A \cap B) \cap C$,

(A$_3$) $(A \cap B) \cup B = B$, $(A \cup B) \cap B = B$,

(A$_4$) $A \cap (B \cup C) = (A \cap B) \cup (A \cap C)$, $A \cup (B \cap C) = (A \cup B) \cap (A \cup C)$,

(A$_5$) $(A \cap -A) \cup B = B$, $(A \cup -A) \cap B = B$.

---

[1] See BENNETT [1], BERNSTEIN [1, 2, 4, 6, 7], BIRKHOFF and BIRKHOFF [1], BRAITHWAITE [1], BYRNE [1, 2, 3], CROISOT [1], DIAMOND [1, 2], FRINK [1], GRAU [1, 2], HAMMER [1], HOBERMAN and MCKINSEY [1], HUNTINGTON [1, 2], KALICKI [1], MILLER [1], MONTAGUE and TARSKI [1], NEWMAN [1], SHEFFER [1], SHOLANDER [1, 2], STABLER [1], STAMM [1], STONE [2, 3], TARSKI [2, 5, 8, 12], WHITEMAN [1, 2]. See also RUDEANU [2].

[2] The set of axioms (A$_1$)—(A$_5$) is not the simplest one. It can be shortened since some axioms are consequences of other ones. For instance, one of the axioms (A$_4$) can be omitted (see e. g. BIRKHOFF [2], p. 133). Many papers quoted in footnote[1] contain much shorter sets of axioms. However, if the set of axioms is short, then it is more difficult to deduce from it various important properties of Boolean operations. To omit these algebraic difficulties we start from the convenient axioms (A$_1$)—(A$_5$).

Observe that every set of axioms for Boolean algebras has to contain at least three variables A, B, C. This follows from the fact that there exists an algebra which is not Boolean but every one of its subalgebras generated by two elements is Boolean. See DIAMOND and MCKINSEY [1].

Thus a *Boolean algebra* is a non-empty set $\mathfrak{A}$ with three operations $A \cup B$, $A \cap B$, $-A$ satisfying axioms (A_1)-(A_5).

Examples. A) By a *field of sets* we shall understand any non-empty class $\mathfrak{A}$ of subsets of a fixed space X, such that $\mathfrak{A}$ is closed with respect to the set-theoretical operations of finite union, intersection and complementation, i.e. such that

(a) if sets A, B are in $\mathfrak{A}$, then their set-theoretical union is in $\mathfrak{A}$;

(b) if sets A, B are in $\mathfrak{A}$, then their set-theoretical intersection is in $\mathfrak{A}$;

(c) if a set A is in $\mathfrak{A}$, then its set-theoretical complement relative to the space X (i.e. the set of all elements of X which do not belong to A) is in $\mathfrak{A}$.

It follows easily from the de Morgan rules for sets that (a) and (c) imply (b) and that (b) and (c) imply (a). Therefore in the definition of a field of sets it suffices to suppose condition (c) and one of conditions (a), (b).

Clearly every field of sets is a Boolean algebra, the Boolean operations $\cup$, $\cap$, $-$ being the set-theoretical union, intersection and complementation respectively.

In particular the class of all subsets of a space X is a field of sets and, therefore, a Boolean algebra.

It is easy to give other examples of fields of sets.

For instance, for every space X, the class composed of all finite subsets of X and of their complements is a field of sets.

Similarly the class of sets of real numbers that are finite unions of (bounded or unbounded) intervals and one-point sets is a field of sets.

For any topological space X, the class of all sets $A \subset X$ that are simultaneously open and closed is a field of sets. Such sets will hereafter be called *open-closed sets*.

Similarly the class of all subsets A of a topological space X such that the boundary of A is nowhere dense is a field[1] of subsets of X. The last remark follows from the fact that the boundary of the union $A \cup B$ and the boundary of the intersection $A \cap B$ are both subsets of the union of the boundaries of A and B, and the boundary of the complement $-A$ is equal to the boundary of A. Hence if the boundaries of A and B are nowhere dense, so are the boundaries of $A \cup B$, $A \cap B$ and $-A$.

B) The following more complex example of a Boolean algebra is one whose elements are sets but the Boolean operations $\cup$, $\cap$, $-$ do not coincide with the set-theoretical ones.

Let $\mathfrak{A}_1$ be the class of all *regular closed subsets* of a topological space X, i.e. subsets that are closures of open subsets of X (or equivalently, are

---
[1] See KURATOWSKI [3], p. 38 and STONE [6].

closures of their interiors). The join $A \cup B$ of sets $A, B \in \mathfrak{A}_1$ is the set-theoretical union of A and B. The meet $A \cap B$ of sets $A, B \in \mathfrak{A}_1$ is the closure of the interior of the set-theoretical intersection of A and B. The Boolean complement $-A$ of a set $A \in \mathfrak{A}_1$ is the closure of the set-theoretical complement of A. One can easily verify that the operations $\cup, \cap, -$, so defined, satisfy the axioms (A_1)—(A_5), i.e. $\mathfrak{A}_1$ is a Boolean algebra.

The class $\mathfrak{A}_2$ of all *regular open subsets* of X (i.e. the class of interiors of closed subsets) is also a Boolean algebra[1] with the following definitions of the Boolean operations. The join $A \cup B$ of two sets $A, B \in \mathfrak{A}_2$ is the interior of the closure of the set-theoretical union of A and B. The meet $A \cap B$ of $A, B \in \mathfrak{A}_2$ is the set-theoretical intersection of A and B. The Boolean complement $-A$ of $A \in \mathfrak{A}_2$ is the interior of the set-theoretical complement of A.

C) The fundamental notion of the theory of probability is *event*. We shall not examine here what events are. We observe only that in the class of all events, three operations corresponding to the logical connectives "or", "and" and "not" are defined: if A, B are events, then "A or B", "A and B" and "not A" are events too. The reader familiar with the theory of probability can easily verify that events form a Boolean algebra, the Boolean operations $\cup, \cap, -$ being defined as "or", "and" and "not" respectively[2].

D) The following example is addressed to readers familiar with mathematical logic.

Let S be the set of all formulas (predicate functions) of a formalized theory based on the two-valued logic. Identify two formulas α, β in S if they are equivalent, i.e. if the formula

$$\alpha \text{ if and only if } \beta$$

is a theorem of the theory. Then S becomes a Boolean algebra $\mathfrak{A}$, the Boolean operations $\cup, \cap, -$ being determined in an obvious way by the logical connectives "or", "and" and "not" respectively. The algebra $\mathfrak{A}$ will be called the *Lindenbaum-Tarski algebra* of the formalized theory in question[3].

Some methods of construction of new Boolean algebras from known Boolean algebras will be described in § 10.

Sometimes it is convenient to denote the complement $-A$ (of an element A of a Boolean algebra $\mathfrak{A}$) by $(-1) \cdot A$. The element A is then

---

[1] See BIRKHOFF [2], p. 176.

[2] The connection between the foundation of the theory of probability and Boolean algebras will be discussed in § 46.

[3] The connection between mathematical logic and Boolean algebras will be discussed in §§ 40 and 41. See also § 18 F.

also denoted by $(+1) \cdot A$. By definition

(1) $(-1) \cdot A = -A , \quad (+1) \cdot A = A .$

Expressions of the form $-A \cup B$, $A \cup -B$, $-A \cup -B$ will be abbrevations for $(-A) \cup B$, $A \cup (-B)$, $(-A) \cup (-B)$ respectively. Similarly for $\cap$.

§ 2. Some consequences of the axioms

Let $\mathfrak{A}$ be a Boolean algebra.

It follows from axioms § 1 (A_1) and (A_2) that the operations $\cup$ and $\cap$ are commutative and associative. Consequently the elements

(1) $A_1 \cup A_2 \cup \cdots \cup A_n , \quad A_1 \cap A_2 \cap \cdots \cap A_n$

are well defined and do not depend on the ordering of the elements $A_1, A_2, \ldots, A_n$. We shall denote them also by

$$\bigcup_{1 \leq i \leq n} A_i \quad \text{and} \quad \bigcap_{1 \leq i \leq n} A_i$$

respectively.

We shall prove that for every $A \in \mathfrak{A}$

(2) $A \cup A = A , \quad A \cap A = A .$

In fact, applying successively (A_3), (A_1), (A_4), (A_4) and (A_3) we obtain

$$A = A \cup (A \cap B)$$
$$= (A \cup A) \cap (A \cup B) = (A \cap (A \cup B)) \cup (A \cap (A \cup B))$$
$$= A \cup A ,$$

and analogously

$$A = A \cap (A \cup B)$$
$$= (A \cap A) \cup (A \cap B) = (A \cup (A \cap B)) \cap (A \cup (A \cap B))$$
$$= A \cap A .$$

Identities (2) are called the *idempotent laws*.

Axioms § 1 (A_4) are called the *distributive laws*.

Axioms § 1 (A_3) are called the *absorption laws*. It follows from the absorption laws that the equalities

(3) $A \cap B = A , \quad A \cup B = B$

are equivalent. In fact, it follows from the first absorption law that $A \cap B = A$ implies $A \cup B = B$. Replacing A by B and conversely in the second absorption law we infer that $A \cup B = B$ implies $A \cap B = A$ by the commutativity law § 1 (A_1). If (3) holds, we write

(4) $A \subset B \quad \text{or} \quad B \supset A$

and say that A is a *subelement* of B, or A *is contained in* B, or B *contains* A.

The relation $\subset$ is called the (Boolean) *inclusion*. Observe that if the Boolean algebra $\mathfrak{A}$ in question is a field of sets, then the Boolean inclusion $\subset$ coincides with the set-theoretical one.

The inclusion $\subset$ is a partial ordering in the Boolean algebra $\mathfrak{A}$, i.e. it has the properties

(5) $$A \subset A \, ;$$

(6) $$\text{if } A \subset B \text{ and } B \subset A \, , \text{ then } A = B \, ;$$

(7) $$\text{if } A \subset B \text{ and } B \subset C \, , \text{ then } A \subset C \, ;$$

A, B, C being arbitrary elements of $\mathfrak{A}$.

In fact, (5) is an immediate consequence of (2). If $A \subset B$ and $B \subset A$, then

$$A = A \cup B = (A \cap B) \cup B = B$$

by § 1 (A_3). Finally, if $A \subset B$ and $B \subset C$, then, by § 1 (A_2),

$$A = A \cap B = A \cap (B \cap C) = (A \cap B) \cap C = A \cap C \, ,$$

i.e. $A \subset C$.

It follows immediately from § 1 (A_5) that

(8) $$A \cap -A \subset B \quad \text{and} \quad B \subset A \cup -A$$

for arbitrary elements A, B.

Replacing B by $B \cap -B$ and by $B \cup -B$ in (8) we obtain

$$A \cap -A \subset B \cap -B \quad \text{and} \quad B \cup -B \subset A \cup -A \, .$$

Substituting here B for A and A for B we obtain also

$$B \cap -B \subset A \cap -A \quad \text{and} \quad A \cup -A \subset B \cup -B \, .$$

This implies by (6) that, for arbitrary $A, B \in \mathfrak{A}$,

(9) $$A \cap -A = B \cap -B \quad \text{and} \quad A \cup -A = B \cup -B \, .$$

The element $A \cap -A$ which, by (9), does not depend on the choice of $A \in \mathfrak{A}$ will be called the *zero element* (or simply the *zero*) of $\mathfrak{A}$ and will be denoted by $\wedge$ or by $\wedge_{\mathfrak{A}}$ if necessary. The element $A \cup -A$ which, by (9), does not depend on the choice of $A \in \mathfrak{A}$ will be called the *unit element* (or simply the *unit*) of $\mathfrak{A}$ and will be denoted by V or by $V_{\mathfrak{A}}$ if necessary. Observe that in the case where the Boolean algebra in question is a field of subsets of a space X, then the zero element of $\mathfrak{A}$ is the empty set, and the unit element of $\mathfrak{A}$ is the whole space X.

By definition, for every $A \in \mathfrak{A}$,

(10) $$A \cap -A = \wedge \, , \quad A \cup -A = V \, .$$

Axiom § 1 (A_5) can be now written in the form

(11) $$\wedge \cup B = B \, , \quad V \cap B = B$$

or in the form

(12) $$\wedge \subset A \, , \quad A \subset V$$

for every $A \in \mathfrak{A}$. This means that the zero element and the unit element are respectively the least element and the greatest element of $\mathfrak{A}$ in the partial ordering $\subset$ of $\mathfrak{A}$.

A Boolean algebra $\mathfrak{A}$ is said to be *degenerate* if it contains only one element. A necessary and sufficient condition for $\mathfrak{A}$ to be degenerate is that $\Lambda = V$, i.e. that the zero and unit of $\mathfrak{A}$ be equal. The necessity is obvious. The sufficiency follows immediately from (12) and (6).

Hence, if the Boolean algebra $\mathfrak{A}$ is non-degenerate (i.e has at least two elements), then $\Lambda \neq V$.

We said at the beginning of § 1 that the operations $\cup, \cap, -$ have the same properties as the corresponding set-theoretical operations. This statement does not follow immediately from the axioms § 1 (A_1)-(A_5). It will be obtained as an immediate consequence of a representation theorem proved in § 8. At present we shall deduce from the axioms only certain properties of $\cup, \cap, -$ analogous to well-known properties of set-theoretical operations on sets.

To simplify our considerations, let us observe that $\cup$ and $\cap$ play a quite symmetrical role in the axioms § 1 (A_1)-(A_5). The set of axioms remains unchanged if we replace everywhere $\cup$ by $\cap$ and $\cap$ by $\cup$. Consequently if, in a true statement about $\cup, \cap, -$, we replace everywhere $\cup$ by $\cap$ and $\cap$ by $\cup$, then we shall obtain also a true statement about $\cup, \cap, -$ called *dual* to the first one. Notice that the replacing of $\cup$ by $\cap$ and $\cap$ by $\cup$ transforms the unit into the zero and the zero into the unit and it transforms $\subset$ into $\supset$ and conversely. Therefore, in order to pass to the dual statement, we should replace everywhere the zero by the unit and conversely, and $\subset$ by $\supset$ and conversely. This general method for construction of dual statements is called the *duality principle*.

First we shall prove that

(13)   if $A \subset C$ and $B \subset D$,   then $A \cup B \subset C \cup D$.

In fact, we have $A \cup C = C$ and $B \cup D = D$. Therefore, by § 1 (A_1) and (A_2), $(A \cup B) \cup (C \cup D) = (A \cup C) \cup (B \cup D) = C \cup D$, i.e. $A \cup B \subset C \cup D$.

By the duality principle, we obtain

(13′)   if $C \subset A$ and $D \subset B$,   then $C \cap D \subset A \cap B$.

The exact proof of (13′) can be obtained from that of (13) by replacing $\cup$ by $\cap$.

It follows immediately from (13) and (2) that

(14)   if $A \subset C$ and $B \subset C$,   then $A \cup B \subset C$,

and, by duality, that

(14′)   if $C \subset A$ and $C \subset B$,   then $C \subset A \cap B$.

We also have

(15) $A \subset A \cup B$, $B \subset A \cup B$

since, by (2) and § 1 (A_2), $A \cup (A \cup B) = (A \cup A) \cup B = A \cup B$. By the duality principle we also obtain

(15′) $A \cap B \subset A , \quad A \cap B \subset B$.

It follows from (14) and (15) that the join $A \cup B$ can be defined in terms of the ordering relation $\subset$ only. In fact, $A \cup B$ is the least element of $\mathfrak{A}$ such that A and B are its subelements. The same remark is true for the meet; by (14′) and (15′), $A \cap B$ is the greatest subelement of A and B simultaneously.

By definition of the inclusion $\subset$, condition (4) implies (3). Hence, by (12), we obtain

(16) $A \cap V = A , \quad A \cup V = V ,$

(16′) $A \cup \wedge = A , \quad A \cap \wedge = \wedge ,$

for every $A \in \mathfrak{A}$.

The complement $-A$ of A is completely characterized by (10), i.e.

(17)   if $A \cap C = \wedge$ and $A \cup C = V$, then $C = -A$.

In fact, by (16), (16′) and § 1 (A_4),

$$C = \wedge \cup C = (A \cap -A) \cup C = (A \cup C) \cap (-A \cup C)$$
$$= V \cap (-A \cup C) = (-A) \cup C ,$$

i.e. $-A \subset C$. On the other hand,

$$C = V \cap C = (A \cup -A) \cap C = (A \cap C) \cup (-A \cap C)$$
$$= \wedge \cup (-A \cap C) = -A \cap C ,$$

i.e. $C \subset -A$. Consequently $C = -A$ by (6).

By (10) and the commutativity law § 1 (A_1),

$$-A \cap A = \wedge \quad \text{and} \quad -A \cup A = V$$

which implies on account of (17) (where A is replaced by $-A$) that

(18) $A = - -A$.

Consequently, if $-A = -B$, then $A = -(-A) = -(-B) = B$. Hence

(19) $A = B$ if and only if $-A = -B$.

Now we shall prove the following identities called the *de Morgan formulas:*

(20) $-(A \cup B) = -A \cap -B , \quad -(A \cap B) = -A \cup -B$.

In fact, the element $C = -A \cap -B$ satisfies the equalities

$$(A \cup B) \cap C = (A \cap -A \cap -B) \cup (B \cap -A \cap -B)$$
$$= \wedge \cup \wedge = \wedge ,$$
$$(A \cup B) \cup C = (A \cup B \cup -A) \cap (A \cup B \cup -B)$$
$$= V \cap V = V$$

on account of the distributive laws and (10), (16), (16′). This proves, by (17), that $C = -(A \cup B)$. The proof of the second identity (20) is by the duality principle.

It follows from (20) that

(21) $A \subset B$ if and only if $-B \subset -A$

since $A \cup B = B$ if and only if $-A \cap -B = -B$.

It follows from (18) and (20) that

(22) $A \cup B = -(-A \cap -B)$, $A \cap B = -(-A \cup -B)$.

Thus the join can be defined by means of the meet and the complement. Similarly, the meet can be defined by means of the join and the complement.

Replacing B by $-A$ in (22) we obtain by (18)

(23) $V = -\wedge$, $\wedge = -V$.

The element $A \cap -B$ will be denoted by $A - B$ and called the *difference* of A and B. Observe that if the Boolean algebra $\mathfrak{A}$ in question is a field of sets, then $A - B$ $(A, B \in \mathfrak{A})$ is the set-theoretical difference of sets A and B, i.e. the set of all points which belong to A but do not belong to B. In every Boolean algebra $\mathfrak{A}$, we have $V - A = -A$.

Notice that

(24) $A \subset B$ if and only if $A - B = \wedge$.

In fact, if $A \subset B$, then $A \cap -B = (A \cap B) \cap -B = A \cap (B \cap -B) = A \cap \wedge = \wedge$ by (16′). Conversely, if $A \cap -B = \wedge$, then, by (16′) and the distributive law, $A = A \cap (B \cup -B) = (A \cap B) \cup (A \cap -B) = (A \cap B) \cup \wedge = A \cap B$, i.e. $A \subset B$.

Since $-(A - B) = -(A \cap -B) = -A \cup B$ by (20) and (18), we infer from (24) and (23) that

(24′) $A \subset B$ if and only if $-A \cup B = V$.

The operation

$$A \to B = -A \cup B,$$

dual to the difference $B - A$, plays an important part in applications of the theory of Boolean algebras to mathematical logic[1]. It will not be examined in detail in this book[2]. Note that, by (24′),

(24″) $A \subset B$ if and only if $A \to B = V$.

Elements $A, B \in \mathfrak{A}$ are said to be *disjoint* provided

$$A \cap B = \wedge.$$

For instance, for arbitrary $A, B \in \mathfrak{A}$, the elements A and $B - A$ are disjoint, i.e.

(25) $A \cap (B - A) = \wedge$

---

[1] For a logical interpretation of →, see p. 194.
[2] For details, see Rasiowa and Sikorski [9].

since $A \cap (B \cap -A) = B \cap (A \cap -A) = \bar{B} \cap \wedge = \wedge$. Observe also that

(26) $$A \cup (B - A) = A \cup B ,$$

since, by the distributive law, $A \cup (B \cap -A) = (A \cup B) \cap (A \cup -A)$
$= (A \cup B) \cap V = A \cup B$.

§ 3. Ideals and filters

A non-empty subset $\varDelta$ of a Boolean algebra $\mathfrak{A}$ is said to be an *ideal*[1]
provided
(a) if $A, B \in \varDelta$, then $A \cup B \in \varDelta$;
(b) if $B \in \varDelta$ and $A \subset B$, then $A \in \varDelta$.

Examples. A) The set of all subelements of a given element $C \in \mathfrak{A}$
is an ideal. This follows immediately from § 2 (14) and (7). Ideals of this
form are called *principal*.

B) If $\mathfrak{A}$ is the field of all subsets of an infinite space X, then the class
of all finite subsets is an example of a non-principal ideal of $\mathfrak{A}$.

C) A real function m defined on a Boolean algebra $\mathfrak{A}$ is said to be a
measure provided
(1) $0 \leq m(A) \leq \infty$ for every $A \in \mathfrak{A}$, and there exists an element
$A_0 \in \mathfrak{A}$ such that $m(A_0) < \infty$;
(2) $m(A \cup B) = m(A) + m(B)$ whenever $A \cap B = \wedge$, $A, B \in \mathfrak{A}$.

The class of all $A \in \mathfrak{A}$ such that $m(A) = 0$ is an ideal. To prove it,
it suffices to show that any measure m has the following properties:

(3) $$m(A \cup B) \leq m(A) + m(B) ;$$

(4) $$\text{if } A \subset B , \quad \text{then} \quad m(A) \leq m(B) ;$$

(5) $$m(\wedge) = 0 .$$

Let A_0 be the element satisfying (1). It follows from (2) that

$$m(A_0) = m(A_0 \cup \wedge) = m(A_0) + m(\wedge) .$$

Since $m(A_0)$ is finite, we obtain (5). If $A \subset B$, then

$$m(A) \leq m(A) + m(B - A) = m(B)$$

on account of (1) and (2) since, by § 2 (25) and (26), B is the join of
disjoint elements A and $B - A$. This proves (4). By (2), (4) and § 2
(25), (26)

$$m(A \cup B) = m(A) + m(B - A) \leq m(A) + m(B)$$

which proves (3).

---

[1] For a detailed study and classification of ideals, see STONE [7]. See also
MAEDA [1], MORI [1], POSPIŠIL [3], TARSKI [3,6].

D) The class of all nowhere dense subsets is an ideal of the field of all subsets of a topological space.

An ideal $\varDelta$ of a Boolean algebra $\mathfrak{A}$ is said to be *proper* if it is a proper subset of $\mathfrak{A}$, i.e. $\varDelta \neq \mathfrak{A}$. The necessary and sufficient condition for the ideal $\varDelta$ to be proper is that $V \notin \varDelta$. The sufficiency is obvious. To prove the necessity let us observe that if $V \in \varDelta$, then by (b) and § 2 (12), $A \in \varDelta$ for every $A \in \mathfrak{A}$, i.e. $\varDelta$ is not proper.

It follows immediately from (b) and § 2 (12) that the zero element of $\mathfrak{A}$ belongs to every ideal of $\mathfrak{A}$. The set composed only of the zero element is an ideal called the *zero ideal*.

It is easy to verify that the intersection of any class of ideals of $\mathfrak{A}$ is an ideal of $\mathfrak{A}$.

For any set $\mathfrak{S}$ of elements of $\mathfrak{A}$ there exist ideals containing $\mathfrak{S}$ (e.g. the whole set $\mathfrak{A}$). The intersection $\varDelta_0$ of all such ideals is the least ideal containing $\mathfrak{S}$. The ideal $\varDelta_0$ is said to be *generated by* $\mathfrak{S}$. It is easy to describe the elements of $\varDelta_0$. If $\mathfrak{S}$ is empty, then $\varDelta_0$ is the zero ideal. Suppose $\mathfrak{S}$ is not empty. Then an element $A \in \mathfrak{A}$ belongs to $\varDelta_0$ if and only if there exists a finite sequence $A_1, \ldots, A_n$ of elements of $\mathfrak{S}$ such that

$$A \subset A_1 \cup \cdots \cup A_n .$$

In fact, by (b), the elements A of this form belong to every ideal $\varDelta$ containing $\mathfrak{S}$ since $A_1 \cup \cdots \cup A_n \in \varDelta$ on account of (a). On the other hand, the elements of this form constitute an ideal containing $\mathfrak{S}$.

In particular, the ideal generated by a given element C is the principal ideal described in Example A). The ideal generated by a given finite set of elements $C_1, \ldots, C_n$ is the principal ideal generated by $C = C_1 \cup \cdots \cup C_n$.

The least ideal $\varDelta_0$ containing a given ideal $\varDelta$ and a given element C is the set of all elements A such that

(6) $\qquad\qquad A \subset B \cup C \quad$ for an element $\quad B \in \varDelta$.

The ideal $\varDelta_0$ generated by $\varDelta$ and C is not proper if and only if

(7) $\qquad\qquad\qquad\qquad -C \in \varDelta$.

In fact, if $-C \in \varDelta$, then also $-C \in \varDelta_0$ and consequently $V = C \cup -C \in \varDelta_0$ by (a), i.e. $\varDelta_0$ is not proper. On the other hand, if $\varDelta_0$ is not proper, then, by (6), there exists an element $B \in \varDelta$ such that $V \subset B \cup C$, i.e. $V = B \cup C$. This implies $-C \subset B$ since $-C = -C \cap (B \cup C) = (-C \cap B) \cup (-C \cap C) = -C \cap B$. Consequently, by (b), $-C \in \varDelta$.

A non-empty subset ∇ of a Boolean algebra $\mathfrak{A}$ is said to be a *filter* provided

(a') if $A, B \in \nabla$, then $A \cap B \in \nabla$;
(b') if $B \in \nabla$ and $A \supset B$, then $A \in \nabla$.

The notion of filters is dual to that of ideal. In fact, the conditions (a′) and (b′) are obtained from (a) and (b) respectively by replacing $\cup, \cap, \subset$ by $\cap, \cup, \supset$.

It follows immediately from § 2 (20) and (21) that if $\varDelta$ is an ideal, then the set of all elements $-A$ where $A \in \varDelta$ is a filter called the *dual* of $\varDelta$. Conversely, if V is a filter, then the set of all elements $-A$ where $A \in V$ is an ideal called the *dual* of V. This natural one-to-one correspondence between ideals and filters shows that it suffices, in practice, to consider ideals only.

Clearly all statements dual to statements proved for ideals are also true for filters.

For instance, if $C \in \mathfrak{A}$ is a given element, then the class of all $A \in \mathfrak{A}$ such that $C \subset A$ is a filter called the *principal* filter *generated* by C. Every filter contains the unit element. The set composed only of the unit element is a filter called the *unit filter* of $\mathfrak{A}$ (of course, the unit filter is dual to the zero ideal). A filter V is *proper* if $V \neq \mathfrak{A}$, i.e. $\wedge \notin V$.

The formulation of other dual statements is left to the reader.

Observe that conditions (a) and (b) can be replaced in the definition of an ideal $\varDelta$ by the following condition:

$$A \cup B \in \varDelta \quad \text{if and only if} \quad A \in \varDelta \quad \text{and} \quad B \in \varDelta.$$

Similarly, conditions (a′) and (b′) can be replaced in the definition of a filter V by the single condition:

$$A \cap B \in V \quad \text{if and only if} \quad A \in V \quad \text{and} \quad B \in V.$$

Example. E) A subset V of a Boolean algebra $\mathfrak{A}$ containing the unit element is a filter if and only if $A \in V$, $A \to B \in V$ imply $B \in V$. By duality, a subset $\varDelta$ of $\mathfrak{A}$ containing the zero element is an ideal if and only if $A \in \varDelta$, $B - A \in \varDelta$ imply $B \in \varDelta$. The proof is left to the reader.

§ 4. Subalgebras

A non-empty subset $\mathfrak{A}_0$ of a Boolean algebra $\mathfrak{A}$ is said to be a *subalgebra* of $\mathfrak{A}$ provided $\mathfrak{A}_0$ is closed under the operations $\cup, \cap, -$, i.e. the following conditions are satisfied:

(a) if $A, B \in \mathfrak{A}_0$, then $A \cup B \in \mathfrak{A}_0$;

(a′) if $A, B \in \mathfrak{A}_0$, then $A \cap B \in \mathfrak{A}_0$;

(b) if $A \in \mathfrak{A}_0$, then $-A \in \mathfrak{A}_0$.

By the de Morgan formulas [see § 2 (22)], condition (b) and one of the conditions (a), (a′) imply the remaining condition (a) or (a′). Consequently if (b) and one of the conditions (a), (a′) are satisfied, then $\mathfrak{A}_0$ is a subalgebra. It follows immediately from the definition that every subalgebra $\mathfrak{A}_0$ is also closed under subtraction, i.e.

(c) if $A, B \in \mathfrak{A}_0$, then $A - B \in \mathfrak{A}_0$.

Each subalgebra $\mathfrak{A}_0$ of any Boolean algebra $\mathfrak{A}$ is also a Boolean algebra under the same operations $\cup$, $\cap$, $-$ restricted to $\mathfrak{A}_0$. The inclusion relation in the Boolean algebra $\mathfrak{A}_0$ is that of $\mathfrak{A}$, restricted to $\mathfrak{A}_0$.

Each subalgebra $\mathfrak{A}_0$ of $\mathfrak{A}$ contains the zero $\wedge$ and the unit $\vee$ of $\mathfrak{A}$. In fact, if $A \in \mathfrak{A}_0$, then $\wedge = A \cap -A \in \mathfrak{A}_0$ and $\vee = A \cup -A \in \mathfrak{A}_0$ by (a), (a') and (b). Clearly the zero and the unit of $\mathfrak{A}$ are also the zero and the unit of $\mathfrak{A}_0$ respectively.

It follows immediately from § 2 (16), (16'), (23) that the set composed only of the zero and the unit of $\mathfrak{A}$ is a subalgebra of $\mathfrak{A}$, viz. the least subalgebra of $\mathfrak{A}$.

The intersection of any number of subalgebras of $\mathfrak{A}$ also is a subalgebra of $\mathfrak{A}$.

For every set $\mathfrak{S}$ of elements of a Boolean algebra $\mathfrak{A}$ there exists a least subalgebra $\mathfrak{A}_0$ of $\mathfrak{A}$ such that $\mathfrak{S}$ is a subset of $\mathfrak{A}_0$. Viz. $\mathfrak{A}_0$ can be defined as the intersection of all subalgebras containing $\mathfrak{S}$. The least subalgebra $\mathfrak{A}_0$ is said to be *generated* by the set $\mathfrak{S}$. It is easy to describe the elements of $\mathfrak{A}_0$. If $\mathfrak{S}$ is empty, then $\mathfrak{A}_0$ is composed of $\wedge$ and $\vee$ only. Suppose $\mathfrak{S}$ is not empty. Then an element $A \in \mathfrak{A}$ belongs to $\mathfrak{A}_0$ if and only if it can be represented in the form

(1)
$$A = (A_{1,1} \cap \cdots \cap A_{1,r_1}) \cup (A_{2,1} \cap \cdots \cap A_{2,r_2}) \cup \cdots$$
$$\cup (A_{s,1} \cap \cdots \cap A_{s,r_s})$$

where, for any m, n, either $A_{m,n} \in \mathfrak{S}$ or $-A_{m,n} \in \mathfrak{S}$.

In fact, the class of elements of the form (1) satisfies condition (a). It follows from the de Morgan formulas [§ 2 (20)] and the distributive laws [§ 1 (A_4)] that the complement of an element A of the form (1) can be also represented in this form. Therefore condition (b) is also satisfied. Hence the elements of the form (1) constitute a subalgebra $\mathfrak{A}_0$ containing $\mathfrak{S}$. On the other hand, each element A of the form (1) belongs to every subalgebra containing $\mathfrak{S}$. Therefore $\mathfrak{A}_0$ is the least subalgebra containing $\mathfrak{S}$.

By duality, we infer that the subalgebra generated by a non-empty set $\mathfrak{S}$ is the set of all elements $A \in \mathfrak{A}$ which can be represented in the form

(2)
$$A = (A_{1,1} \cup \cdots \cup A_{1,r_1}) \cap (A_{2,1} \cup \cdots \cup A_{2,r_2}) \cap \cdots$$
$$\cap (A_{s,1} \cup \cdots \cup A_{s,r_s})$$

where, for any m, n, either $A_{m,n} \in \mathfrak{S}$ or $-A_{m,n} \in \mathfrak{S}$.

In particular, the subalgebra generated by an element $A \in \mathfrak{A}$ is composed of $\wedge, \vee, A$ and $-A$ only.

If $\mathfrak{A}_0$ is a subalgebra of $\mathfrak{A}$ and $A_0 \in \mathfrak{A}$, then the subalgebra generated by $\mathfrak{A}_0$ and A_0 is composed of all elements $A \in \mathfrak{A}$ which can be represented in the form

(3)
$$A = (A_1 \cap A_0) \cup (A_2 - A_0) \text{ where } A_1, A_2 \in \mathfrak{A}_0.$$

The proof is similar to that of (1). It suffices to verify that the join of two elements of the form (3) and the complement of an element of this form is also an element of this form.

§ 5. Homomorphisms, isomorphisms

Let $\mathfrak{A}$ and $\mathfrak{A}'$ be Boolean algebras. A mapping h of $\mathfrak{A}$ into $\mathfrak{A}'$ is said to be a *homomorphism*[1] provided it preserves join, meet and complement, i.e.

(a) $$h(A \cup B) = h(A) \cup h(B) ,$$

(a') $$h(A \cap B) = h(A) \cap h(B) ,$$

(b) $$h(-A) = -h(A) .$$

It follows from the de Morgan formulas [§ 2 (22)] that (b) and one of the conditions (a), (a') imply the other one. Consequently, if (b) and one of the conditions (a), (a') are satisfied, then h is a homomorphism.

It follows immediately from the definition that the homomorphism h also preserves subtraction, i.e.

(c) $$h(A - B) = h(A) - h(B) .$$

The homomorphism h also transforms the zero and the unit of $\mathfrak{A}$ onto the zero and unit of $\mathfrak{A}'$ respectively:

(d) $$h(\wedge_{\mathfrak{A}}) = \wedge_{\mathfrak{A}'} , \quad h(\vee_{\mathfrak{A}}) = \vee_{\mathfrak{A}'}$$

since $h(\wedge_{\mathfrak{A}}) = h(A \cap -A) = h(A) \cap -h(A) = \wedge_{\mathfrak{A}'}$ and dually for $\vee$.

Conversely, if a mapping h satisfies (a), (a') and (d), then it is a homomorphism. In fact, we have then

$$h(A) \cap h(-A) = h(A \cap -A) = \wedge_{\mathfrak{A}} ,$$

$$h(A) \cup h(-A) = h(A \cup -A) = \vee_{\mathfrak{A}'} ,$$

and this implies [see § 2 (17)] that $h(-A) = -h(A)$.

Similarly condition (c), the last of the conditions (d) and one of the conditions (a), (a') imply together that h is a homomorphism.

Any homomorphism h also preserves the inclusion, i.e.

(1)   if $A \subset B$, then $h(A) \subset h(B)$

for if $B = A \cup B$, then $h(B) = h(A \cup B) = h(A) \cup h(B)$.

If h is a homomorphism of $\mathfrak{A}$ into $\mathfrak{A}'$, then the class $h(\mathfrak{A})$ of all elements $h(A) \in \mathfrak{A}'$ ($A \in \mathfrak{A}$) is a subalgebra of $\mathfrak{A}'$.

A one-to-one homomorphism h is called an *isomorphism*. If there exists an isomorphism h of $\mathfrak{A}$ onto $\mathfrak{A}'$, then the Boolean algebras $\mathfrak{A}$ and $\mathfrak{A}'$ are said to be *isomorphic*. In this case h^{-1} is an isomorphism of $\mathfrak{A}'$ onto $\mathfrak{A}$.

---

[1] For examination of a notion more general than that of homomorphism, see HALMOS [4], WRIGHT [2, 3, 4].

In order that a one-to-one mapping h of $\mathfrak{A}$ onto $\mathfrak{A}'$ be an isomorphism it is necessary and sufficient that both h and h^{-1} preserve the inclusion, i.e.

(2) $A \subset B$ if and only if $h(A) \subset h(B)$.

In fact, (2) implies (d). Since join and meet can be defined by means of the inclusion relation only (see § 2, pp. 9), (2) implies also (a) and (a′).

If h is a homomorphism of $\mathfrak{A}$ into $\mathfrak{A}'$ and $\varDelta'$ is an ideal of $\mathfrak{A}'$, then the set $\varDelta = h^{-1}(\varDelta')$ of all elements $A \in \mathfrak{A}$ such that $h(A) \in \varDelta'$ is an ideal. In particular the set $h^{-1}(\bigwedge_{\mathfrak{A}'})$ of all elements $A \in \mathfrak{A}$ such that $h(A) = \bigwedge_{\mathfrak{A}'}$ is an ideal.

A homomorphism h of $\mathfrak{A}$ into $\mathfrak{A}'$ is an isomorphism if and only if $h^{-1}(\bigwedge_{\mathfrak{A}'})$ contains only the zero of $\mathfrak{A}$, i.e. if

(3) $h(A) = \bigwedge_{\mathfrak{A}'}$ implies $A = \bigwedge_{\mathfrak{A}}$.

In fact, if h is one-to-one, then, by (d), (3) is satisfied. Conversely, if (3) is satisfied and $h(A) = h(B)$, then $A - B = \bigwedge_{\mathfrak{A}}$ and $B - A = \bigwedge_{\mathfrak{A}}$ since $h(A - B) = h(A) - h(B) = \bigwedge_{\mathfrak{A}'}$ and $h(B - A) = h(B) - h(A) = \bigwedge_{\mathfrak{A}'}$. Hence we infer [see § 2 (24)] that $A \subset B$ and $B \subset A$ which implies $A = B$. Thus h is one-to-one.

By duality, if V' is a filter of $\mathfrak{A}'$ and h is a homomorphism of $\mathfrak{A}$ into $\mathfrak{A}'$, then $h^{-1}(V')$ is a filter of $\mathfrak{A}$. In particular, $h^{-1}(V_{\mathfrak{A}'})$ is a filter of $\mathfrak{A}$. The homomorphism h is an isomorphism if and only if $h^{-1}(V_{\mathfrak{A}'})$ contains only the element $V_{\mathfrak{A}}$.

If there exists a homomorphism from $\mathfrak{A}$ onto $\mathfrak{A}'$, then $\mathfrak{A}'$ is said to be a *homomorphic image* of $\mathfrak{A}$.

Examples. A) Suppose that $\mathfrak{A}$ and $\mathfrak{A}'$ are fields of subsets of spaces X and X' respectively. Let φ be a mapping of the space X' into the space X such that

$$\varphi^{-1}(A) \in \mathfrak{A}'   \text{ for every set }   A \in \mathfrak{A} .$$

Then the mapping h defined by the formula

$$h(A) = \varphi^{-1}(A)   \text{ for }   A \in \mathfrak{A}$$

is a homomorphism of $\mathfrak{A}$ into $\mathfrak{A}'$. We say then that h is the homomorphism *induced* by the point mapping φ.

B) All two-element Boolean algebras are isomorphic, the isomorphism being given by the mappings which transform zero onto zero and unit onto unit. Any two-element Boolean algebra is isomorphic to the field of all subsets of a one-element space.

C) Let $\mathfrak{A}_1$ and $\mathfrak{A}_2$ be respectively the Boolean algebras of all regular closed subsets of a topological space X and of all regular open subsets of X [see § 1 B)]. The mapping which, with every set $A \in \mathfrak{A}_1$, associates its interior is an isomorphism of $\mathfrak{A}_1$ onto $\mathfrak{A}_2$.

§ 6. Maximal ideals and filters

Let $\mathfrak{A}$ be a Boolean algebra.

A proper ideal (filter) of $\mathfrak{A}$ is said to be *maximal*[1] provided it is not a proper subset of a proper ideal (filter) of $\mathfrak{A}$.

A necessary and sufficient condition for a proper ideal $\varDelta$ (filter V) to be maximal is that, for every $A \in \mathfrak{A}$ either A or $-A$ belongs to $\varDelta$ (to V).

We shall prove this characterization only for ideals; the proof for filters is dual. To prove the sufficiency suppose that the condition in question is satisfied and that $\varDelta$ is a proper subset of an ideal $\varDelta_0$, i.e. there exists an element $A \in \varDelta_0$ such that $A \notin \varDelta$. Hence $-A \in \varDelta$ and consequently also $-A \in \varDelta_0$, which implies that $V = A \cup -A \in \varDelta_0$, i.e. $\varDelta_0$ is not proper. To prove the necessity suppose that $\varDelta$ is maximal. If $A \notin \varDelta$, then $\varDelta$ is a proper subset of the ideal $\varDelta_0$ generated by $\varDelta$ and A. Since $\varDelta$ is maximal, $\varDelta_0$ is not proper. It follows from § 3 (7) that $-A \in \varDelta$.

Every proper ideal (filter) contains at most one of the elements A, $-A$, since if it contains both of them, then it also contains $V = A \cup -A$ ($\varLambda = A \cap -A$) and is not proper. Thus an ideal (filter) is maximal if and only if, for every $A \in \mathfrak{A}$, it contains exactly one of the elements $A, -A$.

Examples. A) If $\mathfrak{A}$ is a field of subsets of a non-empty space X and $x_0 \in X$, then the class $\varDelta$ of all sets $A \in \mathfrak{A}$ such that $x_0 \notin A$ is a maximal ideal of $\mathfrak{A}$ since, for every $A \in \mathfrak{A}$, either A or $-A$ does not contain the point x_0. By the same argument, the class of all sets $A \in \mathfrak{A}$ such that $x_0 \in A$ is a maximal filter of $\mathfrak{A}$. This ideal and this filter are said to be *determined* by the point x_0.

If the field $\mathfrak{A}$ contains all one-point subsets of X, then a maximal filter V is determined by x_0 if and only if $(x_0) \in V$.

B) Let X be any infinite space and let $\mathfrak{A}$ be the field of all sets $A \subset X$ such that either A or $X - A$ is finite. The class of all finite (infinite) sets $A \in \mathfrak{A}$ is a maximal ideal (maximal filter) of $\mathfrak{A}$. This ideal (filter) is not determined by any point $x_0 \in X$.

By a *two-valued homomorphism* of a Boolean algebra $\mathfrak{A}$ we shall understand any homomorphism of $\mathfrak{A}$ into a two-element Boolean algebra.

A measure m on a Boolean algebra $\mathfrak{A}$ [see § 3 C)] is said to be *two-valued* provided it assumes exactly two values: the number 0 and the number 1. We have then

$$m(A) = 0 \text{ or } 1 \quad \text{for every} \quad A \in \mathfrak{A}$$

and in particular

$$m(\varLambda) = 0 \quad \text{and} \quad m(V) = 1 .$$

---

[1] Maximal ideals (filters) are also called *prime* ideals (filters).

The first of these two equalities is a particular case of § 3 (5). By § 3 (4), $0 \leq m(A) \leq m(V)$. The hypothesis that $m(V)$ is equal to zero would imply that the measure m vanishes identically in contradiction to the hypothesis that the two-valued measure assumes both of the values 0 and 1.

Since $m(A) + m(-A) = m(A \cup -A) = m(V) = 1$, we have

(1) $m(-A) = 1 - m(A)$ for every $A \in \mathfrak{A}$.

Observe else that

(2) $m(A \cap B) = m(A) \cdot m(B)$ for arbitrary $A, B \in \mathfrak{A}$.

In fact, the equality (2) holds if either $m(A)$ or $m(B)$ is equal to 0. If $m(A) = 1 = m(B)$, then, by (1), $m(-A) = 0 = m(-B)$ and consequently $m(-A \cup -B) = 0$ [see § 3 (3) and (1)]. Therefore, by (1), $m(A \cap B) = 1 - m(-A \cup -B) = 1$, which completes the proof of (2).

There is a natural one-to-one correspondence between maximal ideals, maximal filters, two-valued homomorphisms and two-valued measures.

In fact, if Δ is a maximal ideal, then the dual of Δ is a maximal filter (see § 3), the formula

(3) $h(A) = \begin{cases} \wedge & \text{if } A \in \Delta \\ V & \text{if } A \notin \Delta \end{cases}$

defines a two valued homomorphism h, and the formula

(4) $m(A) = \begin{cases} 0 & \text{if } A \in \Delta \\ 1 & \text{if } A \notin \Delta \end{cases}$

defines a two-valued measure m.

Similarly, if V is a maximal filter, then the dual of V (see § 3), i.e. the set Δ of all $-A$ where $A \in V$, is a maximal ideal, and (3) and (4) define respectively a two-valued homomorphism and a two-valued measure corresponding to V.

On the other hand, if h is a two-valued homomorphism, then the set Δ of all A such that $h(A) = \wedge$ is a maximal ideal and the set V of all A such that $h(A) = V$ is a maximal filter (dual to Δ). Similarly, if m is a two-valued measure, then the set Δ of all A such that $m(A) = 0$ is a maximal ideal and the set V of all A such that $m(A) = 1$ is a maximal filter dual to Δ [see (1) and (2)].

This natural correspondence permits us to translate theorems on maximal ideals into theorems on maximal filters or two-valued homomorphisms or two-valued measures, and conversely.

A degenerate Boolean algebra $\mathfrak{A}$ does not contain any maximal ideal (and consequently, it does not contain any maximal filter, and there exists no two-valued homomorphism or two-valued measure on it). In fact, the only ideal in $\mathfrak{A}$ is the zero ideal and this ideal is then non-proper.

The following theorem[1] shows that every non-degenerate Boolean algebra $\mathfrak{A}$ has many maximal ideals, maximal filters, two-valued homomorphisms and two-valued measures.

6.1 (i) *For every proper ideal $\varDelta_0$ there exists a maximal ideal containing $\varDelta_0$.*

(ii) *For every proper filter ∇_0 there exists a maximal filter containing ∇_0.*

(iii) *For every proper ideal $\varDelta_0$ (proper filter ∇_0) there exists a two-valued homomorphism h such that $h(A) = \Lambda$ for $A \in \varDelta_0$ (such that $h(A) = \nabla$ for $A \in \nabla_0$).*

(iv) *For every proper ideal $\varDelta_0$ (proper filter ∇_0) there exists a two-valued measure m such that $m(A) = 0$ for $A \in \varDelta_0$ (such that $m(A) = 1$ for $A \in \nabla_0$).*

No known proof of this theorem is effective[2], i.e. every proof is based on the well ordering principle or on other statements equivalent to the axiom of choice.

By the natural correspondence between maximal ideals, maximal filters, two-valued homomorphisms and two-valued measures, it suffices to prove only one of the four parts of 6.1, e.g. (i).

Observe first that if $\{\varDelta_\alpha\}$ is an increasing transfinite sequence of ideals of $\mathfrak{A}$, then the union of all $\varDelta_\alpha$ is also an ideal of $\mathfrak{A}$. If all $\varDelta_\alpha$ are proper ideals (i.e. they do not contain the unit), then their union is also a proper ideal (since it does not contain the unit).

Now let $\{A_\alpha\}_{\alpha < \beta}$ be a transfinite sequence formed of all elements of $\mathfrak{A}$. We define an increasing sequence $\{\varDelta_\alpha\}_{\alpha < \beta}$ of ideals of $\mathfrak{A}$ by transfinite induction as follows:

$\varDelta_0$ is the ideal mentioned in (i). For $0 < \alpha < \beta$, let $\varDelta_\alpha$ be the ideal generated by A_α and the union of all $\varDelta_\gamma$, $\gamma < \alpha$, provided it is proper; otherwise, let $\varDelta_\alpha$ be the union of all $\varDelta_\gamma$ where $\gamma < \alpha$.

The union $\varDelta$ of all ideals $\varDelta_\alpha$ $(\alpha < \beta)$ is a proper ideal containing $\varDelta_0$. We shall prove that $\varDelta$ is maximal, i.e. that for every α, either A_α or $-A_\alpha$ belongs to $\varDelta$.

If $A_\alpha \in \varDelta_\alpha$, then of course $A_\alpha \in \varDelta$. If $A_\alpha \notin \varDelta_\alpha$, then $\varDelta_\alpha$ is the union of all $\varDelta_\gamma$ where $\gamma < \alpha$ and the ideal generated by $\varDelta_\alpha$ and A_α is not proper. Hence, by § 3 (7), $-A_\alpha$ belongs to $\varDelta_\alpha$, and consequently to $\varDelta$.

Theorem 6.1 can be generalized in various directions. A generalization of part (iii) will be given in § 33 (theorem 33.1). Part (iv) is a particular case of the following theorem in measure theory[3]: Every measure m_0 defined on a subalgebra $\mathfrak{A}_0$ of $\mathfrak{A}$ can be extended to a measure m on the whole algebra $\mathfrak{A}$ in such a way that the set of values of m is contained in the closure of the set of values of m_0. In particular, every two-valued

---

[1] This fundamental theorem is due to STONE [5]. See also TARSKI [1], ULAM [2].

[2] The problem of the effectivity of theorem 6.1 will be examined in § 47.

[3] See HORN and TARSKI [1], ŁOŚ and MARCZEWSKI [1], TARSKI [11].

measure m_0 on a subalgebra $\mathfrak{A}_0$ can be extended to a two-valued measure on the whole $\mathfrak{A}$. To deduce (iv) from this theorem, it suffices to assume that $\mathfrak{A}_0$ is the subalgebra generated by Δ_0 (i.e. the set of all elements A and $-A$ where $A \in \Delta_0$) and to define m_0 by $m_0(A) = 0$ and $m_0(-A) = 1$ for every $A \in \Delta_0$.

§ 7. Reduced and perfect fields of sets

A field $\mathfrak{F}$ of subsets of a space X is said to be *reduced* provided any two different points x, y in X are separated by a set A in $\mathfrak{F}$, i.e. there exists a set $A \in \mathfrak{F}$ such that $x \in A$ and $y \notin A$.

Examples. A) The field of all subsets of X is reduced. The degenerate field composed only of the empty set (i.e. the field of all subsets of the empty space) is reduced.

B) If X contains more than one element, then the field composed only of the empty set and the whole space X is not reduced.

C) If a topological space X is zero-dimensional and Hausdorff, then the field of all open-closed sets $A \subset X$ is reduced. The converse statement is not true, in general. If the field of all open-closed subsets of a topological space X is reduced, then X is said to be *totally disconnected*. Clearly every totally disconnected space is Hausdorff.

Every field $\mathfrak{F}$ of subsets of a space X is isomorphic to a reduced field $\mathfrak{F}'$. To obtain $\mathfrak{F}'$ it suffices to identify points in X which are not separated by any set $A \in \mathfrak{F}$. More exactly, for every $x \in X$, let x' denote the set of all $y \in X$ which are not separated from x by any set $A \in \mathfrak{F}$. For every set $A \in \mathfrak{F}$, let A' be the set of all x' where $x \in A$. The class $\mathfrak{F}'$ of all sets A' ($A \in \mathfrak{F}$) is a reduced field of subsets of the space X', and the mapping

$$h(A) = A'$$

is an isomorphism of $\mathfrak{F}$ onto $\mathfrak{F}'$.

A field $\mathfrak{F}$ of subsets of a space X is said to be *perfect* if every maximal filter (or, equivalently, every maximal ideal) of $\mathfrak{F}$ is determined by a point of X [see § 6 A)].

Examples. D) Every finite field (i.e. a field composed of a finite number of subsets) is perfect. For the intersection A_0 of all sets A belonging to a maximal filter V also belongs to V (since it is an intersection of finite number of elements of V) and every point x_0 in A_0 determines V.

E) The field of sets defined in § 6 B) is not perfect.

F) The field of all subsets of an infinite space X is not perfect. In fact, let Δ_0 be the ideal composed of all finite subsets of X. By 6.1, Δ_0 is a subset of a maximal ideal Δ. The ideal Δ is not determined by any point in X.

G) Generalizing E) and F), we see that if X is infinite and if a field $\mathfrak{F}$ of subsets of X contains all one-element sets (and consequently all finite subsets), then $\mathfrak{F}$ is not perfect.

H) The field $\mathfrak{F}$ of all open-closed subsets of a compact topological space X is perfect. In fact, if V is a maximal filter in $\mathfrak{F}$, then the intersection of all sets $A \in V$ is not empty since the sets $A \in V$ are closed and the intersection of any finite number of them belongs to V and, therefore, it is not empty. Every point x_0 belonging to the intersection of all sets $A \in V$ determines V.

If a field $\mathfrak{F}$ of subsets of a space X is perfect and reduced, then the natural one-to-one correspondence between maximal ideals, maximal filters, two-valued homomorphisms and two-valued measures can be extended also over points of X. In fact, every point of X determines uniquely a maximal filter (and therefore a maximal ideal, a two-valued homomorphism and a two-valued measure), and, conversely, every maximal filter (and, therefore, every maximal ideal, two-valued homomorphism and two-valued measure) is determined by a point in X. Different points x, y determine different maximal filters (maximal ideals, two valued homomorphisms, two valued measures). In fact, let $A \in \mathfrak{F}$ be a set such that $x \in A$ and $y \notin A$. Then A belongs to the maximal filter determined by x, but does not belong to the maximal filter determined by y. This proves that the filters determined by x and y are different.

It follows from Examples C) and H) that if X is a compact totally disconnected space, then the field $\mathfrak{F}$ of all open-closed subsets of X is reduced and perfect. The converse is also true:

7.1 *If $\mathfrak{F}$ is a perfect reduced field of subsets of a space X, then we can define a topology in X so that X becomes a compact totally disconnected space and $\mathfrak{F}$ becomes the field of all open-closed subsets of the topological space X.*

For this purpose, let us assume that $\mathfrak{F}$ is an open basis for the topological space X. In other words, a set $G \subset X$ is said to be open if and only if it is the union of some sets belonging to $\mathfrak{F}$. Of course, every set $A \in \mathfrak{F}$ is open. It is also closed in this topology since $X - A$ belongs to $\mathfrak{F}$ and therefore is open. $\mathfrak{F}$ being reduced, the space X is totally disconnected. To prove that X is compact, it suffices to show that if X is the union of a class $\mathfrak{S}$ of open subsets, then there is a finite sequence $A_1, \ldots, A_n \in \mathfrak{S}$ such that

(1) $$A_1 \cup \cdots \cup A_n = X .$$

We can assume that sets in $\mathfrak{S}$ belong to $\mathfrak{F}$ (for if not, we can replace $\mathfrak{S}$ by the class of all sets $A \in \mathfrak{F}$ contained in some set $B \in \mathfrak{S}$).

Suppose that (1) does not hold for any sequence $A_1, \ldots, A_n \in \mathfrak{S}$. This means that the ideal Δ_0 generated by $\mathfrak{S}$ (see § 3) is proper. By 6.1, Δ_0 can be extended to a maximal ideal Δ. Since $\mathfrak{F}$ is perfect, there exists a point $x_0 \in X$ which determines Δ, i.e.

$$A \in \Delta \quad \text{if and only if} \quad x_0 \notin A .$$

Consequently $x_0 \notin A$ for every $A \in \mathfrak{S}$ in contradiction to the hypothesis that $\mathfrak{S}$ covers X.

Now we shall prove that if a set $A \subset X$ is open-closed, then $A \in \mathfrak{F}$. Indeed, A is the union of a class $\mathfrak{R}$ of sets in $\mathfrak{F}$ since A is open. Since A is a closed subset of the compact space X, there exists a finite sequence $A_1, \ldots, A_n \in \mathfrak{R} \subset \mathfrak{F}$ such that $A = A_1 \cup \cdots \cup A_n$. Hence $A \in \mathfrak{F}$.

Let us observe that the topology in X is uniquely determined by $\mathfrak{F}$ and the conditions of theorem 7.1. In fact, suppose that we can also introduce the required topology in another way. Let X_0 denote the space X with this new topology. Since each set $A \in \mathfrak{F}$ is open in X_0, each set open in X is open in X_0 too. This means that the identity mapping of X_0 onto X is continuous. Since X_0 is compact, we infer that the identity mapping of X_0 onto X is a homeomorphism, i.e. the two topologies coincide.

Notice that if two perfect reduced fields $\mathfrak{F}$ and $\mathfrak{F}'$ (of subsets of spaces X and X' respectively) are isomorphic, then the spaces X and X' topologized in the way described in 7.1 are homeomorphic. In fact, let h be the isomorphism of $\mathfrak{F}$ onto $\mathfrak{F}'$. For every $x \in X$, the class Δ of all $A \in \mathfrak{F}$ such that $x \in A$ is a maximal filter. The isomorphism h transforms the filter Δ onto a maximal filter $\Delta' = h(\Delta)$ determined by a point $x' \in X'$. Hence, for every $A \in \mathfrak{F}$,

$$x \in A \quad \text{if and only if} \quad x' \in h(A) .$$

Hence it follows that the one-to-one mapping φ (of X onto X') defined by

$$\varphi(x) = x'$$

has the properties:

$$\varphi(A) = h(A) \in \mathfrak{F}' \quad \text{for every} \quad A \in \mathfrak{F} ,$$
$$\varphi^{-1}(A') = h^{-1}(A') \in \mathfrak{F} \quad \text{for every} \quad A' \in \mathfrak{F}' .$$

Consequently $\varphi(G)$ is open in X' for every set G open in X, and $\varphi^{-1}(G')$ is open in X for every set G' open in X'. This proves that φ is a homeomorphism of X onto X'.

Examples. I) Suppose $\mathfrak{F}$ is a perfect (but not necessarily reduced) field of subsets of X. Defining open subsets as in the proof of 7.1, we turn X into a compact topological space (which, in general, is not a T_0-space). The proof of compactness is the same as that in the proof of 7.1.

J) If $\mathfrak{F}$ is a perfect field of subsets of X, and $\{A_t\}_{t \in T}$ is an infinite indexed set of disjoint non-empty sets in $\mathfrak{F}$, then the set-theoretical union A of all A_t does not belong to $\mathfrak{F}$.

In fact, consider X as a topological space with the topology defined in I). Suppose $A \in \mathfrak{F}$. Then the open sets A_t cover the closed subset A of the compact space X. Hence it follows that A is the union of a finite number of A_t in contradiction to the hypothesis that T is infinite and the A_t are disjoint and non-empty.

§ 8. A fundamental representation theorem

As it was said in § 1, elements of a Boolean algebra are the analogues of subsets of a given space. It follows from the considerations in § 6 and § 7 that maximal filters (or, equivalently, maximal ideals, two-valued homomorphisms, two-valued measures) are the Boolean analogues of points of the space. This remark will be useful in the proof of the following two theorems.

8.1 *Let X be a set of maximal filters of a Boolean algebra $\mathfrak{A}$. For every $A \in \mathfrak{A}$, let $h(A)$ denote the set of all maximal filters $V \in X$ such that $A \in V$. Then the class $\mathfrak{F}$ of all sets $h(A)$ (where $A \in \mathfrak{A}$) is a reduced field of subsets of X and h is a homomorphism of $\mathfrak{A}$ onto $\mathfrak{F}$.*

If, for every $A \neq \wedge$, there is a maximal filter $V \in X$ such that $A \in V$, then h is an isomorphism of $\mathfrak{A}$ onto the field $\mathfrak{F}$.

By the definition of h,

(1) $\qquad\qquad A \in V$ if and only if $V \in h(A)$.

Since $A \cap B \in V$ if and only if $A \in V$ and $B \in V$ (see the end of § 3), we have

$$h(A \cap B) = h(A) \cap h(B)$$

where "$\cap$" on the right side denotes the set-theoretical intersection. Since every filter $V \in X$ is maximal, we infer that (see the beginning of § 6)

$$A \in V \quad \text{if and only if} \quad -A \notin V .$$

This implies by (1) that

$$h(-A) = -h(A)$$

where "$-$" on the right side denotes the set-theoretical complement relative to the space X. Thus h is a homomorphism of $\mathfrak{A}$ into the field of all subsets of X. Consequently the class $\mathfrak{F} = h(\mathfrak{A})$ is a field of subsets of X.

If V_1 and V_2 are different points in X, i.e. different maximal filters, then there is an element $A \in \mathfrak{A}$ which belongs to only one of them, say

$$A \in V_1 \quad \text{and} \quad A \notin V_2 .$$

Consequently, by (1),

$$V_1 \in h(A) \in \mathfrak{F} \quad \text{and} \quad V_2 \notin h(A)$$

which proves that $\mathfrak{F}$ is reduced.

If the condition mentioned in the second part of 8.1 is satisfied, then $h(A)$ is not empty for $A \neq \wedge$. This proves that h is an isomorphism [see § 5 (3)].

8.2 *Every Boolean algebra $\mathfrak{A}$ is isomorphic to a perfect reduced field of sets, i.e. to the field of all open-closed subsets of a compact totally disconnected space*[1].

Assume in 8.1 that X is the set of all maximal filters of $\mathfrak{A}$. Let h and $\mathfrak{F}$ have the same meaning as in 8.1.

If $A \neq \wedge$, then the principal filter generated by A is proper, therefore, by 6.1, it can be extended to a maximal filter V. By definition, $A \in V \in X$. It follows from 8.1 that h is an isomorphism of $\mathfrak{A}$ onto the reduced field $\mathfrak{F}$. We shall prove that $\mathfrak{F}$ is perfect.

Let V' be a maximal filter of $\mathfrak{F}$. By isomorphism, the class V of all elements $A \in \mathfrak{A}$ such that $h(A) \in V'$ is a maximal filter in $\mathfrak{A}$. Let $B \in \mathfrak{F}$, i.e. $B = h(A)$ for an element $A \in \mathfrak{A}$. We have

$$B \in V' \quad \text{if and only if} \quad A \in V,$$

i.e. by (1)

$$B \in V' \quad \text{if and only if} \quad V \in B.$$

This proves that the filter V' is determined by the point $V \in X$, i.e. that $\mathfrak{F}$ is perfect.

The final remark of 8.2 follows immediately from the part just proved and theorem 7.1.

A compact totally disconnected space X is said to be the *Stone space* of a Boolean algebra $\mathfrak{A}$ provided $\mathfrak{A}$ is isomorphic to the (perfect reduced) field of all open-closed subsets of X. It follows from the remarks at the end of § 7 that all Stone spaces of $\mathfrak{A}$ are homeomorphic. Conversely if X and X' are homeomorphic, and X is a Stone space of $\mathfrak{A}$, so is X'. In fact if φ is a homeomorphism of X onto X', then $h(A) = \varphi(A)$ $(A \in \mathfrak{F})$ is an isomorphism of the field $\mathfrak{F}$ of all both open and closed subsets of X onto the field $\mathfrak{F}'$ of all open-closed subsets of X'. If $\mathfrak{F}$ is isomorphic to $\mathfrak{A}$, so is $\mathfrak{F}'$. Hence the Stone space of a Boolean algebra $\mathfrak{A}$ is determined by $\mathfrak{A}$ uniquely up to homeomorphism.

---

[1] This fundamental representation theorem is due to STONE [1, 4, 5, 6, 10]. The representation theorem 8.2 and theorem 6.1 on the existence of maximal ideals and filters are investigated in many papers. See AUMANN [2], DILWORTH [1], DUNFORD and SCHWARTZ [1] (p. 41), DUNFORD and STONE [1], ENGELKING and KURATOWSKI [1], ENOMOTO [1], FRINK [1], ISEKI [1], KAKUTANI [1], LIVENSON [1], MORI [1], NOLIN [1], STABLER [2], TARSKI [1]. Another representation theorem for Boolean algebras was given by HAIMO [2].

Obviously, in the definition of the Stone space of $\mathfrak{A}$ given in the proof of 8.2 we could use maximal ideals (or two-valued homomorphisms, two-valued measures) instead of maximal filters. The definition of $h(A)$ then requires an obvious modification.

We proved in § 2 that the Boolean operations $\cup, \cap, -$ and the Boolean inclusion $\subset$ have some properties of the corresponding set-theoretical operations and the set-theoretical inclusion. It follows from 8.2 that, roughly speaking, they have all the properties of their set-theoretical analogues. In fact, each Boolean algebra $\mathfrak{A}$ is isomorphic to a field of sets and this isomorphism preserves all properties of the (finite) set-theoretical operations and of the set-theoretical inclusion.

It follows from 8.2 that the difference between fields of sets and the more general notion of Boolean algebras is not essential from the point of view of finite set-theoretical operations and their Boolean analogues. We shall show in Chapter II that the difference is essential if some infinite operations are taken into consideration.

Theorem 8.2 also points out the significance of the notion of compact totally disconnected spaces for the theory of Boolean algebras.

Examples. A) If $\mathfrak{A}$ is the field of all open-closed subsets of a compact totally disconnected space X, then X is the Stone space of $\mathfrak{A}$.

B) If a Boolean algebra $\mathfrak{A}$ is finite, then the Stone space X of $\mathfrak{A}$ is a finite Hausdorff space (and conversely). Then $\mathfrak{F}$ is the class of all subsets of X. If X has n elements, then $\mathfrak{A}$ has 2^n elements. Therefore if two finite Boolean algebras have the same number of elements, they are isomorphic.

In particular, any one-point space is the Stone space of the two-element Boolean algebra. The empty set is the Stone space of the degenerate Boolean algebra.

C) The Stone space of a Boolean algebra $\mathfrak{A}$ is metrizable if and only if $\mathfrak{A}$ is at most enumerable. This follows from the theorem that a compact Hausdorff space is metrizable if and only if it has an enumerable open basis[1], i.e. if there exists an enumerable sequence of open sets such that every open set is the union of sets from this sequence. If $\mathfrak{A}$ is enumerable, then the field $\mathfrak{F}$ (of all open-closed subsets of the Stone space X of $\mathfrak{A}$) is an enumerable open basis of X, therefore X is metrizable. Conversely, if X has an open basis $\{G_n\}$, every set $A \in \mathfrak{F}$ is the union of a subsequence of $\{G_n\}$. Since A is a closed subset of a compact space, A is the union of a finite subsequence of $\{G_n\}$. This proves that the cardinality of $\mathfrak{F}$ is $\leq \aleph_0$. Consequently the cardinality of $\mathfrak{A}$ is $\leq \aleph_0$.

---

[1] This theorem is due to URYSOHN. See e.g. ALEXANDROFF and HOPF [1], p. 88.

It is known[1] that the number of topological types of totally dis-connected compact metric spaces is $2^{\aleph_0}$. This implies that there are $2^{\aleph_0}$ isomorphism types of enumerable Boolean algebras, i.e. that all enu-merable Boolean algebras can be divided into $2^{\aleph_0}$ classes in such a way that two Boolean algebras are in the same class if and only if they are isomorphic.

D) Let X_0 be an infinite space and let $\mathfrak{A}$ be the Boolean algebra composed of all finite subsets of X_0 and of their complements with respect to X_0 [see § 6 B)]. Let x_0 be any point which does not belong to X_0, and let $X = X_0 \cup (x_0)$. The mapping

$$h(A) = \begin{cases} A \text{ if } A \in \mathfrak{A} \text{ is finite} \\ A \cup (x_0) \text{ if } A \in \mathfrak{A} \text{ is infinite} \end{cases}$$

is an isomorphism of $\mathfrak{A}$ onto a field $\mathfrak{F}$ of subsets of X. Consider X as a topological space, the class $\mathfrak{F}$ being assumed as the open basis of X. Then X is compact, totally disconnected, and therefore it is the Stone space of $\mathfrak{A}$. In topology X is called the one-point compactification of the discrete space X_0.

E) Let $\mathfrak{A}$ be the least field of subsets of the unit interval $0 \leq x < 1$ containing all intervalls $0 \leq x < a$ $(0 < a \leq 1)$, i.e. the class of all finite unions of left-closed right-open subintervals of this interval. The Stone space of $\mathfrak{A}$ is the set X obtained from the closed unit interval $0 \leq x \leq 1$ by splitting every interior point x into two parts, x^- and x^+. We consider X as an ordered set with the natural order:

$$0 < x^- < x^+ < y^- < y^+ < 1 \quad \text{whenever} \quad 0 < x < y < 1 .$$

The set X with the topology determined by this order is compact and totally disconnected[2]. The Boolean algebra $\mathfrak{A}$ is isomorphic to the field $\mathfrak{F}$ of all open-closed subsets of X (associate, with every interval $a \leq x < b$, the set composed of a^+, b^- and all x^-, x^+ where $a < x < b$!), therefore X is the Stone space of $\mathfrak{A}$.[3]

F) The reader familiar with the Čech-Stone compactification can easily verify that if $\mathfrak{A}$ is the field of all subsets of a space X_0 (considered as a topological space with the discret topology), then the Čech-Stone compactification $\beta(X_0)$ is the Stone space of $\mathfrak{A}$.

G) Let $\mathfrak{F}$ be a field of subsets of X. For every $A \in \mathfrak{F}$, let $g(A)$ be the set of all maximal filters V such that $A \in V$ and V is not determined by any point in X. Then the mapping

$$h(A) = A \cup g(A) \qquad\qquad (A \in \mathfrak{F})$$

is an isomorphism of $\mathfrak{F}$ onto a perfect field of subsets of the space $X' = X \cup g(X)$.

---

[1] Mostowski [1]. See also Mazurkiewicz and Sierpiński [1].

[2] Alexandroff and Urysohn [1].

[3] For a similar construction of the Stone space of the field of all sets $A \cup B$ where $A \in \mathfrak{A}$ and B is finite, see Marczewski [12, 13]. See also Semadeni [3], p. 79.

The fundamental representation theorem 8.2 enables us to translate Boolean notions into the topological language of the corresponding Stone spaces. This will be done often in the sequel. Here we shall give a topological interpretation of ideals and filters.

Let X be the Stone space of a Boolean algebra $\mathfrak{A}$, and let h be the isomorphism of $\mathfrak{A}$ onto the field of all open-closed subsets of X.

For every open set $G \subset X$, the class of all $A \in \mathfrak{A}$ such that $h(A) \subset G$ is an ideal called the *ideal corresponding to G*. Conversely, every ideal Δ corresponds to an open subset of X, viz. to the union of all sets $h(A)$ where $A \in \Delta$.

For every closed set $F \subset X$, the class of all $A \in \mathfrak{A}$ such that $F \subset h(A)$ is a filter called the *filter corresponding to F*. Conversely, every filter V corresponds to a closed subset of X, viz. to the intersection of all sets $h(A)$ where $A \in V$.

Thus we have a one-to-one correspondence between ideals and open subsets, and between filters and closed subsets of the Stone space. Consequently the study of ideals and filters in $\mathfrak{A}$ can be reduced to the study of open and closed subsets of X, respectively. This correspondence also permits us to classify ideals and filters from the point of view of topological properties of the corresponding open and closed sets. For instance, it is natural to distinguish the class of ideals corresponding to regular or dense open sets, the class of filters corresponding to regular or nowhere dense closed sets, etc. Such a classification[1] is not the subject of investigation of this book. We observe only that the zero ideal corresponds to the empty set and proper ideals in $\mathfrak{A}$ correspond to proper open subsets of X. Similarly, the unit filter corresponds to the whole space X and proper filters in $\mathfrak{A}$ correspond to non-empty closed subsets of X. Maximal filters correspond to one-point sets and maximal ideals correspond to the complements of one-point sets.

§ 9. Atoms

An element $a \neq \Lambda$ of a Boolean algebra $\mathfrak{A}$ is said to be an *atom*[2] of $\mathfrak{A}$ provided that for every $A \in \mathfrak{A}$, the inclusion

(1) $$A \subset a$$

implies that

(2) $$\text{either} \quad A = \Lambda \quad \text{or} \quad A = a .$$

The notion of an atom is a Boolean analogue of a one-point set. In fact, if $\mathfrak{A}$ is a field of sets, then every one-point set in $\mathfrak{A}$ is an atom of $\mathfrak{A}$.

If a is an atom of a Boolean algebra $\mathfrak{A}$, then, for every element $B \in \mathfrak{A}$,

(3) $$\text{either} \quad a \subset B \quad \text{or} \quad a \cap B = \Lambda .$$

To prove this, it suffices to assume $A = a \cap B$ in (1) and (2).

---

[1] Discussed in the paper STONE [7].

[2] For a generalization of the notion of atom, see PIERCE [5].

Conversely, if $a \neq \wedge$ and a has the property (3), then a is an atom.

Observe that an element $a \in \mathfrak{A}$ is an atom if and only if the principal filter V generated by a (i.e. the class of all $A \in \mathfrak{A}$ such that $a \subset A$) is a maximal filter. In fact, (3) implies that, for every $B \in \mathfrak{A}$, either B or $-B$ belongs to V, i.e. V is maximal. On the other hand, if V is maximal, then $a \neq \wedge$ and for every $B \in \mathfrak{A}$, either $a \subset B$ or $a \subset -B$. Since $a \subset -B$ implies $a \cap B = \wedge$, condition (3) is satisfied, i.e. a is an atom.

In other words, an element a is an atom of $\mathfrak{A}$ if and only if the principal ideal generated by $-a$ (i.e. the class of all $A \in \mathfrak{A}$ such that $a \cap A = \wedge$) is a maximal ideal.

A Boolean algebra $\mathfrak{A}$ is said to be *atomic* provided, for every element $A \neq \wedge$ $(A \in \mathfrak{A})$, there exists an atom $a \subset A$. A Boolean algebra $\mathfrak{A}$ is said to be *atomless* if it has no atom.

9.1 *For every* $A \in \mathfrak{A}$, *let* $h(A)$ *be the set of all atoms* a *of* $\mathfrak{A}$ *such that* $a \subset A$. *The mapping* h *is a homomorphism of* $\mathfrak{A}$ *into the field of all subsets of the set* $X = h(V)$ *of all atoms.*

If $\mathfrak{A}$ *is atomic, then* h *is an isomorphism.*

Theorem 9.1 follows immediately from theorem 8.1 where X is the set of all principal maximal filters.

Examples. A) Let h be an isomorphism of $\mathfrak{A}$ onto the field $\mathfrak{F}$ of all open-closed subsets of the Stone space X of $\mathfrak{A}$. Then an element $a \in \mathfrak{A}$ is an atom of $\mathfrak{A}$ if and only if $h(a)$ is a one-point subset of X. Of course, a one-point set (x) belongs to $\mathfrak{F}$ if and only if x is an isolated point of X.

$\mathfrak{A}$ is atomic if and only if the set of all isolated points is dense in X. $\mathfrak{A}$ is atomless if and only if X is dense in itself, i.e. if X has no isolated points.

B) The field of all open-closed subsets of the Cantor set X of real numbers (i.e. the set of all real numbers of the form

$$\frac{2\alpha_1}{3} + \frac{2\alpha_2}{3^2} + \cdots$$

where $\alpha_n = 0$ or 1) is an atomless Boolean algebra, since X is dense in itself.

C) It is known[1] that all metrizable dense in themselves totally disconnected compact spaces are homeomorphic, viz. they are homeomorphic with the Cantor set defined in B). This implies that all enumerable atomless Boolean algebras are isomorphic, since they are isomorphic to the field of all open-closed subsets of the Cantor set [see § 8 C)].

D) Every finite Boolean algebra is atomic. If it has 2^n elements [see § 8 B)], then n is the number of atoms.

---

[1] See e.g. KURATOWSKI [4], p. 58.

§ 10. Quotient algebras

Let $\varDelta$ be an ideal of a Boolean algebra $\mathfrak{A}$. For arbitrary elements $A, B \in \mathfrak{A}$ we write

$$A \sim B$$

if and only if

$$A - B \in \varDelta \quad \text{and} \quad B - A \in \varDelta .$$

The relation $\sim$ is an equivalence relation, i.e. it is reflexive, symmetrical and transitive:

(1) $$A \sim A ;$$

(2) $$\text{if} \quad A \sim B , \quad \text{then} \quad B \sim A ;$$

(3) $$\text{if} \quad A \sim B \quad \text{and} \quad B \sim C , \quad \text{then} \quad A \sim C .$$

Property (1) follows from the fact that $\varDelta$ contains the zero element $\bigwedge$. Property (2) is obvious. Property (3) follows from the inclusions:

$$A - C \subset (A - B) \cup (B - C) , \quad C - A \subset (C - B) \cup (B - A) .$$

We also have

(4) $$\text{if} \quad A_1 \sim A_2 , \quad \text{then} \quad -A_1 \sim -A_2 ;$$

(5) $$\text{if} \quad A_1 \sim A_2 \quad \text{and} \quad B_1 \sim B_2 ,$$

$$\text{then} \quad A_1 \cup B_1 \sim A_2 \cup B_2 \quad \text{and} \quad A_1 \cap B_1 \sim A_2 \cap B_2 .$$

In fact, $(-A_1) - (-A_2) = A_2 - A_1 \in \varDelta$ and similarly $(-A_2) - (-A_1) \in \varDelta$, which proves (4). To prove the first part of (5) it suffices to notice that $(A_1 \cup B_1) - (A_2 \cup B_2) \subset (A_1 - A_2) \cup (B_1 - B_2) \in \varDelta$, and similarly for $(A_2 \cup B_2) - (A_1 \cup B_1)$. The second part of (5) follows from the inclusion $(A_1 \cap B_1) - (A_2 \cap B_2) \subset (A_1 - A_2) \cup (B_1 - B_2) \in \varDelta$ and the similar inclusion for $(A_2 \cap B_2) - (A_1 \cap B_1)$.

It follows from (1), (2) and (3) that the relation $\sim$ decomposes the set $\mathfrak{A}$ into disjoint classes (the equivalence classes of the relation $\sim$) in such a way that two elements A_1, A_2 are in the same class if and only if $A_1 \sim A_2$. The class containing an element $A \in \mathfrak{A}$ will be denoted by $[A]$ (or by $[A]_\varDelta$ if necessary). By definition, the following conditions are equivalent

$$A \sim B , \quad A \in [B] , \quad [A] = [B] .$$

The set of all classes $[A]$ $(A \in \mathfrak{A})$ will be denoted by $\mathfrak{A}/\varDelta$. This set is a Boolean algebra under the following definition of Boolean operations

(6) $$[A] \cup [B] = [A \cup B] , \quad [A] \cap [B] = [A \cap B] , \quad -[A] = [-A] .$$

Of course, $\cup, \cap, -$ on the left side of these equalities do not denote the set-theoretical operations on $[A]$, $[B]$ interpreted as sets, but they denote the Boolean operations on $[A]$, $[B]$ interpreted as elements of the Boolean algebra $\mathfrak{A}/\varDelta$.

It follows immediately from (4) and (5) that the elements $[A] \cup [B]$, $[A] \cap [B]$, $-[A]$ do not depend on the choice of the representatives $A, B \in \mathfrak{A}$ of the elements $[A]$, $[B] \in \mathfrak{A}/\Delta$. The proof that the Boolean operations so defined in $\mathfrak{A}/\Delta$ satisfy the axioms (A_1)-(A_5) of § 1 is based directly on the identities (6) which assert that [ ] is commutative with $\cup, \cap, -$. For instance we verify that

$$[A] \cup [B] = [A \cup B] = [B \cup A] = [B] \cup [A]$$

and analogously for the remaining axioms.

It follows from the definition (6) that

(7) $[A] - [B] = [A - B]$.

since $[A] - [B] = [A] \cap -[B] = [A] \cap [-B] = [A \cap -B] = [A - B]$.

The forming of quotient algebras $\mathfrak{A}/\Delta$ in the theory of Boolean algebras is similar to the analogous operations in the theory of groups. The connection between quotient algebras and homomorphisms is also the same as in the theory of groups. In fact, on the one hand, it follows immediately from (6) that the mapping

$$h(A) = [A]$$

is a homomorphism of $\mathfrak{A}$ onto $\mathfrak{A}/\Delta$, called the *natural homomorphism*. On the other hand, if h is a homomorphism of $\mathfrak{A}$ onto a Boolean algebra $\mathfrak{A}'$ and Δ is the ideal of all $A \in \mathfrak{A}$ such that $h(A) = \wedge$, then the formula

$$g([A]) = h(A)$$

defines an isomorphism g of $\mathfrak{A}/\Delta$ onto $\mathfrak{A}'$.

The elements $[\wedge]$ and $[\vee]$ are the zero and unit of $\mathfrak{A}/\Delta$ since the natural homomorphism transforms zero onto zero and onto unit [see § 5 (d)]. Hence we infer that

(8) $[A]$ is the zero of $\mathfrak{A}/\Delta$ if and only if $A \in \Delta$.

Observe also that

(9) $[A] \subset [B]$ if and only if $A - B \in \Delta$.

In fact, this inclusion holds if and only if the element $[A] - [B]$, i.e. the element $[A - B]$, is the zero of $\mathfrak{A}/\Delta$ which is equivalent, by (8), to $A - B \in \Delta$.

Examples. A) For every element E of a Boolean algebra $\mathfrak{A}$, the set of all subelements of E can be considered as a Boolean algebra, denoted by $\mathfrak{A} | E$. The join and meet in $\mathfrak{A} | E$ are the same as in $\mathfrak{A}$. The complement of A in $\mathfrak{A} | E$ is defined as the meet of E and of the complement of A in the Boolean algebra $\mathfrak{A}$. The easy verification of axioms § 1 (A_1)-(A_5) is left to the reader.

Let $\varDelta$ be the principal ideal (in the Boolean algebra $\mathfrak{A}$) generated by the element $-E \in \mathfrak{A}$ (i.e. $\varDelta$ is the class of all $A \in \mathfrak{A}$ such that $A \cap E = \varLambda$). The algebra $\mathfrak{A}|E$ is isomorphic to $\mathfrak{A}/\varDelta$, viz. the mapping $h(A) = [A]$ ($A \in \mathfrak{A}|E$) is an isomorphism of $\mathfrak{A}|E$ onto $\mathfrak{A}/\varDelta$.

B) Let $\mathfrak{F}$ be a field of subsets of a space X, and let E be any subset of X. The class $\varDelta$ of all sets $A \in \mathfrak{F}$ such that the intersection $A \cap E$ is empty (i.e. the class of all subsets of $X - E$ which belong to $\mathfrak{F}$) is an ideal of $\mathfrak{F}$. The Boolean algebra $\mathfrak{F}/\varDelta$ is isomorphic to the field (denoted by $\mathfrak{F}|E$) of all sets $A \cap E$ where $A \in \mathfrak{F}$. The isomorphism is given by the mapping

$$h(E \cap A) = [A] \qquad\qquad (A \in \mathfrak{F}) .$$

In fact, this mapping h is one-to-one since $A_1 \cap E = A_2 \cap E$ if and only if $A_1 - A_2 \subset -E$ and $A_2 - A_1 \subset -E$, i.e. $[A_1] = [A_2]$. It is evident that h preserves the Boolean operations.

Notice that if $E \in \mathfrak{F}$, then the field $\mathfrak{F}|E$ coincides with the Boolean algebra $\mathfrak{F}|E$ defined in Example A).

C) Let $\varDelta$ be an ideal in a Boolean algebra $\mathfrak{A}$, let $E \in \mathfrak{A}$ and let $\mathfrak{A}' = \mathfrak{A}/\varDelta$, $\mathfrak{A}_0 = \mathfrak{A}|E$, $\varDelta_0 = \varDelta \cap \mathfrak{A}_0$. Then $\varDelta_0$ is an ideal in $\mathfrak{A}_0$ and the quotient algebra $\mathfrak{A}_0/\varDelta_0$ is isomorphic to $\mathfrak{A}'|[E]_\varDelta$. Namely the formula

$$h([A]_{\varDelta_0}) = [A]_\varDelta \qquad\qquad (A \in \mathfrak{A}_0)$$

defines an isomorphism from $\mathfrak{A}_0/\varDelta_0$ onto $\mathfrak{A}'|[E]_\varDelta$.

D) Let $\mathfrak{F}$ be the field of all subsets (of a topological space X) with a nowhere dense boundary (see p. 4), i.e. the class of all sets $G \cup N$ where G is open and N is nowhere dense. Let $\varDelta$ be the ideal of all nowhere dense subsets. The Boolean algebra $\mathfrak{F}/\varDelta$ is isomorphic to each of the Boolean algebras $\mathfrak{A}_1$ and $\mathfrak{A}_2$ of all regular closed subsets or of all regular open subsets, respectively [see § 1 B)][1].

This follows from the fact that, for every set $A \in \mathfrak{F}$, there exists exactly one set $A_1 \in \mathfrak{A}_1$ and exactly one set $A_2 \in \mathfrak{A}_2$ such that $[A] = [A_1] = [A_2]$.

It is easy to determine the Stone space of $\mathfrak{A}/\varDelta$ by means of the Stone space X of $\mathfrak{A}$. Let h be the isomorphism of $\mathfrak{A}$ onto the field $\mathfrak{F}$ of all open-closed subsets of X, and let D be the union of all sets $h(A)$ where $A \in \varDelta$. The set D being open, the set $E - X \quad D$ is closed, therefore it is a compact totally disconnected space, and $\mathfrak{F}|E$ is the field of all open-closed subsets of E. Observe that $A \in \varDelta$ if and only if $h(A) \subset D$, i. e. $h(A) \cap E = \varLambda$. Since the mapping h_0 defined by the formula

$$h_0([A]) = h(A) \cap E$$

is an isomorphism of $\mathfrak{A}/\varDelta$ onto $\mathfrak{F}|E$, the space E is the Stone space of $\mathfrak{A}/\varDelta$.

---

[1] STONE [6].

Thus Stone spaces of quotient algebras $\mathfrak{A}/\varDelta$ are (up to homeo-morphisms) closed subsets of the Stone space of $\mathfrak{A}$, and conversely. Every homomorphic image of $\mathfrak{A}$ being isomorphic to a suitable quotient algebra $\mathfrak{A}/\varDelta$, we infer that Stone spaces of homomorphic images of $\mathfrak{A}$ are homeomorphic images of closed subsets of the Stones space of $\mathfrak{A}$, and conversely a Boolean algebra $\mathfrak{A}'$ is a homomorphic image of $\mathfrak{A}$ if and only if its Stone space is homeomorphic to a closed subset of the Stone space of $\mathfrak{A}$.

If V is a filter of $\mathfrak{A}$, then by the Boolean algebra $\mathfrak{A}/V$ we shall under-stand, by duality, the Boolean algebra $\mathfrak{A}/\varDelta$ where $\varDelta$ is the ideal dual to V. Thus two elements $A, B \in \mathfrak{A}$ determine the same element of $\mathfrak{A}/V$ if and only if

$$A \to B \in V \quad \text{and} \quad B \to A \in V .$$

The factor algebra $\mathfrak{A}/\varDelta$ (or $\mathfrak{A}/V$) is a two-element Boolean algebra if and only if $\varDelta$ (V) is a maximal ideal (filter). Then the natural homo-morphism is two-valued.

The factor algebra $\mathfrak{A}/\varDelta$ ($\mathfrak{A}/V$) is degenerate if and only if $\varDelta$ (V) is not proper.

If $\varDelta$ is the zero ideal (if V is the unit filter), then $[A] = [B]$ if and only if $A = B$. Thus the equivalence class $[A]$ contains only the element A. Identifying $[A]$ with A we identify $\mathfrak{A}/\varDelta$ with $\mathfrak{A}$ and this identification preserves Boolean operations.

Let $\varDelta$ be an ideal of a Boolean algebra $\mathfrak{A}$, and let $\varDelta'$ be an ideal of the factor algebra $\mathfrak{A}' = \mathfrak{A}/\varDelta$. The class $\varDelta''$ of all elements $A \in \mathfrak{A}$ such that $[A]_\varDelta \in \varDelta'$ is an ideal of $\mathfrak{A}$. It is easy to verify that the formula

(10) $h([A]_{\varDelta''}) = [[A]_\varDelta]_{\varDelta'}$

defines an isomorphism of $\mathfrak{A}/\varDelta''$ onto the Boolean algebra $(\mathfrak{A}/\varDelta)/\varDelta'$ (i.e. onto $\mathfrak{A}'/\varDelta'$).

By duality, if V is a filter of $\mathfrak{A}$ and V' is a filter of $\mathfrak{A}' = \mathfrak{A}/V$, then the class V'' of all $A \in \mathfrak{A}$ which determine in $\mathfrak{A}'$ elements belonging to V' is a filter and the Boolean algebras $\mathfrak{A}/V''$ and $(\mathfrak{A}/V)/V'$ (i.e. $\mathfrak{A}'/V'$) are isomorphic.

§ 11. Induced homomorphisms between fields of sets

Let $\mathfrak{F}$ and $\mathfrak{F}'$ be fields of subsets of spaces X and X' respectively. We recall [see § 5 A)] that a homomorphism h of $\mathfrak{F}$ into $\mathfrak{F}'$ is said to be *induced* by a mapping φ of X' into X if

(1) $h(A) = \varphi^{-1}(A) \quad \text{for every set} \quad A \in \mathfrak{F} .$

Examples. A) If $\mathfrak{F}' = \mathfrak{F}|X'$ [see § 10 B)] where X' is a subset of X, then the homomorphism

$$h(A) = A \cap X'$$

of $\mathfrak{F}$ onto $\mathfrak{F}'$ is induced by the identity mapping φ of X' into X:

$$\varphi(x') = x' \quad \text{for} \quad x' \in X' .$$

B) The homomorphism h defined in § 8 D) is not induced by any point mapping φ. For suppose that such a mapping φ exists. Since the one-point set $A_0 = (\varphi(x_0))$ is finite, the set $h(A_0)$ does not contain x_0. On the other hand, $x_0 \in \varphi^{-1}(A_0)$. Hence $h(A_0) \neq \varphi^{-1}(A_0)$. Contradiction.

The following condition is necessary and sufficient for a homomorphism h of a field $\mathfrak{F}$ (of subsets of X) into a field $\mathfrak{F}'$ (of subsets of X') to be induced by a point mapping: if a maximal filter V' of $\mathfrak{F}'$ is determined by a point x' in X', then the maximal filter $V = h^{-1}(V')$ [i.e. the set of all $A \in \mathfrak{F}$ such that $h(A) \in V'$] is determined by a point x in X.

In fact, if this condition is satisfied, then the formula

$$\varphi(x') = x$$

defines a mapping (of X' into X) inducing the homomorphism h since, by definition, for every $x' \in X'$ and every $A \in \mathfrak{F}$

$$\varphi(x') \in A \quad \text{if and only if} \quad x' \in h(A)$$

which implies

$$h(A) = \varphi^{-1}(A) \quad \text{for every} \quad A \in \mathfrak{F} .$$

Conversely, if a mapping φ of X' into X induces the homomorphism h, and V' is a maximal filter determined by $x' \in X'$, then $V = h^{-1}(V')$ is determined by the point $x = \varphi(x')$ since, for every $A \in \mathfrak{F}$, the following statements are equivalent:

$$A \in V, \quad h(A) \in V', \quad x' \in h(A), \quad x' \in \varphi^{-1}(A), \quad x \in A .$$

This proves the following theorem.

11.1. *If the field $\mathfrak{F}$ is perfect, then every homomorphism h of $\mathfrak{F}$ into an arbitrary field of sets $\mathfrak{F}'$ is induced by a point mapping.*

Conversely, if the isomorphism of $\mathfrak{F}$ onto the field of all open-closed subsets of the Stone space of $\mathfrak{F}$ is induced by a point mapping, then $\mathfrak{F}$ is perfect[1].

It follows also from the above consideration that a mapping φ of X' into X induces a given homomorphism h of $\mathfrak{F}$ into $\mathfrak{F}'$ if and only if, for every $x' \in X'$, $\varphi(x')$ determines the maximal filter $h^{-1}(V')$ where V' denotes the filter determined by x'. Hence we infer that if $\mathfrak{F}$ is reduced, then there exists at most one mapping φ inducing h. Consequently if $\mathfrak{F}$ is perfect and reduced (in particular if $\mathfrak{F}$ is the field of open-closed subsets of a compact totally disconnected space), then every homomorphism h of $\mathfrak{F}$ into an arbitrary field $\mathfrak{F}'$ is induced by exactly one point mapping φ.

---

[1] Sikorski [6].

Let g be a homomorphism of a Boolean algebra $\mathfrak{A}$ into a Boolean algebra $\mathfrak{A}'$. By 8.2, $\mathfrak{A}$ and $\mathfrak{A}'$ are isomorphic to the fields $\mathfrak{F}$ and $\mathfrak{F}'$ of all open-closed subsets of compact totally disconnected spaces X and X' respectively. Let h be the homomorphism of $\mathfrak{F}$ into $\mathfrak{F}'$ which corresponds, by this isomorphism, to the homomorphism g. The homomorphism h is induced by a point mapping φ. The mapping φ is continuous because $\varphi^{-1}(A)$ is open in X' for every $A \in \mathfrak{F}$ and consequently for every open set $A \subset X$. Conversely, if φ is a continuous mapping of X' into X, then φ induces a homomorphism h of $\mathfrak{F}$ into $\mathfrak{F}'$ and, consequently, φ determines a homomorphism g of $\mathfrak{A}$ into $\mathfrak{A}'$. This natural correspondence between homomorphisms g of $\mathfrak{A}$ into $\mathfrak{A}'$ and continuous mappings φ of X' into X is one-to-one.

We have observed in § 8 that the examination of Boolean algebras can be reduced to the examination of compact totally disconnected spaces. Now we see that the examination of homomorphisms between Boolean algebras can be reduced to examination of continuous mappings between totally disconnected compact spaces.

Observe that a homomorphism h of $\mathfrak{A}$ into $\mathfrak{A}'$ is an isomorphism if and only if the inducing continuous mapping φ maps X' onto X. Thus a Boolean algebra $\mathfrak{A}$ is isomorphic to a subalgebra of $\mathfrak{A}'$ if and only if the Stone space of $\mathfrak{A}$ is a continuous image of the Stone space of $\mathfrak{A}'$.

A homomorphism h (of $\mathfrak{A}$ into $\mathfrak{A}'$) maps $\mathfrak{A}$ onto $\mathfrak{A}'$ if and only if the inducing mapping φ is one-to-one, i.e. it is a homeomorphism of X' onto a closed subset of X. Thus we obtain, for the second time, the statement from § 10 that a Boolean algebra $\mathfrak{A}'$ is a homomorphic image of a Boolean algebra $\mathfrak{A}$ if and only if the Stone space of $\mathfrak{A}'$ is homeomorphic to a closed subset of the Stone space of $\mathfrak{A}$.

Examples. C) An isomorphism h of a Boolean algebra $\mathfrak{A}$ onto itself is called an *automorphism*. Of course, the identity mapping of $\mathfrak{A}$ onto itself is an automorphism of $\mathfrak{A}$. There exist Boolean algebras which have no other automorphism. This follows from the fact that there exist compact totally disconnected spaces[1] X such that the identity mapping of X onto X is the only homeomorphism of X onto X. The field $\mathfrak{F}$ of open-closed subsets of such a space X is a Boolean algebra without proper automorphisms. In fact, every automorphism of the field $\mathfrak{F}$ (of all open-closed subsets of any compact totally disconnected space X) onto itself is induced by a homeomorphism of X onto X, and conversely.

---

[1] An example of such a space was given by KATĚTOV [1] by means of the β-compactification technique. Two similar examples of linearly ordered compact spaces with this property were given independently and simultaneously by JÓNSSON [1] and RIEGER [6].

D) A Boolean algebra $\mathfrak{A}$ is said to be *superatomic*[1] if every homomorphic image of $\mathfrak{A}$ is atomic. Thus $\mathfrak{A}$ is superatomic if and only if the Stone space of $\mathfrak{A}$ is scattered, i.e. it does not contain any non-void dense-in-itself set. Since a compact Hausdorff space is scattered if and only if each of its continuous images is scattered[2], we infer that $\mathfrak{A}$ is superatomic if and only if every subalgebra of $\mathfrak{A}$ is atomic[3].

Note that there exists a superatomic uncountable Boolean algebra with a countable set of atoms because there exists a non-metrizable compact scattered space with a countable dense set of isolated points[4].

The Cantor set (see p. 28) is a continuous image of any non-scattered compact totally disconnected space. Hence it follows that every non-superatomic Boolean algebra contains a subalgebra isomorphic to the Boolean algebra described in § 9 B)[5].

§ 12. Theorems on extending to homomorphisms

We recall the notation introduced in § 1 (1):
$$(-1) \cdot A = -A \quad \text{and} \quad (+1) \cdot A = A$$
for every element A of a Boolean algebra $\mathfrak{A}$.

According to the definition from § 4 (p. 14), a set $\mathfrak{S}$ of elements of a Boolean algebra $\mathfrak{A}$ is said to *generate* $\mathfrak{A}$ (or to be a set of *generators* of $\mathfrak{A}$) provided the subalgebra generated by $\mathfrak{S}$ is identical with $\mathfrak{A}$, i.e. each element $A \in \mathfrak{A}$ is of the form

(1) $$A = \mathsf{U}_{1 \leq p \leq s} \cap_{1 \leq q \leq r_p} \varepsilon_{p,q} A_{p,q}$$

where $A_{p,q} \in \mathfrak{S}$ and $\varepsilon_{p,q} = \pm 1$ [see § 4 (1)].

Let $\mathfrak{A}'$ be another Boolean algebra and let f be a mapping of the set $\mathfrak{S}$ of generators of $\mathfrak{A}$ into $\mathfrak{A}'$. We shall discuss the problem under what conditions the mapping f can be extended to a homomorphism h of $\mathfrak{A}$ into $\mathfrak{A}'$.

First we observe that if such an extension h exists, then it is unique. In fact, if h coincides with f on $\mathfrak{S}$, then, by § 5 (a), (a'), (b), the image $h(A)$ of any element A of the form (1) satisfies the equation

(2) $$h(A) = \mathsf{U}_{1 \leq p \leq s} \cap_{1 \leq q \leq r_p} \varepsilon_{p,q} f(A_{p,q}) .$$

Therefore, if the extension h exists, it is given by (2).

It follows immediately from the uniqueness of extending to a homomorphism that if h_1 and h_2 are two homomorphisms of $\mathfrak{A}$ into $\mathfrak{A}'$ and if they coincide on a set $\mathfrak{S}$ generating $\mathfrak{A}$, then $h_1(A) = h_2(A)$ for every $A \in \mathfrak{A}$.

---

[1] For an investigation of this notion, see DAY [1].
[2] PEŁCZYŃSKI and SEMADENI [1], RUDIN [1].
[3] This remark is due to C. GOFFMAN.
[4] ALEXANDROFF and URYSOHN [1]. See also SEMADENI [1], p. 20.
[5] PEŁCZYŃSKI and SEMADINI [1].

For a characterization of scattered compact spaces (i.e. of superatomic Boolean algebras) in terms of measure theory, see RUDIN [2], PEŁCZYŃSKI and SEMADENI [1].

In particular, if h is a homomorphism of $\mathfrak{A}$ into itself, and $\mathfrak{S}$ generates $\mathfrak{A}$, then

(3)   if $h(A) = A$ for every $A \in \mathfrak{S}$, then $h(A) = A$ for every $A \in \mathfrak{A}$.

The following lemma will be useful in the sequel.

12.1. *If f is a one-to-one mapping of a set $\mathfrak{S}$ generating a Boolean algebra $\mathfrak{A}$ onto a set $\mathfrak{S}'$ generating a Boolean algebra $\mathfrak{A}'$, and if f can be extended to a homomorphism h of $\mathfrak{A}$ into $\mathfrak{A}'$ and f^{-1} can be extended to a homomorphism g of $\mathfrak{A}'$ into $\mathfrak{A}$, then h is an isomorphism of $\mathfrak{A}$ onto $\mathfrak{A}'$ and $g = h^{-1}$.*

In fact, we have

$$g(h(A)) = A$$

for every $A \in \mathfrak{S}$ which, by (3), implies that this equation holds also for every $A \in \mathfrak{A}$. Analogously

$$h(g(A')) = A'$$

for every $A' \in \mathfrak{S}'$ and consequently for every $A' \in \mathfrak{A}'$. This proves that $g = h^{-1}$. Thus h is an isomorphism of $\mathfrak{A}$ onto $\mathfrak{A}'$.

The following theorem gives a complete answer to the problem formulated on the beginning of this section.

12.2 *A mapping f of a set $\mathfrak{S}$ of generators of a Boolean algebra $\mathfrak{A}$ into a Boolean algebra $\mathfrak{A}'$ can be extended to a homomorphism h of $\mathfrak{A}$ into $\mathfrak{A}'$ if and only if*

(4) $\varepsilon_1 A_1 \cap \cdots \cap \varepsilon_n A_n = \bigwedge_{\mathfrak{A}}$ *implies* $\varepsilon_1 f(A_1) \cap \cdots \cap \varepsilon_n f(A_n) = \bigwedge_{\mathfrak{A}'}$

for every sequence $A_1, \ldots, A_n \in \mathfrak{S}$ and for every sequence $\varepsilon_1, \ldots, \varepsilon_n$ of numbers $-1, 1$.[1]

The necessity of (4) follows immediately from § 5 (a'), (b), (d).

To prove the sufficiency, suppose that (4) holds. If (1) and

(1') $A = \bigcup_{1 \leq k \leq s'} \bigcap_{1 \leq l \leq r'_k} \varepsilon'_{k,l} A'_{k,l} \quad (A'_{k,l} \in \mathfrak{S},\ \varepsilon'_{k,l} = -1 \text{ or } 1)$

are two representations of the same element A by means of elements in $\mathfrak{S}$, then

(5) $\bigcup_{1 \leq p \leq s} \bigcap_{1 \leq q \leq r_p} \varepsilon_{p,q} f(A_{p,q}) = \bigcup_{1 \leq k \leq s'} \bigcap_{1 \leq l \leq r'_k} \varepsilon'_{k,l} f(A'_{k,l})$.

In fact, (1) and (1') imply, by the de Morgan law and the distributive law, that

$$\bigcup_{\{l_k\}} \bigcup_{1 \leq p \leq s} \bigcap_{1 \leq q \leq r_p} \bigcap_{1 \leq k \leq s'} (\varepsilon_{p,q} A_{p,q} \cap -\varepsilon'_{k,l_k} A'_{k,l_k})$$

$$= (\bigcup_{1 \leq p \leq s} \bigcap_{1 \leq q \leq r_p} \varepsilon_{p,q} A_{p,q}) \cap -(\bigcup_{1 \leq k \leq s'} \bigcap_{1 \leq l \leq r'_k} \varepsilon'_{k,l} A'_{k,l}) = \bigwedge_{\mathfrak{A}}$$

where the join $\bigcup_{\{l_k\}}$ is extended over all sequences of integers l_k such that $1 \leq l_k \leq r'_k$ $(1 \leq k \leq s')$. It follows from (4) that

$$(\bigcup_{1 \leq p \leq s} \bigcap_{1 \leq q \leq r_p} \varepsilon_{p,q} f(A_{p,q})) \cap -(\bigcup_{1 \leq k \leq s'} \bigcap_{1 \leq l \leq r'_k} \varepsilon'_{k,l} f(A'_{k,l}))$$

$$= \bigcup_{\{l_k\}} \bigcup_{1 \leq p \leq s} \bigcap_{1 \leq q \leq r_p} \bigcap_{1 \leq k \leq s'} (\varepsilon_{p,q} f(A_{p,q}) \cap -\varepsilon'_{k,l_k} f(A'_{k,l_k})) = \bigwedge_{\mathfrak{A}'}.$$

---

[1] SIKORSKI [14].

Similarly we prove that

$$(\bigcup_{1 \leq k \leq s'} \bigcap_{1 \leq l \leq r_k'} \varepsilon_{k,l}' f(A_{k,l}')) \cap -(\bigcup_{1 \leq p \leq s} \bigcap_{1 \leq q \leq r_p} \varepsilon_{p,q} f(A_{p,q})) = \wedge_{\mathfrak{A}'}$$

which completes the proof of (5).

Thus the equality (2) defines uniquely a mapping h of $\mathfrak{A}$ into $\mathfrak{A}'$. If $A \in \mathfrak{S}$, then, of course, $A = A$ is a representation of A in the form (1). Consequently $h(A) = f(A)$, i.e. h is an extension of f. It is easy to verify that h is a homomorphism.

12.3. *Let f be a one-to-one mapping of a set $\mathfrak{S}$ generating a Boolean algebra $\mathfrak{A}$ onto a set $\mathfrak{S}'$ generating a Boolean algebra $\mathfrak{A}'$. The mapping f can be extended to an isomorphism of $\mathfrak{A}$ onto $\mathfrak{A}'$ if and only if, for every sequence $A_1, \ldots, A_n \in \mathfrak{S}$ and for every sequence $\varepsilon_1, \ldots, \varepsilon_n$ of numbers ± 1,*

$$\varepsilon_1 A_1 \cap \cdots \cap \varepsilon_n A_n = \wedge_{\mathfrak{A}} \quad \text{if and only if} \quad \varepsilon_1 f(A_1) \cap \cdots \cap \varepsilon_n f(A_n) = \wedge_{\mathfrak{A}'}{}^1.$$

The necessity of this condition follows from § 5 (a'), (b), (d), (3). If this condition is satisfied, then, by 12.2, f can be extended to a homomorphism of $\mathfrak{A}$ into $\mathfrak{A}'$, and f^{-1} can be extended to a homomorphism of $\mathfrak{A}'$ into $\mathfrak{A}$. By 12.1, f can be extended to an isomorphism of $\mathfrak{A}$ onto $\mathfrak{A}'$.

12.4 *For every $t \in T$, let $\mathfrak{A}_t$ be a subalgebra of an algebra $\mathfrak{A}$ and let h_t be a homomorphism of $\mathfrak{A}_t$ into a Boolean algebra $\mathfrak{A}'$. Suppose that the set-theoretical union of all $\mathfrak{A}_t$ generates $\mathfrak{A}$. Then, in order that there be a homomorphism h of $\mathfrak{A}$ into $\mathfrak{A}'$ which is a common extension of all the h_t, i.e.*

$$h(A) = h_t(A) \quad \text{for} \quad A \in \mathfrak{A}_t,$$

it is necessary and sufficient that, for every sequence $t_1, \ldots, t_n$ of distinct elements of T, and for arbitrary elements $A_1 \in \mathfrak{A}_{t_1}, \ldots, A_{t_n} \in \mathfrak{A}_{t_n}$,

(6) $A_1 \cap \cdots \cap A_n = \wedge_{\mathfrak{A}} \quad \text{implies} \quad h_{t_1}(A_1) \cap \cdots \cap h_{t_n}(A_n) = \wedge_{\mathfrak{A}'}.{}^2$

The necessity is obvious.

To prove the sufficiency, let us observe that if $A \in \mathfrak{A}_{t_1}$ and $A \in \mathfrak{A}_{t_2}$ $(t_1 \neq t_2)$, then $A \cap -A = \wedge, -A \cap A = \wedge$ and, by (6),

$$h_{t_1}(A) \cap -h_{t_2}(A) = \wedge, \quad -h_{t_1}(A) \cap h_{t_2}(A) = \wedge$$

which implies $h_{t_1}(A) = h_{t_2}(A)$. Thus the equality

$$f(A) = h_t(A) \quad \text{for} \quad A \in \mathfrak{A}_t$$

defines uniquely a mapping f of the set-theoretical union $\mathfrak{S}$ of all $\mathfrak{A}_t$ into $\mathfrak{A}'$. It follows from (6) that f satisfies (4). The homomorphism h mentioned in 12.2 satisfies all the conditions mentioned in 12.4.

A set $\mathfrak{S}$ of elements of a Boolean algebra $\mathfrak{A}$ is said to be *dense (in $\mathfrak{A}$)* if, for every non-zero element $A \in \mathfrak{A}$, there exists an element $B \in \mathfrak{S}$ such that $\wedge \neq B \subset A$.

12.5. *Suppose $\mathfrak{S}$ is a dense subset of non-zero elements of a Boolean algebra $\mathfrak{A}$ which generates $\mathfrak{A}$, and $\mathfrak{S}'$ is a dense subset of non-zero elements*

---

[1] KURATOWSKI and POSAMENT [1].
[2] SIKORSKI [14].

of a Boolean algebra $\mathfrak{A}'$ *which generates* $\mathfrak{A}'$. *Then every mapping* f *from* $\mathfrak{S}$
onto $\mathfrak{S}'$ *such that*

(7) $\qquad\qquad A_1 \subset A_2 \quad \text{if and only if} \quad f(A_1) \subset f(A_2) \qquad (A_1, A_2 \in \mathfrak{S})$

can be extended uniquely to an isomorphism from $\mathfrak{A}_1$ *onto* $\mathfrak{A}_2$.[1]

Observe that (7) implies the following property of f:

(8) $\quad A_1 \cap A_2 = \wedge_{\mathfrak{A}} \quad \text{if and only if} \quad f(A_1) \cap f(A_2) = \wedge_{\mathfrak{A}'} \quad (A_1, A_2 \in \mathfrak{S})\,.$

For if $A_1 \cap A_2 \neq \wedge$, there is a $B \in \mathfrak{S}_1$ such that $B \subset A_1$ and $B \subset A_2$.
Consequently $f(B) \subset f(A_1)$ and $f(B) \subset f(A_2)$. Moreover $f(B) \neq \wedge$. This
proves that $f(A_1) \cap f(A_2) \neq \wedge$. Analogously we prove that $f(A_1) \cap$
$\cap f(A_2) \neq \wedge$ implies $A_1 \cap A_2 \neq \wedge$.

To prove 12.5 it suffices to show that the mapping f satisfies the
condition of theorem 12.3. Suppose that for some $A_1, \ldots, A_n$,
$B_1, \ldots B_m$ in $\mathfrak{S}$,

(9) $\qquad\qquad A_1 \cap \cdots \cap A_n \cap -B_1 \cap \cdots \cap -B_m \neq \wedge\,.$

Then there exists an element $A \in \mathfrak{S}$ such that A is a subelement of the
element (9), i.e.

$\quad A \subset A_i \quad \text{for} \quad i = 1, \ldots, n \quad \text{and} \quad A \cap B_j = \wedge \quad \text{for} \quad j = 1, \ldots, m\,.$

By (7) and (8),

$f(A) \subset f(A_i) \quad \text{for} \quad i = 1, \ldots, n \quad \text{and} \quad f(A) \cap f(B_j) = \wedge \quad \text{for} \quad j = 1, \ldots, m\,.$
Since $f(A) \neq \wedge$, we infer that

(10) $\qquad\quad f(A_1) \cap \cdots \cap f(A_n) \cap -f(B_1) \cap \cdots \cap -f(B_m) \neq \wedge\,.$

Similarly we prove that (10) implies (9).

Examples. A) A Boolean algebra is atomic if and only if the set of
all atoms is dense.

B) Any dense set $\mathfrak{S}$ of non-zero elements of a Boolean algebra $\mathfrak{A}$
is partially ordered by the Boolean inclusion $\subset$ and satisfies the following
condition:

(a) If $A, B \in \mathfrak{S}$ and $A \not\subset B$, then there exists an element $C \subset A$,
$C \in \mathfrak{S}$, such that no element $D \in \mathfrak{S}$ satisfies simultaneously $D \subset C$ and
$D \subset B$.

In other words, if A is not a subelement of B, then there exists a
$C \subset A$ $(C \in \mathfrak{S})$ which is disjoint from B.

Conversely, it can be proved[2] that if a set $\mathfrak{S}$ partially ordered by a
relation $\subset$ satisfies condition (a), then there exists a Boolean algebra $\mathfrak{A}$
such that $\mathfrak{S}$ is a dense subset of $\mathfrak{A}$ and the Boolean inclusion in $\mathfrak{A}$ is an
extension of the partial ordering $\subset$.

---

[1] SIKORSKI [32].
[2] BÜCHI [1]. For another proof see SIKORSKI [31].

§ 13. Independent subalgebras. Products

An indexed set $\{\mathfrak{A}_t\}_{t \in T}$ of subalgebras of a Boolean algebra $\mathfrak{A}$ is said to be *independent*[1] provided

(1) $$A_1 \cap \cdots \cap A_n \neq \wedge$$

for every finite sequence of non-zero elements A_i chosen from subalgebras with different indices, i.e. for arbitrary elements $A_1, \ldots, A_n$ such that

$$\wedge \neq A_j \in \mathfrak{A}_{t_j}, \quad t_j \neq t_k \quad \text{for} \quad j \neq k \quad (j, k = 1, \ldots, n) \, .$$

Notice that (1) implies that no element $A \in \mathfrak{A}$ different from $\wedge$ and $\vee$ belongs to two subalgebras $\mathfrak{A}_t$ with different indices. More generally, if $A_1 \in \mathfrak{A}_{t_1}$, $A_2 \in \mathfrak{A}_{t_2}$ and $t_1 \neq t_2$, then the inclusion $A_1 \subset A_2$ does not hold except in the case where $A_1 = \wedge$ or $A_2 = \vee$. This follows from the fact that $A_1 \subset A_2$ is equivalent to $A_1 \cap -A_2 = \wedge$.

13.1 *If $\{\mathfrak{A}_t\}_{t \in T}$ is an independent indexed set of subalgebras of a Boolean algebra $\mathfrak{A}$ and, for every $t \in T$, h_t is a homomorphism of $\mathfrak{A}_t$ into a Boolean algebra $\mathfrak{A}'$, then there is a homomorphism h of the subalgebra $\mathfrak{A}_0 \subset \mathfrak{A}$ generated by the union of all $\mathfrak{A}_t$ ($t \in T$) into $\mathfrak{A}'$, such that*

$$h(A) = h_t(A) \quad \text{for} \quad A \in \mathfrak{A}_t, t \in T \, .$$

In fact, if $A_1 \cap \cdots \cap A_n = \wedge$ and $A_j \in \mathfrak{A}_{t_j}, t_j \neq t_k$ for $j \neq k$, then there exists an index j_0 such that $A_{j_0} = \wedge$. Hence $h_{t_{j_0}}(A_{j_0}) = \wedge$ and consequently $h_{t_1}(A_1) \cap \cdots \cap h_{t_n}(A_n) = \wedge$. By 12.4 all the homomorphisms h_t have a common extension which is a homomorphism of $\mathfrak{A}_0$ into $\mathfrak{A}'$.

13.2. *If $\{\mathfrak{A}_t\}_{t \in T}$ and $\{\mathfrak{A}'_t\}_{t \in T}$ are independent indexed sets of subalgebras of Boolean algebras $\mathfrak{A}$ and $\mathfrak{A}'$ respectively, and h_t ($t \in T$) is an isomorphism of $\mathfrak{A}_t$ onto $\mathfrak{A}'_t$, then the isomorphisms h_t have a common extension h which is an isomorphism of the subalgebra $\mathfrak{A}_0 \subset \mathfrak{A}$, generated by the union of all $\mathfrak{A}_t$, onto the subalgebra $\mathfrak{A}'_0 \subset \mathfrak{A}'$ generated by the union of all $\mathfrak{A}'_t$.*

By 13.1, we can extend the isomorphisms h_t to a homomorphism h of $\mathfrak{A}_0$ into $\mathfrak{A}'_0$. By the same argument we can extend all the isomorphisms h_t^{-1} to a homomorphism h' of $\mathfrak{A}'_0$ into $\mathfrak{A}_0$. By 12.1, this implies that h is an isomorphism of $\mathfrak{A}_0$ onto $\mathfrak{A}'_0$.

The following operation of forming products of fields of sets yields an important example of an independent indexed set of subalgebras.

For every $t \in T$, let $\mathfrak{F}_t$ be a (non-degenerate) field of subsets of a non-empty space X_t. Let X be the Cartesian product of all the spaces X_t, i.e. the set of all $x = \{x_t\} = \{x_t\}_{t \in T}$ where $x_t \in X_t$ for every $t \in T$. For every

---

[1] The notion of independence (and of $\mathfrak{m}$-independence — see § 37) is a slight generalization of the notion of set-theoretical independence of sets introduced by FICHTENHOLZ and KANTOROVITCH [1] HAUSDORFF [1],. MARCZEWSKI [5, 6, 8, 14, 16] examined this notion from the point of view of applications to measure theory. See also KAPPOS [5].

The theorems mentioned in § 13 were proved by SIKORSKI [11, 13, 14].

set $A \subset X_t$, let A^* be the set of all points $x \in X$ whose t^{th} co-ordinate x_t belongs to A, and let $\mathfrak{F}_t^*$ be the field (of subsets of X) composed of all sets A^* where $A \in \mathfrak{F}_t$. The field $\mathfrak{F}$ (of subsets of X) generated by the union of all classes $\mathfrak{F}_t^*$ ($t \in T$) is called the *field product* of the indexed set $\{\mathfrak{F}_t\}_{t \in T}$ of fields of sets.

The operation of forming the field product is often used in measure theory. One proves in measure theory that, if m_t are given measures on $\mathfrak{F}_t$ and $m_t(X_t) = 1$, then there is exactly one measure m on the field product $\mathfrak{F}$ such that

(2) $$m(A_1^* \cap \cdots \cap A_n^*) = m_{t_1}(A_1) \cdot \cdots \cdot m_{t_n}(A_n)$$

for every finite sequence $A_j \in \mathfrak{F}_{t_j}$, $t_j \neq t_k$ for $j \neq k$ ($j, k = 1, \ldots, n$). The measure m is said to be the *product* of the measures m_t.

$\{\mathfrak{F}_t^*\}_{t \in T}$ is an independent indexed set of subalgebras of $\mathfrak{F}$. In fact, if sets $A_j^* \in \mathfrak{F}_{t_j}^*$ ($t_j \neq t_k$ for $j \neq k$) are non-empty, then the sets A_j are also non empty, i.e. there are points $x_j \in A_j$. Every point $x = \{x_t\} \in X$ such that $x_{t_j} = x_j$ belongs to $A_1^* \cap \cdots \cap A_n^*$.

Observe that $\mathfrak{F}_t^*$ is isomorphic to $\mathfrak{F}_t$. Viz. the mapping

(3) $$g_t(A) = A^* \quad \text{for} \quad A \in \mathfrak{F}_t$$

is an isomorphism of $\mathfrak{F}_t$ onto $\mathfrak{F}_t^*$.

The above consideration suggests the following generalization of the notion of the field products.

Let $\{\mathfrak{A}_t\}_{t \in T}$ be an indexed set of non-degenerate Boolean algebras. By a *Boolean product*[1] (or simply *product*) of $\{\mathfrak{A}_t\}_{t \in T}$ we shall understand any pair

(4) $$\{\{i_t\}_{t \in T}, \mathfrak{B}\}$$

such that

(a) $\mathfrak{B}$ is a Boolean algebra;
(b) for every $t \in T$, i_t is an isomorphism of $\mathfrak{A}_t$ into $\mathfrak{B}$;
(c) the indexed set $\{i_t(\mathfrak{A}_t)\}_{t \in T}$ of subalgebras of $\mathfrak{B}$ is independent;
(d) the union of all the subalgebras $i_t(\mathfrak{A}_t)$, $t \in T$, generates $\mathfrak{B}$.

If (4) and

(4') $$\{\{i_t'\}_{t \in T}, \mathfrak{B}'\}$$

are two pairs satisfying (a) and (b), then we say that (4') is *isomorphic* to (4) if there exists an isomorphism h of $\mathfrak{B}$ onto $\mathfrak{B}'$ such that

(5) $$i_t' = h i_t \quad \text{for every} \quad t \in T.$$

Note that an isomorphism h of $\mathfrak{B}$ onto $\mathfrak{B}'$ satisfies (5) if and only if
(5') h is a common extension of all the isomorphisms $i_t' i_t^{-1}$ (from $i_t(\mathfrak{A}_t)$ onto $i_t'(\mathfrak{A}_t)$), $t \in T$.

If (4') is isomorphic to (4), then (4) is also isomorphic to (4') (take the isomorphism h^{-1}!). Therefore we can say simply that (4) and (4') are isomorphic.

---

[1] SIKORSKI [13]. Another equivalent definition was given by KAPPOS [2, 5] and RIDDER [1].

It follows easily from (5') and 13.2 that any two products (4), (4') of $\{\mathfrak{A}_t\}_{t \in T}$ are isomorphic. On the other hand, if (4) is a product of $\{\mathfrak{A}_t\}_{t \in T}$ and (4') is isomorphic to (4), then (4') is also a product of $\{\mathfrak{A}_t\}_{t \in T}$. Thus the Boolean product of $\{\mathfrak{A}_t\}_{t \in T}$ is determined by $\{\mathfrak{A}_t\}_{t \in T}$ uniquely up to isomorphism.

The Boolean product of any indexed set $\{\mathfrak{A}_t\}_{t \in T}$ of non-degenerate Boolean algebras always exists. In fact, let h_t be an isomorphism of $\mathfrak{A}_t$ onto the field $\mathfrak{F}_t$ of all open-closed subsets of the Stone space X_t of $\mathfrak{A}_t$. Let $\mathfrak{F}$ be the field product of all the fields $\mathfrak{F}_t$, $t \in T$, and let

(6) $$h_t^*(A) = h_t(A)^* \qquad \text{for all } A \in \mathfrak{A}_t,$$

$(t \in T)$. Then

(7) $$\{\{h_t^*\}_{t \in T}, \mathfrak{F}\}$$

is a Boolean product of $\{\mathfrak{A}_t\}_{t \in T}$. Note that $\mathfrak{F}$ is the field of all open-closed subsets of the Cartesian product X of all the spaces X_t, and X is compact and totally disconnected since the X_t are. Thus X is a Stone space of $\mathfrak{F}$, i.e. the product of the Stone spaces of the Boolean algebras $\mathfrak{A}_t$ is a Stone space of the Boolean product of the $\mathfrak{A}_t$.

The following theorem can be considered as another equivalent definition of a Boolean product.

13.3. $\{\{i_t\}_{t \in T}, \mathfrak{B}\}$ *is a Boolean product of* $\{\mathfrak{A}_t\}_{t \in T}$ *if and only if conditions* (a), (b), (d) *and the following condition* (e) *are satisfied:*

(e) *if* h_t *is any homomorphism from* $\mathfrak{A}_t$ *into a Boolean algebra* $\mathfrak{C}$, $t \in T$, *then there exists a homomorphism* h *from* $\mathfrak{B}$ *into* $\mathfrak{C}$ *such that*

(8) $$h_t = h i_t \qquad \text{for every } t \in T.$$

First observe that (8) is equivalent to the following statement:

(e') *the homomorphism* h *is a common extension of all the homomorphisms* $h_t i_t^{-1}$ *(from* $i_t(\mathfrak{A}_t)$ *into* $\mathfrak{C}$*),* $t \in T$.

It follows from 13.1 that every Boolean product of $\{\mathfrak{A}_t\}_{t \in T}$ has property (e).

On the other hand, it follows from 12.1 that if (4) and (4') satisfy (a), (b), (d), (e), then (4) and (4') are isomorphic. Now let (4) be any pair satisfying (a), (b), (d), (e), and let (4') be a Boolean product of $\{\mathfrak{A}_t\}_{t \in T}$. Thus (4) and (4') are isomorphic. Consequently (4) is also a Boolean product of $\{\mathfrak{A}_t\}_{t \in T}$.

Note that if (4) is a Boolean product of $\{\mathfrak{A}_t\}_{t \in T}$, then the set of all elements

(9) $$\bigcap_{t \in T'} i_t(A_t) \quad \text{where } A_t \in \mathfrak{A}_t \text{ for } t \in T', \text{ and } T' \text{ is a finite subset of } T$$

is dense in $\mathfrak{B}$. This can be proved easily in the case where (4) is identical with (7) and may be extended to arbitrary Boolean products by isomorphism.

Observe that the theorem on forming the product of measures is also true for Boolean algebras: if, for every $t \in T$, m_t is a measure on $\mathfrak{A}_t$ such that $m_t(V) = 1$, and (4) is a Boolean product of $\{\mathfrak{A}_t\}_{t \in T}$, then there exists exactly one measure m on $\mathfrak{B}$ such that (2) holds, where now A_j^* denotes the image of A_j given by the isomorphism i_{t_j}; i.e. $A_j^* = i_{t_j}(A_j)$ for $j = 1, 2, \ldots, n$.

If $\{\{i_t\}_{t \in T}, \mathfrak{B}\}$ is a Boolean product of $\{\mathfrak{A}_t\}_{t \in T}$, $\mathfrak{A}_t$ is a subalgebra of $\mathfrak{B}$ and i_t is the identity mapping for every $t \in T$, then $\mathfrak{B}$ will also be called a *Boolean product* of $\{\mathfrak{A}_t\}_{t \in T}$.

§ 14. Free Boolean algebras

A set $\mathfrak{S}$ of generators of a Boolean algebra $\mathfrak{A}$ is said to be a set of *free generators* of $\mathfrak{A}$ if every mapping f of $\mathfrak{S}$ into an arbitrary Boolean algebra $\mathfrak{A}'$ can be extended to a homomorphism h of $\mathfrak{A}$ into $\mathfrak{A}'$. Of course, free generators are always different from the zero and the unit [see § 5 (d)].

A Boolean algebra $\mathfrak{A}$ is said to be *free (with* $\mathfrak{n}$ *free generators)* provided it contains a set $\mathfrak{S}$ (of power $\mathfrak{n}$) of free generators of $\mathfrak{A}$.

For instance, every four-element Boolean algebra $\mathfrak{A}$ is a free Boolean algebra with one free generator. Viz. either of the two elements distinct from $\wedge$ and V can be taken as a free generator. In fact, let A be such an element. The algebra $\mathfrak{A}$ is composed of elements $\wedge$, V, A and $-A$. If $f(A) = A' \in \mathfrak{A}'$, then the mapping h defined by the equalities

$$h(\wedge) = \wedge, \quad h(V) = V, \quad h(A) = A', \quad h(-A) = -A'$$

is a homomorphism and an extension of f.

14.1. *In order that* $\mathfrak{A}$ *be a free Boolean algebra with* $\mathfrak{n}$ *free generators it is necessary and sufficient that* $\mathfrak{A}$ *be a Boolean product of an indexed set of* $\mathfrak{n}$ *four-element Boolean algebras.*

Suppose that $\mathfrak{A}$ is a free Boolean algebra and elements A_t ($t \in T$, $\overline{\overline{T}} = \mathfrak{n}$) are free generators of $\mathfrak{A}$. The set composed of $\wedge$, V, A_t and $-A_t$ is a four-element subalgebra $\mathfrak{A}_t$ of $\mathfrak{A}$, and $\mathfrak{A}$ is generated by the union of all $\mathfrak{A}_t$. Let h_t be a homomorphism of $\mathfrak{A}_t$ into a Boolean algebra $\mathfrak{A}'$ and let

$$f(A_t) = h_t(A_t) \quad \text{for} \quad t \in T.$$

The elements A_t being free generators, the mapping f can be extended to a homomorphism h which is also a common extension of all the h_t. By 13.3, $\mathfrak{A}$ is a Boolean product of $\{\mathfrak{A}_t\}_{t \in T}$.

On the other hand, suppose that $\mathfrak{A}$ is a Boolean product of $\mathfrak{n}$ four-element Boolean algebras, i.e. $\mathfrak{A}$ contains $\mathfrak{n}$ four-element subalgebras $\mathfrak{A}_t$ such that the union of all $\mathfrak{A}_t$ generates $\mathfrak{A}$ and any homomorphisms h_t from $\mathfrak{A}_t$ into any Boolean algebra $\mathfrak{A}'$ can be extended to a homomorphism h of $\mathfrak{A}$ into $\mathfrak{A}'$ (see 13.3 (e')). The subalgebra $\mathfrak{A}_t$ is composed of elements $\wedge$, V, A_t and $-A_t$. We shall prove that all the elements A_t are free generators of $\mathfrak{A}$. In fact, they generate $\mathfrak{A}$. Suppose that f is any mapping of the class $\mathfrak{S}$ of all the A_t into a Boolean algebra $\mathfrak{A}'$. Then the mapping

h_t defined by

$$h_t(\Lambda) = \Lambda, \quad h_t(V) = V, \quad h_t(A_t) = f(A_t), \quad h_t(-A_t) = -f(A_t)$$

is a homomorphism of $\mathfrak{A}_t$ into $\mathfrak{A}'$. The homomorphism h which is a common extension of all the h_t is also an extension of f.

It follows from the investigation in § 13 and 14.1 that two free Boolean algebras with the same number of free generators are isomorphic. Of course, an algebra isomorphic to a free Boolean algebra is also free with the same number of free generators. Thus free Boolean algebras are determined by the number of their free generators uniquely up to isomorphism.

An indexed set $\{A_t\}_{t \in T}$ of elements of a Boolean algebra $\mathfrak{A}$ is said to be *independent* provided

$$\varepsilon_1 A_{t_1} \cap \varepsilon_2 A_{t_2} \cap \cdots \cap \varepsilon_n A_{t_n} \neq \Lambda$$

for every sequence $t_j \in T$, $t_j \neq t_k$ for $j \neq k$, and for every sequence of numbers $\varepsilon_j = -1$ or 1. All elements A_t are then different from Λ and V.

It is easy to observe that $\{A_t\}_{t \in T}$ is independent if and only if the indexed set $\{\mathfrak{A}_t\}_{t \in T}$, where $\mathfrak{A}_t$ is the four-element algebra generated by A_t, is independent. Hence we obtain the theorem:

14.2. *In order that a Boolean algebra $\mathfrak{A}$ be free (with $\mathfrak{n}$ free generators) it is necessary and sufficient that it be generated by an independent indexed set of $\mathfrak{n}$ elements. These elements are then free generators.*

Let T_0 be a set with cardinality $\mathfrak{n}$. For every $t \in T_0$, let H_t be the Hausdorff space H composed only of the numbers -1 and 1. The Cartesian product $\mathscr{D}_{\mathfrak{n}} = P_{t \in T_0} H_t = H^{T_0}$ is a totally disconnected compact space called the *Cantor space* (or, more exactly, the *Cantor $\mathfrak{n}$-space*)[1]. For every fixed point $t \in T_0$, let D_t denote the set of all points in $\mathscr{D}_{\mathfrak{n}}$ whose t^{th} co-ordinate is equal to 1 (i.e. the set of all $\{a_\tau\}_{\tau \in T_0} \in \mathscr{D}_{\mathfrak{n}}$ such that $a_t = 1$). Let $\mathfrak{F}_{0,\mathfrak{n}}$ be the field of all open-closed subsets of $\mathscr{D}_{\mathfrak{n}}$. It follows from the definition of the topology in the Cartesian product $\mathscr{D}_{\mathfrak{n}}$ that $\mathfrak{F}_{0,\mathfrak{n}}$ is the smallest field containing all the sets $D_t, t \in T_0$. The sets D_t are independent since if $t_1, \ldots, t_n$ are different indices in T_0, then every point $\{a_\tau\}_{\tau \in T_0} \in \mathscr{D}_{\mathfrak{n}}$ with $a_{t_j} = \varepsilon_j$ $(j = 1, \ldots, n)$ belongs to $\varepsilon_1 \cdot D_{t_1} \cap \cdots \cap \varepsilon_n \cdot D_{t_n}$ $(\varepsilon_j = \pm 1$ for $j = 1, \ldots, n)$. This implies, by 14.2, that

14.3. *The field $\mathfrak{F}_{0,\mathfrak{n}}$ of all open-closed subsets of the Cantor $\mathfrak{n}$-space $\mathscr{D}_{\mathfrak{n}}$ is a free Boolean algebra with $\mathfrak{n}$ free generators $D_t, t \in T_0$.*

Theorem 14.3 can also be deduced from 14.1 and remarks on p. 41. In fact, H_t is the Stone space of the four element Boolean algebra. Hence it follows that $\mathscr{D}_{\mathfrak{n}}$ is the Stone space for the product of $\mathfrak{n}$ replicas of the

---

[1] If $\mathfrak{n} = \aleph_0$, then $\mathscr{D}_{\mathfrak{n}}$ is homeomorphic to the Cantor set of real numbers defined in § 9 B).

four-element Boolean algebra, i.e. that $\mathfrak{F}_{0,n}$ is a free Boolean algebra with n generators.

If $\mathfrak{A}'$ is a free Boolean algebra with a set $\mathfrak{S}'$ of free generators, then, for every Boolean algebra $\mathfrak{A}$ generated by a subset $\mathfrak{S}$ with $\overline{\overline{\mathfrak{S}}} \leq \overline{\overline{\mathfrak{S}'}}$, there exists a homomorphism h of $\mathfrak{A}'$ onto $\mathfrak{A}$ such that h maps $\mathfrak{S}'$ onto $\mathfrak{S}$. This follows immediately from the existence of a mapping of $\mathfrak{S}'$ onto $\mathfrak{S}$ and from the definition of free algebra. Hence, every Boolean algebra is a homomorphic image of a free Boolean algebra with a sufficiently large number of free generators.

The above remark can be also expressed as follows:

14.4. *Every Boolean algebra $\mathfrak{A}$ with a set $\mathfrak{S}$ of generators with $\overline{\overline{\mathfrak{S}}} \leq n$ is isomorphic to the quotient algebra $\mathfrak{F}_{0,n}/\Delta$ for a suitable ideal Δ. Moreover, we may assume that this isomorphism maps the set of all elements $[D_t]$ onto the set $\mathfrak{S}$ of generators of $\mathfrak{A}$.*

Incidentally we have proved that every compact totally disconnected space is homeomorphic to a closed subset of the Cantor space $\mathscr{D}_n$ whenever n is sufficiently large (see remarks on p. 31—32).

Examples. A) A finite Boolean algebra is free if and only if it has 2^{2^n} elements $(n = 1, 2, \ldots)$. It then has n free generators and 2^n atoms and is isomorphic to the field of all subsets of a 2^n-element space. This follows directly from 14.3 [see also § 8 B) and § 9 D)].

The two-element Boolean algebra may be considered as the free algebra with the empty set of free generators.

B) A countable Boolean algebra is free if and only if it is atomless. This follows from the fact that $\mathscr{D}_{\aleph_0}$ is homeomorphic with the Cantor set examined in § 9 B). See also § 8 C) and § 9 C).

C) If an infinite Boolean algebra has at least one atom, then it is not free.

D) For every infinite set X of power 2^n there exists a reduced perfect field $\mathfrak{F}$ of subsets of X such that $\overline{\overline{\mathfrak{F}}} = n$.

In fact, the existence of such a field $\mathfrak{F}$ is invariant under one-to-one transformations of X. On the other hand, the set $\mathscr{D}_n$ has cardinality 2^n, $\mathfrak{F}_{0,n}$ is a reduced perfect field of subsets of $\mathscr{D}_n$, and $\overline{\overline{\mathfrak{F}_{0,n}}} = n$.

E) For every infinite cardinal m, the Cantor space $\mathscr{D}_{2^m}$ contains a dense subset of power m.[1]

In fact, we have $\mathscr{D}_{2^m} = \mathsf{P}_{t \in T_0} H_t$ (see p. 43) where $\overline{\overline{T}}_0 = 2^m$. There exists a reduced field $\mathfrak{F}$ of subsets of T_0 such that $\overline{\overline{\mathfrak{F}}} = m$. For every set $A \in \mathfrak{F}$, let $a_A = \{a_t\}_{t \in T_0}$ be a point in $\mathscr{D}_{2^m}$ defined as follows:

$$a_t = 1 \quad \text{if} \quad t \in A, \quad \text{and} \quad a_t = -1 \quad \text{if} \quad t \in -A.$$

The set X of all the points a_A, where $A \in \mathfrak{F}$, has power m. The field $\mathfrak{F}$

---

[1] This theorem is a particular case of a general theorem (due to HEWITT [2]) on Cartesian products of topological spaces. The case $m = \aleph_0$ of this theorem was independently found by MARCZEWSKI [9].

being reduced, for any distinct points $t_1, \ldots, t_n$ in T_0 and for every sequence $\varepsilon_1, \ldots, \varepsilon_n$ of numbers -1 and 1 there exists a set $A \in \mathfrak{F}$ such that

$$t_i \in A \quad \text{if} \quad \varepsilon_i = 1 \quad \text{and} \quad t_i \in -A \quad \text{if} \quad \varepsilon_i = -1 \,.$$

This implies that $a_A \in \varepsilon_1 \cdot D_{t_1} \cap \cdots \cap \varepsilon_n \cdot D_{t_n}$. Since sets of the form $\varepsilon_1 \cdot D_{t_1} \cap \cdots \cap \varepsilon_n \cdot D_{t_n}$ constitute an open basis in $\mathscr{D}_2\mathfrak{m}$, the set X is dense.

F) The field $\mathfrak{F}$ of all subsets of a set X of power $\mathfrak{m} \geqq \aleph_0$ contains a subfield isomorphic to $\mathfrak{F}_{0, 2}\mathfrak{m}$ (consequently X contains $2^\mathfrak{m}$ independent subsets[1]).

It suffices to prove this in the case where X is a dense subset of $\mathscr{D}_2\mathfrak{m}$, $\overline{\overline{X}} = \mathfrak{m}$. Then the mapping

$$h(A) = A \cap X \quad \text{for} \quad A \in \mathfrak{F}_{0, 2}\mathfrak{m}$$

is an isomorphism of $\mathfrak{F}_{0, 2}\mathfrak{m}$ into $\mathfrak{F}$.

G) The Stone space X_0 of the field $\mathfrak{F}$ of all subsets of a set of power $\mathfrak{m} \geqq \aleph_0$ has power $2^{2^\mathfrak{m}}$.[2]

In fact, it is easy to prove that the set of all maximal filters in $\mathfrak{F}$ has cardinality $\leqq 2^{2^\mathfrak{m}}$. Hence $\overline{\overline{X}}_0 \leqq 2^{2^\mathfrak{m}}$. On the other hand, it follows from F) that there exists a continuous mapping of X_0 onto $\mathscr{D}_2\mathfrak{m}$ (see the remarks on p. 34). Hence $\overline{\overline{X}}_0 \geqq \overline{\overline{\mathscr{D}_2\mathfrak{m}}} = 2^{2^\mathfrak{m}}$.

§ 15. Induced homomorphisms between quotient algebras

Let $\mathfrak{F}$ and $\mathfrak{F}'$ be fields of subsets of spaces X and X' respectively, and let $\varDelta$ and $\varDelta'$ be ideals of $\mathfrak{F}$ and $\mathfrak{F}'$ respectively.

Generalizing the definition from § 11 [see also § 5 A)] we say that a homomorphism h of $\mathfrak{F}/\varDelta$ into $\mathfrak{F}'/\varDelta'$ is induced[3] by a point mapping φ of X' into X if

(1) $\qquad\qquad h([A]_\varDelta) = [\varphi^{-1}(A)]_{\varDelta'} \quad \text{for every} \quad A \in \mathfrak{F} \,.$

Of course, condition (1) implies that

(2) $\qquad\qquad \varphi^{-1}(A) \in \mathfrak{F}' \quad \text{for every} \quad A \in \mathfrak{F} \,.$

It also implies that

(3) $\qquad\qquad \varphi^{-1}(A) \in \varDelta' \quad \text{for every} \quad A \in \varDelta$

since the homomorphism h transforms the zero of $\mathfrak{F}/\varDelta$ into the zero of $\mathfrak{F}'/\varDelta'$. Conversely, if a mapping φ satisfies (2) and (3), then formula (1) defines a homomorphism h of $\mathfrak{F}/\varDelta$ into $\mathfrak{F}'/\varDelta'$.

---

[1] This theorem is a particular case of more general theorems proved by HAUS-DORFF [1] and TARSKI [6]. For $\mathfrak{m} = \aleph_0$ see also FICHTENHOLZ and KANTOROVITCH [1].

[2] GILLMAN and JERISON [1], NOVÁK [1], POSPÍŠIL [5, 6]. See also TARSKI [6], HAUSDORFF [1].

[3] SIKORSKI [6]. Theorems 15.1—15.2 were proved by SIKORSKI [6, 18]. The first theorem of this kind was that of VON NEUMANN [1] on induced measure-preserving isomorphisms.

In particular, if $\varDelta$ is the zero ideal, then, identifying $\mathfrak{F}/\varDelta$ with $\mathfrak{F}$ (see the end of § 10), we say that a homomorphism h of $\mathfrak{F}$ into $\mathfrak{F}'/\varDelta'$ is induced by a mapping φ of X' into X if

(4) $\qquad\qquad h(A) = [\varphi^{-1}(A)]_{\varDelta'}$ for every $A \in \mathfrak{F}$.

Condition (2) is then also satisfied. Conversely, if φ is a mapping of X' into X, satisfying condition (2), then (4) defines a homomorphism h of $\mathfrak{F}$ into $\mathfrak{F}'/\varDelta'$.

If both $\varDelta$ and $\varDelta'$ are zero ideals, then $\mathfrak{F}/\varDelta$ and $\mathfrak{F}'/\varDelta'$ can be identified with $\mathfrak{F}$ and $\mathfrak{F}'$ respectively (see the end of § 10) and, in this case, the above definition of induced homomorphism coincides with that from § 11 and § 5 A).

The sufficient condition given in the first part of 11.1 is, in general, insufficient if homomorphisms of $\mathfrak{F}$ into arbitrary quotient algebras $\mathfrak{F}'/\varDelta'$ are taken under consideration.

Example. A) Let X be a closed subset of a totally disconnected compact space X', let $\mathfrak{F}$ and $\mathfrak{F}'$ be the fields of all open-closed subsets of the spaces X and X' respectively, and let $\varDelta'$ be the ideal of all $B \in \mathfrak{F}'$ which do not intersect X. Every set $A \in \mathfrak{F}$ is of the form

$$A = A' \cap X$$

where $A' \in \mathfrak{F}'$. The set A' in this equality is not uniquely determined by A, but the element $[A']_{\varDelta'}$ is uniquely determined by A, and the mapping h defined by

$$h(A) = [A']_{\varDelta'}$$

is an isomorphism of $\mathfrak{F}$ onto $\mathfrak{F}'/\varDelta'$ [see § 10 B]. The isomorphism h is induced by a point mapping if and only if X is a *retract* of X',[1] i.e. if there exists a continuous mapping ψ (called the *retraction of X' onto X*) of X' into X such that

$$\psi(x) = x \quad \text{for every} \quad x \in X.$$

In fact, on the one hand, a continuous mapping ψ of X' into X is a retraction of X' onto X if and only if

(5) $\qquad\qquad \psi^{-1}(A) \cap X = A \quad \text{for every} \quad A \in \mathfrak{F}$.

On the other hand, a mapping ψ (of X' into X) induces h if and only if it is continuous and satisfies (5).

There exist compact totally disconnected spaces X' containing closed subsets X such that X is not the retract of X' (or where X is not even a continuous image of X'). In this case, the homomorphism h defined above is not induced by any point mapping. Observe that such spaces are always non-metrizable since every closed subset X of any compact totally disconnected metric space X' is a retract of X'.[2]

---

[1] Sikorski [6, 18].
[2] Sierpiński [4].

The space X defined in § 8 E) is an example of such a singular space. It is compact and totally disconnected, therefore it can be (homeomorphically) imbedded in the Cantor space $X' = \mathscr{D}_\mathfrak{n}$ if $\mathfrak{n}$ is sufficiently large. On the other hand, X is not a continuous image of $\mathscr{D}_\mathfrak{n}$ for any cardinal $\mathfrak{n}$.[1]

The Stone space of the Boolean algebra of all subsets of a countable set X (i.e. the Čech-Stone compactification of the discrete space X — see § 8 F)) is another example of a compact totally disconnected space which is not a continuous image of $\mathscr{D}_\mathfrak{n}$ for any cardinal $\mathfrak{n}$.

Consider now the case where $\mathfrak{F} = \mathfrak{F}_{0,\,\mathfrak{n}}$ is the field of all open-closed subsets of the Cantor space $\mathscr{D}_\mathfrak{n}$, and $\varDelta$ is an ideal of it. $\mathfrak{F}'$ denotes an arbitrary field of subsets of a set X', and $\varDelta'$ is any ideal of $\mathfrak{F}'$.

15.1. *Every homomorphism h of $\mathfrak{F}_{0,\,\mathfrak{n}}/\varDelta$ into any quotient algebra $\mathfrak{F}'/\varDelta'$ is induced by a point mapping.*

Assume all notation from § 14, p. 43. For every $t \in T_0$, let $A'_t \in \mathfrak{F}'$ be a set such that

(6) $$h([D_t]_\varDelta) = [A'_t]_{\varDelta'} .$$

Let

$$\varphi_t(x') = \begin{cases} 1 & \text{if } x' \in A'_t \\ -1 & \text{if } x' \in -A'_t \end{cases}$$

and let

$$\varphi(x') = \{\varphi_t(x')\}_{t \in T_0} \in \mathscr{D}_\mathfrak{n} \quad \text{for} \quad x' \in X' .[2]$$

We have

(7) $$\varphi^{-1}(D_t) = A'_t \in \mathfrak{F}'$$

and consequently $\varphi^{-1}(A) \in \mathfrak{F}'$ for every $A \in \mathfrak{F}_{0,\,\mathfrak{n}}$. Thus the formula

$$h_1(A) = [\varphi^{-1}(A)]_{\varDelta'} \qquad (A \in \mathfrak{F}_{0,\mathfrak{n}})$$

defines a homomorphism h_1 of $\mathfrak{F}_{0,\,\mathfrak{n}}$ into $\mathfrak{F}'/\varDelta'$. The formula

$$h_2(A) = h([A]_\varDelta) \qquad (A \in \mathfrak{F}_{0,\,\mathfrak{n}})$$

also defines a homomorphism h_2 of $\mathfrak{F}_{0,\,\mathfrak{n}}$ into $\mathfrak{F}'/\varDelta'$. It follows from (6) and (7), that the homomorphisms h_1 and h_2 coincide on the set of all generators D_t of $\mathfrak{F}_{0,\,\mathfrak{n}}$. Hence it follows (see the remark in § 12, p. 35) that $h_1 = h_2$, which proves (1).

15.2. *Let $\mathfrak{F} = \mathfrak{F}_{0,\,\mathfrak{n}} | X$ be the field of all open-closed subsets of a closed subset X of the Cantor $\mathfrak{n}$-space $\mathscr{D}_\mathfrak{n}$. In order that every homomorphism h of $\mathfrak{F}$ into any quotient algebra $\mathfrak{F}'/\varDelta'$ be induced by a point mapping, it is necessary and sufficient that X be a retract of $\mathscr{D}_\mathfrak{n}$.*

---

[1] Šanin [1].

[2] The mapping φ is called the *characteristic function* of $\{A'_t\}_{t \in T}$. The notion of characteristic functions was examined by MARCZEWSKI [2, 10] in the case where T is the set of positive integers. For the generalization to uncountable sets T, see STONE [11]. MARCZEWSKI's method of characteristic functions is used several times in this book.

In this case, X is also a retract of every compact totally disconnected space containing X as a subspace.

To prove the sufficiency, suppose that ψ is a retraction of $\mathscr{D}_n$ onto X, i.e.

$$\psi^{-1}(A) \in \mathfrak{F}_{0,n} \quad \text{and} \quad \psi^{-1}(A) \cap X = A \quad \text{for every} \quad A \in \mathfrak{F}.$$

The mapping h_0 given by

$$h_0(B) = h(B \cap X) \quad \text{for} \quad B \in \mathfrak{F}_{0,n}$$

is a homomorphism of $\mathfrak{F}_{0,n}$ into $\mathfrak{F}'/\Delta'$, therefore, by 15.1, it is induced by a mapping φ_0 of X' into $\mathscr{D}_n$, i.e.

$$h(B \cap X) = [\varphi_0^{-1}(B)]_{\Delta'} \quad \text{for every} \quad B \in \mathfrak{F}_{0,n}.$$

Assuming $B = \psi^{-1}(A)$ we obtain

$$h(A) = h(\psi^{-1}(A) \cap X) = [\varphi_0^{-1}(\psi^{-1}(A))]_{\Delta'} \quad \text{for} \quad A \in \mathfrak{F}.$$

This proves that the superposition

$$\varphi(x') = \psi(\varphi_0(x')) \quad \text{for} \quad x' \in X'$$

induces the homomorphism h.

The necessity and the final part of 15.2 follow from the argument in Example A).

It follows from 15.2 that the following three conditions are equivalent for the field $\mathfrak{F}$ of all open-closed subsets of any compact totally disconnected space X:

(a) every homomorphism of $\mathfrak{F}$ into $\mathfrak{F}'/\Delta'$ is induced by a point mapping;

(b) X is homeomorphic to a retract of $\mathscr{D}_n$;

(c) X is an absolute retract in the domain of totally disconnected compact spaces, i.e. X is a retract of every totally disconnected compact space $X' \supset X$.

It follows from the investigation in § 11 that if a field $\mathfrak{F}$ has property (a), then $\mathfrak{F}$ is perfect (the case where Δ' is the zero ideal). However, if $\mathfrak{F}$ is perfect, then the existence of a mapping φ inducing a given homomorphism h of $\mathfrak{F}$ into $\mathfrak{F}'/\Delta'$ is equivalent to the existence of a homomorphism h' of $\mathfrak{F}$ into $\mathfrak{F}'$ such that $h(A) = [h'(A)]$ for every $A \in \mathfrak{F}$. In fact, if h' exists, then h' is induced by a mapping φ which also induces h. Conversely, if h is induced by a mapping φ, then $h'(A) = \varphi^{-1}(A)$ satisfies the condition mentioned above.

Thus the problem of the validity of (a) can be reduced to the following question. Under what condition has a given Boolean algebra $\mathfrak{A}$ the following property?

(a') For every homomorphism h of $\mathfrak{A}$ into any algebra $\mathfrak{A}'/\Delta'$ (where $\mathfrak{A}'$ is a Boolean algebra and Δ' is an ideal of $\mathfrak{A}'$) there exists a homomorphism

h' of $\mathfrak{A}$ into $\mathfrak{A}'$ such that

(8) $\qquad\qquad h(A) = [h'(A)]_{\Delta'}$ for every $A \in \mathfrak{A}$.

To give the answer to this question we introduce the following definition which is an algebraic analogue of the topological notion of retract examined above: a subalgebra $\mathfrak{A}_0$ of a Boolean algebra $\mathfrak{A}$ is said to be *retract* of $\mathfrak{A}$ provided there exists a homomorphism g (called a *retract homomorphism*) of $\mathfrak{A}$ onto $\mathfrak{A}_0$ such that $g(A) = A$ for $A \in \mathfrak{A}_0$.

15.3. *For every Boolean algebra* $\mathfrak{A}$, *property* (a') *is equivalent to each of the following properties:*

(b') $\mathfrak{A}$ *is isomorphic to a retract of a free Boolean algebra;*

(c') $\mathfrak{A}$ *is an absolute Boolean retract, i.e., for every Boolean algebra* $\mathfrak{A}'$, *such that* $\mathfrak{A}$ *is a homomorphic image of* $\mathfrak{A}'$, $\mathfrak{A}$ *is isomorphic to a retract of* $\mathfrak{A}'$.[1]

Theorem 15.3 can be deduced from the equivalence of (a), (b), (c) just proved since we can restrict our investigation to the case where $\mathfrak{A}$ is a perfect field and $\mathfrak{A}'$ is a field of sets. Then (a) is equivalent to (a'). Since a Boolean algebra $\mathfrak{B}$ is isomorphic to a retract of a Boolean algebra $\mathfrak{B}'$ if and only if the Stone space of $\mathfrak{B}$ is homeomorphic to a retract of the Stone space of $\mathfrak{B}'$, the properties (b') and (c') are equivalent to (b) and (c), respectively.

Theorem 15.3 can also be proved directly as follows [and this proof, by the mentioned equivalence of (a), (b), (c) with (a'), (b'), (c') respectively, can be considered as the second proof of 15.2].

First observe that if $\mathfrak{A}$ is free, then $\mathfrak{A}$ has property (a'). In fact, let $\mathfrak{S}$ be the set of free generators of $\mathfrak{A}$. For every $A \in \mathfrak{S}$, let $f(A) \in \mathfrak{A}'$ be such that $[f(A)] = h(A)$. The mapping f can be extended to a homomorphism h' and this extension h' satisfies (8).

On the other hand, any retract $\mathfrak{A}_0$ of a Boolean algebra $\mathfrak{A}$ with property (a') also has this property. In fact, let g be the retract homomorphism of $\mathfrak{A}$ onto $\mathfrak{A}_0$. By (a'), there exists a homomorphism h' such that $hg(A) = [h'(A)]$ for every $A \in \mathfrak{A}$. In particular, $h(A) = [h'(A)]$ for every $A \in \mathfrak{A}_0$.

Since property (a') is invariant under isomorphisms, the two above statements prove that (b') implies (a').

Since every Boolean algebra is a homomorphic image of a free Boolean algebra (see 14.4), (c') implies (b').

Suppose that $\mathfrak{A}$ has property (a') and that $\mathfrak{A}$ is a homomorphic image of a Boolean algebra $\mathfrak{A}'$, i.e. there exists an isomorphism h of $\mathfrak{A}$ onto $\mathfrak{A}'/\Delta'$ for a suitable ideal Δ'. The homomorphism h' satisfying (8) is an isomorphism of $\mathfrak{A}$ onto a subalgebra $\mathfrak{A}_0'$ of $\mathfrak{A}'$, and the mapping g given by

$$g(A') = h'h^{-1}([A']) \quad \text{for} \quad A' \in \mathfrak{A}'$$

is a retract homomorphism of $\mathfrak{A}'$ onto $\mathfrak{A}_0'$. Thus (a') implies (c').

---

[1] HALMOS [10]. For a generalization, see SEMADENI [5].

§ 16. Direct unions

Let $\{\mathfrak{A}_t\}_{t \in T}$ be an indexed set of Boolean algebras. Consider their Cartesian product $\mathfrak{A}$, i.e. the set of all indexed sets $A = \{A_t\} = \{A_t\}_{t \in T}$ such that $A_t \in \mathfrak{A}_t$ for every $t \in T$. The set $\mathfrak{A}$ is also a Boolean algebra with the following definition of Boolean operations

$$\{A_t\} \cup \{B_t\} = \{A_t \cup B_t\},$$
$$\{A_t\} \cap \{B_t\} = \{A_t \cap B_t\},$$
$$-\{A_t\} = \{-A_t\}.$$

The easy verification that the operations $\cup, \cap, -$ so defined in $\mathfrak{A}$ satisfy the axioms § 1 (A_1)-(A_5) is left to the reader.

The Boolean algebra $\mathfrak{A}$ defined above is called the *direct union* of $\{\mathfrak{A}_t\}_{t \in T}$.

Of course, $\{\wedge\}$ and $\{\vee\}$ (i.e. the elements of the Cartesian product $\mathfrak{A}$, whose all coordinates are equal to the zero elements or the unit elements of $\mathfrak{A}_t$ respectively) are the zero element and the unit element of $\mathfrak{A}$ respectively. The Boolean inclusion $\{A_t\} \subset \{B_t\}$ holds if and only if $A_t \subset B_t$ for every $t \in T$. We also have

$$\{A_t\} - \{B_t\} = \{A_t - B_t\}.$$

If $\mathfrak{n}_t$ is the cardinal of $\mathfrak{A}_t$, then the cardinal $\mathfrak{n}$ of the direct union of $\{\mathfrak{A}_t\}_{t \in T}$ is

(1) $$\mathfrak{n} = \Pi_{t \in T} \mathfrak{n}_t.$$

If $\mathfrak{A}_t$ are fields of subsets of disjoint spaces X_t respectively, then the direct union of $\{\mathfrak{A}_t\}_{t \in T}$ is isomorphic to the field of all subsets B of the union X of all X_t $(t \in T)$, such that $B \cap X_t \in \mathfrak{A}_t$ for every $t \in T$. The isomorphism h is given by the formula

$$h(\{A_t\}) = \mathsf{U}_{t \in T} A_t.$$

In particular, if the $\mathfrak{A}_t$ are two-element Boolean algebras, i.e. fields of all subsets of the one-element spaces $X_t = (t)$ respectively, then the direct union of $\{\mathfrak{A}_t\}_{t \in T}$ is isomorphic to the field of all subsets of the set T.

The fundamental representation theorem 8.2 can now be expressed in the following way: every non-degenerate Boolean algebra is isomorphic to a subalgebra of a direct union of two-element Boolean algebras.

More exactly, let T be the set of all two-valued homomorphisms of a Boolean algebra $\mathfrak{A}$ onto a fixed two-element Boolean algebra $\mathfrak{A}_0$ and let $\mathfrak{A}_t = \mathfrak{A}_0$. The mapping

$$h(A) = \{t(A)\}_{t \in T}$$

is an isomorphism of $\mathfrak{A}$ into the direct union of $\{\mathfrak{A}_t\}_{t \in T}$. The isomorphism h is, in fact, identical with the isomorphism h defined in the proof of 8.2 provided the two-valued homomorphisms are identified with the corresponding maximal filters.

Examples. A) A Boolean algebra $\mathfrak{A}$ is isomorphic to the direct union of Boolean algebras $\mathfrak{A}_1, \ldots, \mathfrak{A}_n$ if and only if there exist disjoint elements $A_1, \ldots, A_n \in \mathfrak{A}$ such that

$$A_1 \cup \cdots \cup A_n = V \,,$$

and

$$\mathfrak{A} \mid A_j \text{ is isomorphic to } \mathfrak{A}_j \qquad (j = 1, \ldots, n) \,.$$

Namely, if h_j is an isomorphism of $\mathfrak{A} \mid A_j$ onto $\mathfrak{A}_j$, then

$$h(A) = \{h_1(A \cap A_1), \ldots, h_n(A \cap A_n)\} \qquad (A \in \mathfrak{A})$$

is an isomorphism of $\mathfrak{A}$ onto the direct union of $\mathfrak{A}_1, \ldots, \mathfrak{A}_n$.

An analogous problem for an infinite indexed set of Boolean algebras will be discussed in § 22 E).

B) For every $t \in T$, let X_t be a Stone space of a Boolean algebra $\mathfrak{A}_t$. Suppose that the spaces X_t are disjoint. The union X of the X_t will be considered as a topological space, the class of all open-closed subsets of the spaces X_t being assumed as an open basis. In other words, a set $A \subset X$ is open if $A \cap X_t$ is open in X_t for every t. In particular, every set X_t is both open and closed in X.

The reader familiar with the Čech-Stone compactification can verify that the Čech-Stone compactification $\beta(X)$ of X is a Stone space of the direct union of $\{\mathfrak{A}_t\}_{t \in T}$.[1]

§ 17. Connection with algebraic rings

We recall that an (algebraic) *ring* is a non empty set $\mathfrak{A}$ of elements (denoted by the letters $A, B, C, \ldots$) with two operations; addition $A + B$ and multiplication $A \cdot B$ satisfying the following axioms:

(R₁) $$A + B = B + A \,;$$

(R₂) $$A + (B + C) = (A + B) + C \,;$$

(R₃) given A and C, there exists exactly one element B such that

$$A + B = C \,;$$

(R₄) $$A \cdot (B \cdot C) = (A \cdot B) \cdot C \,;$$

(R₅) $$A \cdot (B + C) = A \cdot B + A \cdot C \,;$$

(R₆) $$(A + B) \cdot C = A \cdot C + B \cdot C \,.$$

It follows from (R₁)—(R₃) that there exists an element, called the *zero* of $\mathfrak{A}$ and denoted by $\mathbf{0}$, such that

(1) $$A + \mathbf{0} = A \quad \text{for every} \quad A \,.$$

An element $\mathbf{1}$ of $\mathfrak{A}$ is said to be the *unit* of $\mathfrak{A}$ provided

$$A \cdot \mathbf{1} = A = \mathbf{1} \cdot A \quad \text{for every} \quad A \,.$$

Every ring contains at most one unit.

---

[1] Dwinger [2].

A ring $\mathfrak{A}$ is said to be *commutative* provided

$$A \cdot B = B \cdot A \quad \text{for every } A, B .$$

A ring $\mathfrak{A}$ is said to be a *Boolean ring* provided $\mathfrak{A}$ contains a unit[1] $\boldsymbol{1}$ and

(2) $$A \cdot A = A \quad \text{for every } A .$$

A simple example of a Boolean ring is given by the ring $\mathfrak{R}$ of all integers modulo 2, i.e. the algebraic ring containing only two elements, the zero $\boldsymbol{0}$ and the unit $\boldsymbol{1}$, with the following definition of addition and multiplication:

$$\boldsymbol{0 + 0 = 0 = 1 + 1}, \quad \boldsymbol{0 + 1 = 1 = 1 + 0},$$
$$\boldsymbol{0 \cdot 0 = 0 \cdot 1 = 1 \cdot 0 = 0}, \quad \boldsymbol{1 \cdot 1 = 1}.$$

Another example of a Boolean ring is given by the ring $\mathfrak{R}^X$ of all mappings of any abstract set X into $\mathfrak{R}$ with the usual definition of addition and multiplication; the sum $f = f_1 + f_2$ (the product $f = f_1 \cdot f_2$) is the mapping defined by the formula $f(x) = f_1(x) + f_2(x)$ $(f(x) = f_1(x) \cdot f_2(x))$. If A is a subset of X, then the function $f \in \mathfrak{R}^X$ given by

$$f(x) = \begin{cases} \boldsymbol{1} & \text{if } x \in A \\ \boldsymbol{0} & \text{if } x \notin A \end{cases}$$

is called the *characteristic function* of the set A. Observe that the correspondence between subsets of X and their characteristic functions is one-to-one. The ring $\mathfrak{R}^X$ is the set of characteristic functions of all subsets of X. The zero (unit) of $\mathfrak{R}^X$ is the function equal identically to $\boldsymbol{0}$ $(\boldsymbol{1})$, i.e. the characteristic function of the empty set (the whole space X).

Subrings of $\mathfrak{R}^X$ are also Boolean rings whenever they contain the unit of $\mathfrak{R}^X$. In particular, if X is a topological space, then the class of all continuous mappings of X into the discrete space $\mathfrak{R}$ is a Boolean ring, This ring is the class of characteristic functions of all open-closed subsets of X.

Every Boolean ring $\mathfrak{A}$ has the following properties:

(3) $$A + A = \boldsymbol{0} \quad \text{for every } A ,$$

(4) $$\text{if } A + B = \boldsymbol{0}, \quad \text{then } B = A ,$$

(5) $$A \cdot B = B \cdot A \quad \text{for every } A \text{ and } B$$

(i.e. every Boolean ring is commutative).

In fact, if follows from (2) that

$$(\boldsymbol{1} + A) \cdot (\boldsymbol{1} + A) = (\boldsymbol{1} + A) ,$$

i.e., by (2),

$$(\boldsymbol{1} + A) + (A + A) = (\boldsymbol{1} + A) .$$

Hence, by (R_3) (uniqueness!) and (1) (where A is replaced by $\boldsymbol{1} + A$) we obtain (3). Property (4) follows from (3) and from the uniqueness

---

[1] Usually the existence of the unit is omitted in the definition of Boolean rings. The definition assumed here is more convenient from the point of view of the applications in this book.

guaranted by (R_3). To prove (5), let us observe that

$$(A + B) \cdot (A + B) = A + B ,$$

i.e., by (R_5) and (2),

$$(A + B) + (A \cdot B + B \cdot A) = (A + B)$$

which, by (R_3) and (1), implies $A \cdot B + B \cdot A = 0$. This proves (5) on account of (4).

17.1. *Every Boolean algebra is a Boolean ring with the following definitions of addition and multiplication:*

(6) $A + B = (A - B) \cup (B - A) ,$

(7) $A \cdot B = A \cap B .$

Conversely, every Boolean ring is a Boolean algebra with the following definitions of join, meet and complement:

(8) $A \cup B = A + B + A \cdot B ,$

(9) $A \cap B = A \cdot B ,$

(10) $-A = 1 + A .$

In both the cases the algebraic zero and unit coincide with the Boolean zero and unit respectively[1].

The first part of 17.1 can be proved by the verification that, if $\mathfrak{A}$ is a Boolean algebra, then the operations $A + B$ (often called the *symmetric difference*[2] of A and B) and $A \cdot B$ defined by (6) and (7) satisfy the axioms $(R_1) - (R_6)$. Another shorter proof can be given by the following argument. We may restrict ourselves to the case where $\mathfrak{A}$ is the field of all open-closed subsets of a totally disconnected compact space X. Identifying subsets of X with their characteristic functions, we may consider $\mathfrak{A}$ as the class of all continuous mappings of X into $\mathfrak{R}$. By this identification, the sum and product defined by (6) and (7) coincide with the usual sum and product of mappings from X to $\mathfrak{R}$, therefore they satisfy the axioms $(R_1) - (R_6)$.

To prove the second part of 17.1, suppose that $\mathfrak{A}$ is a Boolean ring and $A \cup B$, $A \cap B$, $-A$ are defined by (8), (9) and (10). The commutativity of addition and multiplication [see (R_1) and (5)] implies the commutative laws § 1 (A_1). We have

$$A \cup (B \cup C) = A + (B + C + B \cdot C) + A \cdot (B + C + B \cdot C)$$
$$= A + B + C + A \cdot B + A \cdot C + B \cdot C + A \cdot B \cdot C$$

and

$$(A \cup B) \cup C = (A + B + A \cdot B) + C + (A + B + A \cdot B) \cdot C$$
$$= A + B + C + A \cdot B + A \cdot C + B \cdot C + A \cdot B \cdot C$$

---

[1] Theorem 17.1 is due to Stone [2, 5].

There also exist other binary operations with respect to which every Boolean algebra is a ring and which define uniquely the Boolean operations. See Rudeanu [2].

[2] For the investigation of the symmetric difference, see e.g. Helson [1], Marczewski [7].

which proves the first associative law § 1 (A_2). The second follows immediately from (R_4). We have

$$(A \cap B) \cup B = A \cdot B + B + A \cdot B \cdot B$$
$$= B + A \cdot B + A \cdot B = B$$

by (2) and (3), and analogously

$$(A \cup B) \cap B = (A + B + A \cdot B) \cdot B$$
$$= A \cdot B + B + A \cdot B = B$$

which proves the absorption laws § 1 (A_3). Similarly we verify that, on account of (2), (3) and (5),

$$A \cap (B \cup C) = A \cdot (B + C + B \cdot C)$$
$$= A \cdot B + A \cdot C + A \cdot B \cdot C,$$
$$(A \cap B) \cup (A \cap C) = A \cdot B + A \cdot C + A \cdot B \cdot A \cdot C$$
$$= A \cdot B + A \cdot C + A \cdot B \cdot C,$$
$$A \cup (B \cap C) = A + B \cdot C + A \cdot B \cdot C,$$
$$(A \cup B) \cap (A \cup C) = (A + B + A \cdot B)(A + C + A \cdot C)$$
$$= A \cdot A + B \cdot A + A \cdot B \cdot A + A \cdot C + B \cdot C +$$
$$+ A \cdot B \cdot C + A \cdot A \cdot C + B \cdot A \cdot C + A \cdot B \cdot A \cdot C$$
$$= A + B \cdot C + A \cdot B \cdot C,$$

which proves the distributive laws § 1 (A_4). Since

$$A \cap -A = A \cdot (\mathbf{1} + A) = A + A = \mathbf{0}$$

and

$$A \cup -A = A + (\mathbf{1} + A) + A \cdot (\mathbf{1} + A)$$
$$= \mathbf{1} + A + A + A + A = \mathbf{1},$$

the laws § 1 (A_5) are also satisfied.

Examples. A) The ring $\mathfrak{R}$ considered as a Boolean algebra is the two-element Boolean algebra.

B) The ring $\mathfrak{R}^X$ considered as a Bolean algebra is isomorphic to the field of all subsets of X.

Chapter II

Infinite joins and meets

§ 18. Definition

Let $A_1, \ldots, A_n$ be elements of a Boolean algebra $\mathfrak{A}$. The finite join

$$B = A_1 \cup \cdots \cup A_n$$

is the least element containing all the elements $A_1, \ldots, A_n$, i.e. it is characterized uniquely by the following two conditions:

(j_1) $A_i \subset B$ for $i = 1, \ldots, n$;
(j_2) if $A_i \subset B'$ for $i = 1, \ldots, n$, then $B \subset B'$.

Similarly, the finite meet

$$C = A_1 \cap \cdots \cap A_n$$

is the greatest element contained in the elements $A_1, \ldots, A_n$, i.e. it is uniquely characterized by the following two conditions:

(m$_1$) $C \subset A_i$ for $i = 1, \ldots, n$;
(m$_2$) if $C' \subset A_i$ for $i = 1, \ldots, n$, then $C' \subset C$.

These properties of join and meet were proved in § 2 for $n = 2$ (see § 2 (15), (14) and (15'), (14')). By an induction argument we can generalize them for every positive integer n. These properties suggest the following generalization of join and meet for the case of an arbitrary, finite or infinite number of elements.

Let $\mathfrak{S}$ be a non-empty set of elements of a Boolean algebra $\mathfrak{A}$. An element $B \in \mathfrak{A}$ is said to be the *join* of all elements $A \in \mathfrak{S}$ *in the algebra* $\mathfrak{A}$ provided it is the least element in $\mathfrak{A}$ containing all the elements $A \in \mathfrak{S}$, i.e. provided the following two conditions are satisfied:

(J$_1$) $A \subset B$ for all $A \in \mathfrak{S}$;
(J$_2$) if $A \subset B'$ $(B' \in \mathfrak{A})$ for all $A \in \mathfrak{S}$, then $B \subset B'$.

By duality, an element $C \in \mathfrak{A}$ is said to be the *meet* of all elements $A \in \mathfrak{S}$ *in the algebra* $\mathfrak{A}$ provided it is the greatest element contained in all the elements $A \in \mathfrak{S}$, i.e. provided the following two conditions are satisfied:

(M$_1$) $C \subset A$ for all $A \in \mathfrak{S}$;
(M$_2$) if $C' \subset A$ $(C' \in \mathfrak{A})$ for all $A \in \mathfrak{S}$, then $C' \subset C$.

The join B of all $A \in \mathfrak{S}$ will be denoted (if it exists) by

(1) $$\bigcup\nolimits_{A \in \mathfrak{S}}^{\mathfrak{A}} A ,$$

the meet C of all $A \in \mathfrak{S}$ will be denoted (if it exists) by

(1') $$\bigcap\nolimits_{A \in \mathfrak{S}}^{\mathfrak{A}} A .$$

If $\mathfrak{S}$ is the set of all elements of an indexed set $\{A_t\}_{t \in T}$, then instead of (1) and (1') we write respectively

(2) $$\bigcup\nolimits_{t \in T}^{\mathfrak{A}} A_t$$

and

(2') $$\bigcap\nolimits_{t \in T}^{\mathfrak{A}} A_t .$$

Of course, if T is the set of all positive integers, we also write

(3) $$\bigcup\nolimits_{1 \leq n < \infty}^{\mathfrak{A}} A_n$$

and

(3') $$\bigcap\nolimits_{1 \leq n < \infty}^{\mathfrak{A}} A_n$$

instead of (2) and (2') respectively.

The superscript $\mathfrak{A}$ on U and $\mathsf{\cap}$ is, in general, necessary if we examine a subalgebra of $\mathfrak{A}$ simultaneously with the whole algebra $\mathfrak{A}$. In fact, suppose that $\mathfrak{B}$ is a subalgebra of $\mathfrak{A}$ and $\mathfrak{S}$ is a subset of $\mathfrak{B}$. Then, by definition,

(4) $$\mathsf{U}^{\mathfrak{A}}_{A\in\mathfrak{S}}A \subset \mathsf{U}^{\mathfrak{B}}_{A\in\mathfrak{S}}A$$

and

(4') $$\mathsf{\cap}^{\mathfrak{B}}_{A\in\mathfrak{S}}A \subset \mathsf{\cap}^{\mathfrak{A}}_{A\in\mathfrak{S}}A$$

provided these joins or meets exist. Indeed, $\mathsf{U}^{\mathfrak{A}}_{A\in\mathfrak{S}}A$ is the least element in $\mathfrak{A}$, containing all $A\in\mathfrak{S}$, and $\mathsf{U}^{\mathfrak{B}}_{A\in\mathfrak{S}}A$ is the least element in $\mathfrak{B}$, containing all $A\in\mathfrak{S}$. Since $\mathfrak{B}$ is a subset of $\mathfrak{A}$, the relation (4) holds. By duality we obtain (4').

If the set $\mathfrak{S}$ is finite, then the sign $\subset$ can be replaced in (4) and (4') by $=$ since both joins (meets) coincide with the join (meet) of elements of $\mathfrak{S}$ in the sense considered in the first chapter. If the set $\mathfrak{S}$ is infinite, then the sign $\subset$ cannot, in general, be replaced by $=$. This will be shown below (see A)).

However, if we are considering infinite joins and meets with respect to some fixed Boolean algebra $\mathfrak{A}$, then for simplicity we shall often omit the superscript $\mathfrak{A}$ in (1), (1'), (2), (2'), (3), (3').

Observe that if $\mathfrak{B}$ is a subalgebra of $\mathfrak{A}$, $\mathfrak{S}$ is a subset of $\mathfrak{B}$ and the join $\mathsf{U}^{\mathfrak{A}}_{A\in\mathfrak{S}}A$ exists and belongs to $\mathfrak{B}$, then this join is also the join of all $A\in\mathfrak{S}$ in the subalgebra $\mathfrak{B}$, i.e.

(5) $$\mathsf{U}^{\mathfrak{A}}_{A\in\mathfrak{S}}A = \mathsf{U}^{\mathfrak{B}}_{A\in\mathfrak{S}}A \ .$$

By duality, if $\mathfrak{S}$ is a subset of the subalgebra $\mathfrak{B}$ of $\mathfrak{A}$ and the meet $\mathsf{\cap}^{\mathfrak{A}}_{A\in\mathfrak{S}}A$ exists and belongs to $\mathfrak{B}$, then this meet is also the meet of all $A\in\mathfrak{S}$ in the subalgebra $\mathfrak{B}$, i.e.

(5') $$\mathsf{\cap}^{\mathfrak{A}}_{A\in\mathfrak{S}}A = \mathsf{\cap}^{\mathfrak{B}}_{A\in\mathfrak{S}}A \ .$$

The infinite join and meet are invariant under isomorphism *onto*, i.e. if h is an isomorphism of $\mathfrak{A}$ onto a Boolean algebra $\mathfrak{A}'$ and $\mathfrak{S}$ is a subset of $\mathfrak{A}$, then

(6) $$h\big(\mathsf{U}^{\mathfrak{A}}_{A\in\mathfrak{S}}A\big) = \mathsf{U}^{\mathfrak{A}'}_{A\in\mathfrak{S}}h(A) \ ,$$

(6') $$h\big(\mathsf{\cap}^{\mathfrak{A}}_{A\in\mathfrak{S}}A\big) = \mathsf{\cap}^{\mathfrak{A}'}_{A\in\mathfrak{S}}h(A) \ .$$

These equalities should be read as follows: if the join (meet) on one side exists, then the join (meet) on the second side also exists and the equality holds.

To prove (6) and (6') it suffices to recall that the isomorphisms h and h^{-1} preserve the inclusion $\subset$ (see § 5 (2)). Since the infinite joins and meets are defined only by means of $\subset$, h also preserves them.

However, infinite joins and meets are not preserved, in general, by homomorphisms and by isomorphisms *into*.

Let Δ be an ideal of a Boolean algebra $\mathfrak{A}$ and let $\mathfrak{A}' = \mathfrak{A}/\Delta$. An element $A' = [A] \in \mathfrak{A}'$ is the join of an indexed set of elements $A_t' = [A_t]$ $(t \in T)$ in the Boolean algebra $\mathfrak{A}'$ if and only if

(J$_1^*$) $A_t - A \in \Delta$ for every $t \in T$;
(J$_2^*$) if $A_t - A_0 \in \Delta$ $(A_0 \in \mathfrak{A})$ for every $t \in T$, then $A - A_0 \in \Delta$.

This follows immediately from (J$_1$), (J$_2$) and § 10 (9).

By duality, we see that an element $A' = [A] \in \mathfrak{A}/\Delta = \mathfrak{A}'$ is the meet of an indexed set of elements $A_t' = [A_t]$ $(t \in T)$ in the Boolean algebra $\mathfrak{A}'$ if and only if

(M$_1^*$) $A - A_t \in \Delta$ for every $t \in T$;
(M$_2^*$) if $A_0 - A_t \in \Delta$ $(A_0 \in \mathfrak{A})$ for every $t \in T$, then $A_0 - A \in \Delta$.

Examples. A) Let $\mathfrak{A}$ be the field of all subsets of the space X of all non-negative integers, and let $\mathfrak{B}$ be its subalgebra composed of all finite sets of positive integers and their complements in X. Let $A_n = (n)$ $(n = 1, 2, \ldots)$ be the one-point set containing only the number n. Then

$$\mathsf{U}_{1 \leq n < \infty}^{\mathfrak{A}} A_n = X - (0) \,,$$

i.e. the Boolean join of all A_n in the algebra $\mathfrak{A}$ coincides with the set-theoretical union. On the other hand

$$\mathsf{U}_{1 \leq n < \infty}^{\mathfrak{B}} A_n = X = V \,,$$

since X is the only element of $\mathfrak{B}$ which contains all A_n. Similarly, if $B_n = X - A_n$, then

$$\mathsf{\cap}_{1 \leq n < \infty}^{\mathfrak{A}} B_n = (0) \,,$$

i.e. this Boolean meet coincides with the set-theoretical intersection, but

$$\mathsf{\cap}_{1 \leq n < \infty}^{\mathfrak{B}} B_n = \Lambda \,.$$

This example shows that $\subset$ cannot be replaced by $=$ in (4) and (4'). It shows also that, in fields of sets, the set-theoretical union and intersection do not generally coincide with the (infinite) Boolean join and meet, respectively. However it is easy to see that if $\mathfrak{F}$ is a field of sets, $A_t \in \mathfrak{F}$ for every $t \in T$ and the set-theoretical union (intersection) of all A_t belongs to $\mathfrak{F}$, then it is also the Boolean join (meet) of all A_t in the Boolean algebra $\mathfrak{F}$.

B) If a field $\mathfrak{F}$ (of subsets of a space X) contains all one-point subsets of X, then infinite joins and meets always coincide with the set-theoretical unions and intersections, respectively. More exactly, if $A_t \in \mathfrak{F}$ for every $t \in T$, then the join $\mathsf{U}_{t \in T}^{\mathfrak{F}} A_t$ (the meet $\mathsf{\cap}_{t \in T}^{\mathfrak{F}} A_t$) exists if and only if the set-theoretical union (intersection) of all A_t belongs to $\mathfrak{F}$, and then they coincide.

In fact, suppose that the join $A = \bigcup_{t \in T}^{\mathfrak{F}} A_t$ exists. By (J_1) all A_t are subsets of A. If A is not the union of all A_t, then there exists a point $x_0 \in A$ such that $A_t \subset A - (x_0) \in \mathfrak{F}$ for every $t \in T$. By (J_2), this contradicts the hypothesis that A is the Boolean join of all A_t in $\mathfrak{F}$.

The converse statement was proved in Example A). By duality we obtain the proof for meets.

C) Let $\mathfrak{A}$ be any Boolean algebra. A join $A = \bigcup_{t \in T}^{\mathfrak{A}} A_t$ is said to be *essentially infinite* provided $A \neq A_{t_1} \cup \cdots \cup A_{t_n}$ for every finite sequence $t_1, \ldots, t_n \in T$.

If $\mathfrak{A}$ is the field of all open-closed subsets of a compact space X, then the join $A = \bigcup_{t \in T}^{\mathfrak{A}} A_t$ $(A_t \in \mathfrak{A})$ coincides with the set-theoretical union if and only if it is not essentially infinite, i.e. if $A = A_{t_1} \cup \cdots \cup A_{t_n}$ for some $t_1, \ldots, t_n \in T$. In fact, if the open-closed set A is the union of all A_t, then, by the compactness of X, it is the union of some sets $A_{t_1}, \ldots, A_{t_n}$ which yields $A = A_{t_1} \cup \cdots \cup A_{t_n}$. The converse statement is obvious.

By duality we obtain analogous remarks on meets.

D) Let $\mathfrak{F}$ be a field of subsets of a space X. Suppose that all at most enumerable subsets of X belong to $\mathfrak{F}$. Let $\varDelta$ be the ideal of all finite subsets of X, and $\mathfrak{A} = \mathfrak{F}/\varDelta$ [1]. No countable join $\bigcup_{1 \leq n < \infty}^{\mathfrak{A}} A_n$ is essentially infinite.

By (J_1) and (J_2) it suffices to prove that if

(7) $A, A_n \in \mathfrak{A}, \ A_n \subset A$ and $A_1 \cup \cdots \cup A_n \neq A$ for $n = 1, 2, \ldots,$

then there exists an element $A' \in \mathfrak{A}$ such that

$$A' \subset A, \ A' \neq A \quad \text{and} \quad A_n \subset A' \quad \text{for} \quad n = 1, 2, \ldots.$$

Let $A_n = [B_n]$, $A = [B]$ where $B_n, B \in \mathfrak{F}$. By (7) and § 10 (9), the sets $B - (B_1 \cup \cdots \cup B_n)$ are infinite and the sets $B_n - B$ are finite. We can choose a sequence $\{x_n\}$ of distinct elements such that

$$x_n \in B - (B_1 \cup \cdots \cup B_n) \quad \text{for} \quad n = 1, 2, \ldots.$$

Let $B' = B - (x_1, x_2, \ldots) \in \mathfrak{F}$ and $A' = [B']$. Since $B' \subset B$ and $B - B'$ is infinite we have $A' \subset A$ and $A' \neq A$. Since $B_n - B' \subset (B_n - B) \cup \cup (x_1, \ldots, x_n) \in \varDelta$, we have $A_n \subset A'$, q.e.d.

By duality we obtain an analogous statement for meets.

E) Let $\varDelta$ be an ideal of a Boolean algebra $\mathfrak{A}$ and let $\mathfrak{A}' = \mathfrak{A}/\varDelta$.

Observe that, in general, [ ] does not commute with infinite joins and meets $\bigcup_{t \in T}$ and $\bigcap_{t \in T}$, i.e. that the equalities

(8) $\bigcup_{t \in T}^{\mathfrak{A}'} [A_t] = \left[\bigcup_{t \in T}^{\mathfrak{A}} A_t \right],$

(8') $\bigcap_{t \in T}^{\mathfrak{A}'} [A_t] = \left[\bigcap_{t \in T}^{\mathfrak{A}} A_t \right]$

do not hold, in general, even if the joins or meets exist.

---

[1] For investigation of this Boolean algebra, see SIERPIŃSKI [3, 5].

In fact, let $\mathfrak{A}$ be the field of all subsets of an infinite set T, let Δ be the ideal of all finite subsets and let $A_t = (t)$, then

$$\bigcup\nolimits^{\mathfrak{A}'}_{t \in T}[A_t] = \Delta \neq V = \left[\bigcup\nolimits^{\mathfrak{A}}_{t \in T} A_t\right].$$

By passing to complements of A_t, we obtain a counterexample for meets.

F) The example below is given only for readers familiar with mathematical logic.

Let $\mathfrak{A}$ be the Lindenbaum-Tarski algebra of a formalized theory (see § 1 D)). The element (of $\mathfrak{A}$) determined by a formula α will be denoted by $|\alpha|$.

Let $\alpha(\tau)$ be a formula containing at least one free individual variable τ. One proves[1] that

$$\left|\bigcup\nolimits_\tau \alpha(\tau)\right| = \bigcup\nolimits^{\mathfrak{A}}_{t \in T} |\alpha(t)|, \quad \left|\bigcap\nolimits_\tau \alpha(\tau)\right| = \bigcap\nolimits^{\mathfrak{A}}_{t \in T}|\alpha(t)|$$

where $\bigcup_\tau$ and $\bigcap_\tau$ denote on the left side the logical quantifiers "there exists a τ such that ..." and "for every τ, ..." respectively, and T is the (infinite) set of all terms.

G) Let $\mathfrak{S}$ be the set of all atoms of a Boolean algebra $\mathfrak{A}$. The algebra $\mathfrak{A}$ is atomic if and only if $\bigcup_{A \in \mathfrak{S}} A = V$ (we assume here $\bigcup_{A \in \mathfrak{S}} A = \Delta$ if $\mathfrak{S}$ is empty).

§ 19. Algebraic properties of infinite joins and meets. $(\mathfrak{m}, \mathfrak{n})$-distributivity

Let $\mathfrak{A}$ be a Boolean algebra. All joins and meets below are taken in $\mathfrak{A}$.

It follows immediately from the definition that infinite joins and meets are commutative, i.e., for every one-to-one mapping τ of T onto T,

$$(1) \qquad \bigcup\nolimits_{t \in T} A_t = \bigcup\nolimits_{t \in T} A_{\tau(t)}, \quad \bigcap\nolimits_{t \in T} A_t = \bigcap\nolimits_{t \in T} A_{\tau(t)}.$$

Each of these equalities should be read as follows: if one of the joins (meets) exists, so does the second one, and the equality holds.

It is easy to verify that infinite joins and meets are also associative, i.e. if the set T is the union of sets T_s $(s \in S)$, then

$$(2) \qquad \bigcup\nolimits_{s \in S} \bigcup\nolimits_{t \in T_s} A_t = \bigcup\nolimits_{t \in T} A_t, \quad \bigcap\nolimits_{s \in S} \bigcap\nolimits_{t \in T_s} A_t = \bigcap\nolimits_{t \in T} A_t.$$

Each of these equalities should be read as follows: if the joins (meets) on the left side exist, then the join (meet) on the right side also exists and the equality holds.

The *de Morgan laws* are also true for infinite joins and meets:

$$(3) \qquad \bigcup\nolimits_{t \in T} -A_t = -\bigcap\nolimits_{t \in T} A_t, \quad \bigcap\nolimits_{t \in T} -A_t = -\bigcup\nolimits_{t \in T} A_t.$$

---

[1] See e.g. Rasiowa and Sikorski [1, 7]. See also § 40. To perform the substitution $\alpha(t)$, a change of bound variables is sometimes necessary.

The existence of the join (or meet) on one side of either of these equalities implies the existence of the join (meet) on the second side and then the equality holds.

The de Morgan laws (3) follow easily from the definition of infinite joins and meets and from § 2 (21).

If, for every $t \in T$, there is an $s \in S$ such that $A_t \subset B_s$, then

$$(4) \qquad\qquad \mathsf{U}_{t \in T} A_t \subset \mathsf{U}_{s \in S} B_s$$

whenever both joins exist. Similarly, if, for every $t \in T$, there exists an $s \in S$ such that $B_s \subset A_t$, then

$$(4') \qquad\qquad \mathsf{\cap}_{s \in S} B_s \subset \mathsf{\cap}_{t \in T} A_t$$

whenever both meets exist.

In particular we have

$$(5) \quad \mathsf{U}_{t \in T'} A_t \subset \mathsf{U}_{t \in T} A_t \quad \text{and} \quad \mathsf{\cap}_{t \in T} A_t \subset \mathsf{\cap}_{t \in T'} A_t \quad \text{for} \quad T' \subset T,$$

$$(6) \qquad A_{t_0} \subset \mathsf{U}_{t \in T} A_t \quad \text{and} \quad \mathsf{\cap}_{t \in T} A_t \subset A_{t_0} \quad \text{for} \quad t_0 \in T,$$

$$(7) \qquad\quad \mathsf{U}_{t \in T} A_t \subset A \quad \text{if} \quad A_t \subset A \quad \text{for every} \quad t \in T,$$

$$(7') \qquad\quad A \subset \mathsf{\cap}_{t \in T} A_t \quad \text{if} \quad A \subset A_t \quad \text{for every} \quad t \in T,$$

provided the joins or meets in question exist.

We have also

$$(8) \qquad \begin{aligned} A \cup (\mathsf{U}_{t \in T} A_t) &= \mathsf{U}_{t \in T} (A \cup A_t), \\ A \cap (\mathsf{\cap}_{t \in T} A_t) &= \mathsf{\cap}_{t \in T} (A \cap A_t). \end{aligned}$$

The existence of the join (meet) on the left side implies the existence of the join (meet) on the right side.

It is a little more difficult to verify that the following *distributive laws* hold

$$(9) \qquad \begin{aligned} A \cap (\mathsf{U}_{t \in T} A_t) &= \mathsf{U}_{t \in T} (A \cap A_t), \\ A \cup (\mathsf{\cap}_{t \in T} A_t) &= \mathsf{\cap}_{t \in T} (A \cup A_t). \end{aligned}$$

Each of these laws should be read: if the infinite join (meet) on the left side exists, so does the infinite join (meet) on the right side and the equality holds.

To prove the first law (9), suppose that $B = \mathsf{U}_{t \in T} A_t$. We have

$$A \cap A_t \subset A \cap B \quad \text{for every} \quad t \in T$$

since $A_t \subset B$. Suppose now that C is an element such that

$$A \cap A_t \subset C \quad \text{for every} \quad t \in T,$$

that is

$$A - C \subset -A_t \quad \text{for every} \quad t \in T.$$

Hence

$$A - C \subset \mathsf{\cap}_{t \in T} -A_t,$$

i.e., by the de Morgan law,

$$A - C \subset - B,$$

which implies

$$A \cap B \subset C.$$

This proves that $A \cap B$ is the join of all $A \cap A_t$.

By duality we obtain the proof of the second distributive law.

The set-theoretical union and intersection also satisfy the following distributive laws:

(10)
$$\bigcap_{t \in T} \bigcup_{s \in S} A_{t,s} = \bigcup_{\varphi \in S^T} \bigcap_{t \in T} A_{t, \varphi(t)},$$

(10')
$$\bigcup_{t \in T} \bigcap_{s \in S} A_{t,s} = \bigcap_{\varphi \in S^T} \bigcup_{t \in T} A_{t, \varphi(t)},$$

where S^T denotes the set of all mappings of T into S (see p. 2). These laws generally do not hold in Boolean algebras, even under the hypothesis that all the infinite joins and meets in question exist, and even in the case where S is finite[1].

Example. A) Let $\mathfrak{F}$ be the field of all Borel sets of real numbers. Let Δ be the ideal of all at most enumerable sets, and let $\mathfrak{A} = \mathfrak{F}/\Delta$. Let $B_{n,i}$ be the set of all numbers x of the form

$$x = \sum_{j=1}^{\infty} \frac{a_j + 1}{2^{j+1}}$$

where $a_j = -1$ or 1 for $j \neq n$, and $a_n = i$ $(i = \pm 1, n = 1, 2, \ldots)$. The sets $B_{n,i}$ belong to $\mathfrak{F}$ since they are unions of finite numbers of closed subintervals of the closed unit interval U. Let $A_{n,i} = [B_{n,i}]$. We have

$$\bigcap_{1 \leq n < \infty} (A_{n,-1} \cup A_{n,1}) = [U] \neq \wedge = \bigcup_{\varphi \in \Phi} \bigcap_{1 \leq n < \infty} A_{n, \varphi(n)}$$

where Φ denotes the set of all mappings φ of the set T of all positive integers into the set S composed of the numbers -1 and 1 only.

In fact, $A_{n,-1} \cup A_{n,1} = [U]$, and therefore the meet of all these elements is equal to $[U] \neq \wedge$. On the other hand,

(11)
$$\bigcap_{1 \leq n < \infty} A_{n, \varphi(n)} = \wedge$$

for every $\varphi \in \Phi$. In fact, if $[B] \subset A_{n, \varphi(n)} = [B_{n, \varphi(n)}]$ for $n = 1, 2, \ldots$, then all the sets $C_n = B - B_{n, \varphi(n)}$ belong to Δ. Since $B \subset B_{n, \varphi(n)} \cup C_n$, we have $B \subset \bigcap_{1 \leq n < \infty} (B_{n, \varphi(n)} \cup C_n) \subset (\bigcap_{1 \leq n < \infty} B_{n, \varphi(n)}) \cup (\bigcup_{1 \leq n < \infty} C_n) \in \Delta$ since the set $\bigcap_{1 \leq n < \infty} B_{n, \varphi(n)}$ contains only one element, viz. the number

$$\sum_{n=1}^{\infty} \frac{\varphi(n) + 1}{2^{n+1}}.$$

---

[1] For a study of infinite distributivity in Boolean algebras, see CHIN and TARSKI [1], CHRISTENSEN and PIERCE [1], ENOMOTO [1], KERSTAN [1], KOWALSKY [1], MATTHES [1, 2], PIERCE [3, 5, 6], SCOTT [1], SIKORSKI [25, 27], SIKORSKI and TRACZYK [2], SMITH [1], SMITH and TARSKI [1]. See also §§ 20, 21, 24, 25, 29, 34, 35, 36, 38.

Hence $B \in \varDelta$, i.e. $[B] = \wedge$. This proves (11)[1].

Notice that the above argument remains valid if the ideal $\varDelta$ of all enumerable subsets is replaced e.g. by the ideal of all Borel sets of the first category, or by the ideal of all Borel sets of Lebesgue measure zero.

A Boolean algebra $\mathfrak{A}$ is said to be $(\mathfrak{m}, \mathfrak{n})$-*distributive* if the equality (10) holds for every $(\mathfrak{m}, \mathfrak{n})$-indexed set $\{A_{t,s}\}_{t \in T, s \in S}$ of elements in $\mathfrak{A}$ such that all the meets $\bigcap_{t \in T} A_{t, \varphi(t)}$ exist and

(12)   all the joins $\bigcup_{s \in S} A_{t,s}$ and the meet $\bigcap_{t \in T} \bigcup_{s \in S} A_{t,s}$ exist.

By the de Morgan laws, $\mathfrak{A}$ is $(\mathfrak{m}, \mathfrak{n})$-distributive if and only if (10′) holds for every $(\mathfrak{m}, \mathfrak{n})$-indexed set $\{A_{t,s}\}_{t \in T, s \in S}$ such that all the joins $\bigcup_{t \in T} A_{t, \varphi(t)}$ exist and

(12′)   all the meets $\bigcap_{s \in S} A_{t,s}$ and the join $\bigcup_{t \in T} \bigcap_{s \in S} A_{t,s}$ exist.

A Boolean algebra is said to be $\mathfrak{m}$-*distributive* if it is $(\mathfrak{m}, \mathfrak{m})$-distributive. It is said to be *completely distributive* if it is $\mathfrak{m}$-distributive for every $\mathfrak{m}$.

If $\mathfrak{m}$ is finite, then every Boolean algebra is $(\mathfrak{m}, \mathfrak{n})$-distributive by (9). Also every Boolean algebra is $(\mathfrak{m}, 1)$-distributive. To exclude these trivial cases, we shall always assume in the sequel that $\mathfrak{m}$ is infinite and $\mathfrak{n} \geq 2$.

It follows immediately from the definition that if $\mathfrak{A}$ is $(\mathfrak{m}, \mathfrak{n})$-distributive and $\mathfrak{m}' \leq \mathfrak{m}$, $\mathfrak{n}' \leq \mathfrak{n}$, then $\mathfrak{A}$ is $(\mathfrak{m}', \mathfrak{n}')$-distributive. Thus a completely distributive algebra is $(\mathfrak{m}, \mathfrak{n})$-distributive for every $\mathfrak{m}$ and $\mathfrak{n}$.

Let H denote the set composed of numbers 1 and -1.

19.1[2]. *For every Boolean algebra $\mathfrak{A}$, the following conditions are equivalent:*

   (i) $\mathfrak{A}$ *is* $\mathfrak{m}$-*distributive;*
   (ii) $\mathfrak{A}$ *is* $(\mathfrak{m}, 2)$-*distributive;*
   (iii) *if* $\{A_t\}_{t \in T}$ *is an* $\mathfrak{m}$-*indexed set of elements in* $\mathfrak{A}$, *then for every element* $A \neq \wedge$ ($A \in \mathfrak{A}$) *there exists a function* $\varepsilon \in H^T$ *such that*

(13) $A \cap \bigcap_{t \in T} \varepsilon(t) \cdot A_t \neq \wedge$.

Condition (13) should be understood as follows: either the meet of all the elements A, $\varepsilon(t) \cdot A_t$ $(t \in T)$ does not exist or it exists but is not equal to $\wedge$. Note that if all the infinite meets $\bigcap_{t \in T} \varepsilon(t) \cdot A_t$ exist, then the existence of $\varepsilon \in H^T$ for every $A \neq \wedge$ with the property (13) is equivalent to

(14) $\bigcup_{\varepsilon \in H^T}^{\mathfrak{A}} \bigcap_{t \in T} \varepsilon(t) \cdot A_t = \vee$,

by (9).

---

[1] Identity (11) follows also easily from theorem 21.1.
[2] Pierce [3], Smith and Tarski [1].

(i) implies (ii) trivially.

(ii) implies (iii). Suppose that an element $A \in \mathfrak{A}$ has the property that
$$A \cap \bigcap_{t \in T} \varepsilon(t) \cdot A_t = \wedge \quad \text{for every} \quad \varepsilon \in H^T,$$
(i.e. the meet of all the elements A, $\varepsilon(t) \cdot A_t$ is always equal to $\wedge$).

Let t_0 be an element which does not belong to T, let $T' = T \cup (t_0)$, $S = H$ and let
$$A_{t,s} = s \cdot A_t \quad \text{for} \quad t \in T, \ s \in S,$$
$$A_{t_0,s} = A \quad \text{for} \quad s \in S.$$
Then applying (10) to the $(\mathfrak{m}, 2)$-indexed set $\{A_{t,s}\}_{t \in T', s \in S}$ we get
$$A = \bigcap_{t \in T'} \bigcup_{s \in S} A_{t,s} = \wedge.$$

(iii) implies (i). Suppose $\mathfrak{A}$ is not $\mathfrak{m}$-distributive, i.e. there exists an $\mathfrak{m}$-indexed set $\{A_{t,s}\}_{t \in T, s \in S}$ satisfying all the conditions mentioned in the definition of distributivity and such that (10) does not hold. Thus there exists an element $A \neq \wedge$ such that

(15) $$A \subset \bigcup_{s \in S} A_{t,s} \quad \text{for every} \quad t \in T$$
and
(16) $$A \cap \bigcap_{t \in T} A_{t, \varphi(t)} = \wedge \quad \text{for every} \quad \varphi \in S^T.$$
We shall prove that for every $\varepsilon \in H^{T \times S}$,
(17) $$A \cap \bigcap_{(t,s) \in T \times S} \varepsilon(t, s) \cdot A_{t,s} = \wedge.$$
In fact, if there exists a $t \in T$ such that $\varepsilon(t, s) = -1$ for all $s \in S$, then (17) follows from (15). If for every $t \in T$ there exists an $s = \varphi(t)$ such that $\varepsilon(t, s) = 1$, then (17) follows from (16), since every element appearing on the left side of (16) appears also on the left hand side of (17). Property (17) just proved shows that the $\mathfrak{m}$-indexed set $\{A_{t,s}\}_{(t,s) \in T \times S}$ does not have property (iii).

By 19.1 it suffices to examine $(\mathfrak{m}, \mathfrak{n})$-distributivity only for $\mathfrak{n} \geq \mathfrak{m} \geq \aleph_0$. Therefore in all theorems concerning $(\mathfrak{m}, \mathfrak{n})$-distributivity we shall always suppose that $\mathfrak{m}$ and $\mathfrak{n}$ are infinite cardinals.

19.2. *The following three conditions are equivalent for any Boolean algebra* $\mathfrak{A}$:

(d) $\mathfrak{A}$ *is* $(\mathfrak{m}, \mathfrak{n})$-*distributive;*

(d$_1$) *if an* $(\mathfrak{m}, \mathfrak{n})$-*indexed set* $\{A_{t,s}\}_{t \in T, s \in S}$ *of elements in* $\mathfrak{A}$ *satisfies* (12) *and*
(18) $$\bigcap_{t \in T} \bigcup_{s \in S} A_{t,s} \neq \wedge,$$
then there exists a mapping $\varphi \in S^T$ *such that*
(19) $$\bigcap_{t \in T} A_{t, \varphi(t)} \neq \wedge;$$

(d$_2$) *if an* $(\mathfrak{m}, \mathfrak{n})$-*indexed set* $\{A_{t,s}\}_{t \in T, s \in S}$ *of elements in* $\mathfrak{A}$ *satisfies*
(20) $$\bigcup_{s \in S} A_{t,s} = \vee \quad \text{for every} \quad t \in T,$$
then, for every $A \neq \wedge$ $(A \in \mathfrak{A})$, *there exists a mapping* $\varphi \in S^T$ *such that*
(21) $$A \cap \bigcap_{t \in T} A_{t, \varphi(t)} \neq \wedge.$$

Condition (19) should be read as follows: either the meet $\bigcap_{t \in T} A_{t, \varphi(t)}$ does not exist or it exists but is not equal to $\wedge$. Condition (21) should be interpreted in the same way: either the meet of A and all the elements $A_{t, \varphi(t)}$ $(t \in T)$ does not exist, or it exists but is not equal to $\wedge$. Note that if all the meets $\bigcap_{t \in T} A_{t, \varphi(t)}$ exist, then the conclusion in (d_2) is equivalent to the identity

(21') $$\bigcup_{\varphi \in S^T} \bigcap_{t \in T} A_{t, \varphi(t)} = V \; .$$

This follows easily from (9).

(d) implies (d_1) since (10) and (18) imply (19) for some $\varphi \in S^T$.

To deduce (d_2) from (d_1) it suffices to augment the set T with a new element t_0, to assume $A_{t_0, s} = A$ for all $s \in S$, and to apply (d_1) for $\{A_{t, s}\}_{t \in T \cup (t_0), s \in S}$ under the hypothesis that (20) holds.

(d_2) implies (d). In fact, let an $(\mathfrak{m}, \mathfrak{n})$-indexed set $\{A_{t, s}\}_{t \in T, s \in S}$ satisfy (12) and $B = \bigcap_{t \in T} \bigcup_{s \in S} A_{t, s}$. Suppose that (10) does not hold, i.e. there exists an element $A \neq \wedge$ such that

(22) $A \subset B$ and $\bigcap_{t \in T} A_{t, \varphi(t)} \subset B - A$ for every $\varphi \in S^T$.

Augment the set S with a new element s_0 and denote

$$B_{t, s_0} = - B \quad \text{for every} \quad t \in T \; ,$$
$$B_{t, s} = B \cap A_{t, s} \quad \text{for every} \quad t \in T \quad \text{and every} \quad s \in S \; .$$

The $(\mathfrak{m}, \mathfrak{n})$-indexed set $\{B_{t, s}\}_{t \in T, s \in S \cup (s_0)}$ satisfies (20). Applying (d_2) to this indexed set, we infer that there exists a $\varphi \in S^T$ such that $A \cap \bigcap_{t \in T} A_{t, \varphi(t)} \neq \wedge$, contradicting (22).

The definition of $(\mathfrak{m}, \mathfrak{n})$-distributivity on p. 62 is a little complicated since the existence of various infinite joins and meets has to be postulated. Theorem 19.3 below suggests another equivalent definition which avoids this difficulty. In order to formulate it, let us define an indexed (an $\mathfrak{n}$-indexed) set $\{A_s\}_{s \in S}$ to be a *covering* (an $\mathfrak{n}$-*covering*) of a Boolean algebra $\mathfrak{A}$ if $\bigcup_{s \in S} A_s = V_{\mathfrak{A}}$. A covering $\{B_r\}_{r \in R}$ is said to be a *refinement* of a covering $\{A_s\}_{s \in S}$ if, for every $r \in R$, there is an $s \in S$ such that $B_r \subset A_s$.

19.3. *The following conditions are equivalent for every Boolean algebra* $\mathfrak{A}$:

(d) $\mathfrak{A}$ *is* $(\mathfrak{m}, \mathfrak{n})$-*distributive;*

(d_3) *every set of at most* $\mathfrak{m}$ $\mathfrak{n}$-*coverings has a common refinement*[1].

It suffices to prove that (d_3) is equivalent to condition (d_2) of theorem 19.2.

(d_2) implies (d_3). Indeed, suppose that

(23)   for every $t \in T$ $(\overline{\overline{T}} \leq \mathfrak{m})$, $\{A_{t, s}\}_{s \in S}$ is an $\mathfrak{n}$-covering of $\mathfrak{A}$;

i.e. (20) holds. Let $\{B_r\}_{r \in R}$ be the indexed set formed of all the elements B in $\mathfrak{A}$ with the property:

(24)   there exists a $\varphi \in S^T$ such that $B \subset A_{t, \varphi(t)}$ for every $t \in T$.

---
[1] PIERCE [3].

If (d_2) holds, then $\{B_r\}_{r \in R}$ is a covering of $\mathfrak{A}$. For if not, then there would exist an element $A \neq \wedge$ such that

$$(25) \qquad\qquad A \cap B_r = \wedge \quad \text{for every} \quad r \in R .$$

By (d_2), there would then exist a mapping $\varphi \in S^T$ such that (21) holds; i.e. there would exist an element $B \neq \wedge$ such that $B \subset A$ and $B \subset A_{t, \varphi(t)}$ for every $t \in T$. The last property implies that B is one of the elements B_r, which is impossible by (25). By (24) the covering $\{B_r\}_{r \in R}$ is a refinement of each of the n-coverings $\{A_{t, s}\}_{s \in S}$, $t \in T$.

(d_3) implies (d_2). Suppose that an (m, n)-indexed set $\{A_{t, s}\}_{t \in T, s \in S}$ satisfies (20), i.e. that (23) holds. Let a covering $\{B_r\}_{r \in R}$ be a common refinement of each of the n-coverings $\{A_{t, s}\}_{s \in S}$. Thus if $A \neq \wedge$, then there exists an $r \in R$ such that $A \cap B_r \neq \wedge$, and for every $t \in T$ there is an $s = \varphi(t) \in S$ such that $B_r \subset A_{t, \varphi(t)}$. Since the non-zero element $A \cap B_r$ is a subelement of all the elements A, $A_{t, \varphi(t)}$ $(t \in T)$, inequality (21) holds.

§ 20. m-complete Boolean algebras

Let $\mathfrak{A}$ be a Boolean algebra and let m be an infinite cardinal. By the de Morgan laws (see § 19 (3)), the following two conditions are equivalent:

(a) for every m-indexed set $\{A_t\}_{t \in T}$ (of elements of $\mathfrak{A}$) the join $\bigcup_{t \in T} A_t$ exists in $\mathfrak{A}$;

(a') for every m-indexed set $\{A_t\}_{t \in T}$ (of elements of $\mathfrak{A}$) the meet $\bigcap_{t \in T} A_t$ exists in $\mathfrak{A}$.

If one of the conditions mentioned above is satisfied (or, what is the same, if both of the conditions are satisfied), then $\mathfrak{A}$ is said to be an m-*complete Boolean algebra* or a *Boolean* m-*algebra*. If $\mathfrak{A}$ is an m-complete Boolean algebra for every m, then $\mathfrak{A}$ is said to be a *complete Boolean algebra*.

Of course, every Boolean algebra isomorphic to an m-complete (complete) Boolean algebra is also an m-complete (complete) Boolean algebra.

Let $\mathfrak{F}$ be a field of subsets of a space X. By the de Morgan laws for sets, the following two conditions are equivalent:

(b) for every m-indexed set $\{A_t\}_{t \in T}$ of sets in $\mathfrak{F}$, the set-theoretical union of all A_t $(t \in T)$ belongs to $\mathfrak{F}$;

(b') for every m-indexed set $\{A_t\}_{t \in T}$ of sets in $\mathfrak{F}$, the set-theoretical intersection of all A_t $(t \in T)$ belongs to $\mathfrak{F}$.

If one (or both) of these conditions is satisfied, then $\mathfrak{F}$ is said to be an m-*complete field of sets*, or an m-*field of sets*. If $\mathfrak{F}$ is an m-complete field of sets for every m, then $\mathfrak{F}$ is said to be a *complete field of sets*.

Examples. A) The class of all subsets A of a space X such that either $\bar{\bar{A}} \leq \mathfrak{m}$ or $\overline{\overline{X - A}} \leq \mathfrak{m}$ is an $\mathfrak{m}$-field of subsets of X. The class of all subsets of any space X is a complete field of sets.

B) It follows from the remark at the end of § 18 A) that every $\mathfrak{m}$-complete (complete) field of sets is an $\mathfrak{m}$-complete (complete) Boolean algebra and the Boolean joins $\bigcup_{t \in T} A_t$ and meets $\bigcap_{t \in T} A_t$ coincide then with the set-theoretical unions and intersections respectively provided $\bar{\bar{T}} \leq \mathfrak{m}$. The converse statement is, in general, not true; a field of sets $\mathfrak{F}$ can be an $\mathfrak{m}$-complete Boolean algebra, but fail to be an $\mathfrak{m}$-complete field of sets.

For instance, let h be the isomorphism of the field $\mathfrak{A}$ of all subsets of an infinite space X onto the field $\mathfrak{F}$ of all open-closed subsets of the Stone space of $\mathfrak{A}$. Then $\mathfrak{F}$ is a complete Boolean algebra since it is isomorphic to the complete Boolean algebra $\mathfrak{A}$. However $\mathfrak{F}$ is not a σ-field[1] of sets (and consequently, $\mathfrak{F}$ is not an $\mathfrak{m}$-complete field for any infinite cardinal $\mathfrak{m}$). In fact, if $\{A_n\}$ is an infinite sequence of non-empty disjoint subsets of X, then the set-theoretical union of all $h(A_n)$ $(n = 1, 2, \ldots)$ does not belong to $\mathfrak{F}$ since it is not closed (see also § 7 J)).

C) The Boolean algebra $\mathfrak{A}_1$ of all regular closed subsets of a topological space X (see § 1 B)) is a complete Boolean algebra (which, in general, is not a complete field of sets)[2].

In fact, it is easy to verify that, for every indexed set $\{A_t\}$ of regular closed subsets of X, the closure of the set-theoretical union of all A_t is the Boolean join of all A_t in $\mathfrak{A}_1$; the closure of the interior of the set-theoretical intersection of all A_t is the Boolean meet of all A_t in $\mathfrak{A}_1$.

In a similar way we can prove that the Boolean algebra $\mathfrak{A}_2$ of all regular open subsets of X (see § 1 B)) is a complete Boolean algebra[2]. For any indexed set $\{A_t\}$ of regular open subsets, the interior of the closure of the set-theoretical union of all A_t is the Boolean join of all A_t in $\mathfrak{A}_2$; the interior of the set-theoretical intersection of all A_t is the Boolean meet of all A_t in $\mathfrak{A}_2$.

D) If $\mathfrak{A}$ is an $\mathfrak{m}$-complete (complete) Boolean algebra and $E \in \mathfrak{A}$, then $\mathfrak{A}|E$ (see § 10 A)) is also an $\mathfrak{m}$-complete (complete) Boolean algebra. If $\mathfrak{F}$ is an $\mathfrak{m}$-complete (complete) field of subsets of a space X and E is a subset of X, then $\mathfrak{F}|E$ (see § 10 B)) is also an $\mathfrak{m}$-complete (complete) field of sets.

E) Every infinite Boolean σ-algebra $\mathfrak{A}$ contains a subalgebra isomorphic to the field $\mathfrak{F}$ of all subsets of an enumerable set X, and consequently it has power $\geq 2^{\aleph_0}$.

---

[1] The letter σ denotes $\aleph_0$. See Terminology and notation, p. 1.
[2] See TARSKI [3], BIRKHOFF [2] and MACNEILLE [1].

In fact, if $\mathfrak{A}$ is infinite, then $\mathfrak{A}$ contains an indexed set $\{A_x\}_{x \in X}$ of disjoint elements such that $\bigcup_{x \in X} A_x = V$. The mapping which associates the element $\bigcup_{x \in B} A_x \in \mathfrak{A}$, to each set $B \subset X$, is an isomorphism from $\mathfrak{F}$ into $\mathfrak{A}$.

Thus the Stone space X_0 of $\mathfrak{F}$ (i.e. the Čech-Stone compactification $\beta(X)$ of the discrete space X — see § 8 F)) is a continuous image of the Stone space of any infinite Boolean σ-algebra $\mathfrak{A}$. Since X_0 is not a continuous image of the Cantor space $\mathscr{D}_\mathfrak{n}$ for any cardinal $\mathfrak{n}$ (see p. 47), the Stone space of $\mathfrak{A}$ is not a continuous image of $\mathscr{D}_\mathfrak{n}$.[1] Hence it follows that no infinite Boolean σ-algebra $\mathfrak{A}$ is isomorphic to a subalgebra of the free Boolean algebra $\mathfrak{F}_{0,\mathfrak{n}}$, $\mathfrak{n}$ being any cardinal.

F) Let X be a topological space. We recall that a set $A \subset X$ is said to have the *Baire property*[2] if there exists an open set G such that $A - G$ and $G - A$ are of the first category. In other words, A has the Baire property if

$$A = (G - N_1) \cup N_2$$

where G is open, and N_1, N_2 are of the first category.

If A has the Baire property, so has its complement. In fact, let $G_0 = -\mathbf{C}G$ be the complement of the closure of G. Then

$$(-A) - G_0 = \mathbf{C}G - A \subset (\mathbf{C}G - G) \cup (G - A) \, ,$$
$$G_0 - (-A) = A - \mathbf{C}G \subset A - G \, .$$

Hence it follows that $(-A) - G_0$ and $G_0 - (-A)$ are of the first category.

If the sets A_n $(n = 1, 2, \ldots)$ have the Baire property, so has their union A. In fact, let G_n be an open set such that $A_n - G_n$ and $G_n - A_n$ are sets of the first category, $n = 1, 2, \ldots$. Let G be the union of the sets G_n. Since

$$A - G \subset \bigcup_{1 \le n < \infty} (A_n - G_n) \, , \quad G - A \subset \bigcup_{1 \le n < \infty} (G_n - A_n) \, ,$$

the sets $A - G$ and $G - A$ are of the first category.

Thus the class of all sets having the Baire property is a σ-field of subsets of X. This σ-field contains all open sets. By definition, the class of all Borel sets is the smallest σ-field containing all open sets. Hence it follows that every Borel set has the Baire property.

G) Let $\mathfrak{F}$ and $\mathfrak{F}'$ be fields of subsets of spaces X and X' respectively. Let $\mathfrak{F}_1$ and $\mathfrak{F}_1'$ be the smallest complete fields containing $\mathfrak{F}$ and $\mathfrak{F}'$ respectively.

$\mathfrak{F}$ and $\mathfrak{F}'$ are said to be *strongly isomorphic* provided there exists an isomorphism h_1 of $\mathfrak{F}_1$ onto $\mathfrak{F}_1'$ such that $h_1(\mathfrak{F}) = \mathfrak{F}'$. The mapping $h = h_1 | \mathfrak{F}$ is then called a *strong isomorphism*[3] *of* $\mathfrak{F}$ *onto* $\mathfrak{F}'$.

Not every isomorphism between fields of sets is strong. For instance, if X is a totally disconnected compact space, X' is a dense

[1] ENGELKING and PEŁCZYŃSKI [1].
[2] See e.g. KURATOWSKI [3], p. 54—56.
[3] MARCZEWSKI [3].

subset of X, $\mathfrak{F}$ is the field of all open-closed subsets of X, and $\mathfrak{F}' = \mathfrak{F} \mid X'$ is the field of all sets $A \cap X'$ where $A \in \mathfrak{F}$, then

$$h(A) = A \cap X' \quad \text{for} \quad A \in \mathfrak{F}$$

is an isomorphism of $\mathfrak{F}$ onto $\mathfrak{F}'$ but, if $X' \neq X$, h is not a strong isomorphism.

Observe that an isomorphism h of a field $\mathfrak{F}$ onto a field $\mathfrak{F}'$ is strong if and only if both h and h^{-1} are induced by point mappings (see § 11).

Fields $\mathfrak{F}, \mathfrak{F}'$ of subsets of X, X' respectively are said to be *equivalent*[1] if there exists a one-to-one mapping ψ of X onto X' such that the mapping h defined by the formula

$$h(A) = \psi(A) \quad \text{for} \quad A \in \mathfrak{F}$$

is an isomorphism of $\mathfrak{F}$ onto $\mathfrak{F}'$.

Observe that two reduced fields are equivalent if and only if they are strongly isomorphic.

In the general case, two fields $\mathfrak{F}, \mathfrak{F}'$ are strongly isomorphic if and only if the reduced fields $\mathfrak{F}_0, \mathfrak{F}_0'$ obtained from $\mathfrak{F}$ and $\mathfrak{F}'$ respectively by identification of non-separated points (see § 7, p. 20) are equivalent.

H) The direct union of any indexed set of Boolean $\mathfrak{m}$-algebras (of $\mathfrak{m}$-fields of subsets of disjoint spaces) is a Boolean $\mathfrak{m}$-algebra (an $\mathfrak{m}$-field of subsets of the union of those spaces). The same remark is true for complete Boolean algebras (complete fields of sets).

I) Every $\mathfrak{m}$-field of sets is $\mathfrak{m}$-distributive. In fact, the equality § 19 (10) follows easily from the remark at the end of § 18 A). Another proof can be obtained by verification that condition (d_1) [or (d_2)] in 19.2 is satisfied.

Every complete field of sets is completely distributive.

We shall now prove the following simple criterion for $\mathfrak{m}$-completeness of Boolean algebras.

20.1. *If the join* $\bigcup_{t \in T} B_t$ *exists for any* $\mathfrak{m}$-*indexed set* $\{B_t\}_{t \in T}$ *of disjoint elements of a Boolean algebra* $\mathfrak{A}$, *then* $\mathfrak{A}$ *is* $\mathfrak{m}$-*complete*[2].

The proof is by induction on the cardinal $\mathfrak{m}$. Suppose that theorem 20.1 is true for all cardinals $\mathfrak{m}' < \mathfrak{m}$ and that the join of any $\mathfrak{m}$-indexed set of disjoint elements exists in $\mathfrak{A}$. Thus $\mathfrak{A}$ is an $\mathfrak{m}'$-complete Boolean algebra for every infinite cardinal $\mathfrak{m}' < \mathfrak{m}$.

Let $\{A_t\}_{t \in T}$ be any indexed set of elements in $\mathfrak{A}$, $\overline{\overline{T}} = \mathfrak{m}$. It is convenient to assume here that T is the set of all ordinals $t < \alpha$ where α is the least ordinal of power $\mathfrak{m}$. Let

(1) $\qquad B_0 = A_0 \quad \text{and} \quad B_t = A_t - \bigcup_{t' < t} A_{t'} \quad \text{for} \quad 0 < t < \alpha .$

---

[1] MARCZEWSKI [3, 10].

[2] SMITH and TARSKI [1].

We have

$$\bigcup_{t' < t} A_{t'} = \bigcup_{t' < t} B_{t'} \quad \text{for} \quad 0 < t \leq \alpha.$$

The proof is by transfinite induction. It is easy to verify that if this equality holds for an ordinal t, then it also holds for $t + 1$. Suppose that t is a limit ordinal and that the equality holds for all ordinals less than t. Then, by the induction hypothesis, every $A_{t'}$ is a subelement of $\bigcup_{t'' \leq t'} B_{t''}$. This proves that $\bigcup_{t' < t} A_{t'} \subset \bigcup_{t' < t} B_{t'}$. The converse inclusion is also true since $B_{t'} \subset A_{t'}$ for every t'.

Since $\bigcup_{t < \alpha} B_t$ exists, we infer that $\bigcup_{t \in T} A_t$ exists. This proves that $\mathfrak{A}$ is m-complete.

Notice that incidentally we proved the following theorem.

20.2. *If the Boolean algebra $\mathfrak{A}$ is m'-complete for every $\mathfrak{m}' < \mathfrak{m}$, then for every m-indexed set $\{A_t\}_{t \in T}$ of elements in $\mathfrak{A}$ there exists an m-indexed set $\{B_t\}_{t \in T}$ of disjoint elements of $\mathfrak{A}$ such that*

$$B_t \subset A_t \quad \text{for every} \quad t \in T$$

and

$$\bigcup_{t \in T} B_t = \bigcup_{t \in T} A_t.$$

The hypothesis that a Boolean algebra $\mathfrak{A}$ under consideration is m-complete is very useful in investigations of infinite joins and meets of at most m-elements because it is not necessary to postulate additionally that the joins and meets exist.

For instance, if $\mathfrak{m} \leq \mathfrak{n}$, the hypothesis that $\mathfrak{A}$ is n-complete facilitates remarkably the definition of $(\mathfrak{m}, \mathfrak{n})$-distributivity on p. 62: $\mathfrak{A}$ is $(\mathfrak{m}, \mathfrak{n})$-distributive if identity (10) (or (10')) holds for every $(\mathfrak{m}, \mathfrak{n})$-indexed set $\{A_{t, s}\}_{t \in T, s \in S}$ of elements of $\mathfrak{A}$. Also some proofs can be simplified. For instance, in the proof that (d_2) implies (d_3) on p. 64—65 it suffices to take for $\{B_r\}_{r \in R}$ the indexed set $\{\bigcap_{t \in T} A_{t, \varphi(t)}\}_{\varphi \in S^T}$.

We shall now apply theorem 20.2 to formulate a criterion for $(\mathfrak{m}, \mathfrak{n})$-distributivity which is a modification of 19.3. A covering (n-covering) composed of disjoint elements will be called, for brevity, a *partition* (n-*partition*).

20.3. *The following conditions are equivalent for any Boolean algebra $\mathfrak{A}$ which is n'-complete for every infinite cardinal $\mathfrak{n}' < \mathfrak{n}$:*

(d) *$\mathfrak{A}$ is $(\mathfrak{m}, \mathfrak{n})$-distributive;*

(d_4) *every set of at most $\mathfrak{m}$ n-partitions has a common refinement*[1].

By 19.3, it suffices to prove that (d_4) is equivalent to condition (d_3) from 19.3. Since (d_3) directly implies (d_4), we have to prove only the converse implication. Suppose that for every $t \in T$ ($\overline{\overline{T}} \leq \mathfrak{m}$), $\{A_{t, s}\}_{s \in S}$ is an n-covering of $\mathfrak{A}$. By 20.2 there exists a partition $\{B_{t, s}\}_{s \in S}$ such that $B_{t, s} \subset A_{t, s}$. By (d_4), there is a covering $\{B_r\}_{r \in R}$ which is a common

---

[1] Pierce [3].

refinement of all the $\mathfrak{n}$-partitions $\{B_{t,s}\}_{s \in S}$, $t \in T$. Clearly $\{B_r\}$ is also a common refinement of all the $\mathfrak{n}$-coverings $\{A_{t,s}\}_{s \in S}$, $t \in T$.

It follows from 20.3 that in condition (d_2) in 19.2 we can additionally require in (20) that $A_{t,s} \cap A_{t,s'} = \wedge$ for $s \neq s'$. We shall use this remark in the proof of the following theorem.

20.4. *Every $2^{\mathfrak{m}}$-complete $\mathfrak{m}$-distributive Boolean algebra is $(\mathfrak{m}, 2^{\mathfrak{m}})$-distributive*[1].

Let H be the set composed of the numbers 1 and -1, and let $\bar{\bar{T}} = \mathfrak{m} = \bar{\bar{S}}$. It suffices (by the last formulation of (d_2) — see also § 19 (21')) to prove that

$$\bigcup_{\varphi \in (H^S)^T} \bigcap_{t \in T} A_{t, \varphi(t)} = V$$

for any indexed set $\{A_{t,s}\}_{t \in T, s \in H^S}$ of elements of $\mathfrak{A}$, such that

(2) $\bigcup_{s \in H^S} A_{t,s} = V$ for every $t \in T$,

(3) $A_{t,s} \cap A_{t,s'} = \wedge$ for $\varepsilon \neq \varepsilon'$.

For $j \in H$, let $B_{t,s,j}$ be the join of all the elements $A_{t,s}$ such that $\varepsilon(s) = j$. It follows directly from the definition and (3) that

(4) $A_{t,s} = \bigcap_{s \in S} B_{t,s,\varepsilon(s)}$

since the element on the right side contains $A_{t,s}$ as a subelement and it is disjoint from all the $A_{t,s'}$ where $\varepsilon' \neq \varepsilon$.

Since $\mathfrak{A}$ is $\mathfrak{m}$-distributive, we have by (4) and (2)

$$\bigcap_{s \in S} \bigcup_{j \in H} B_{t,s,j} = \bigcup_{\varepsilon \in H^S} \bigcap_{s \in S} B_{t,s,\varepsilon(s)} = \bigcup_{\varepsilon \in H^S} A_{t,s} = V$$

for every $t \in T$. Consequently, by the $\mathfrak{m}$-distributivity of $\mathfrak{A}$,

$$V = \bigcap_{t \in T} \bigcap_{s \in S} \bigcup_{j \in H} B_{t,s,j} = \bigcap_{(t,s) \in T \times S} \bigcup_{j \in H} B_{t,s,j}$$
$$= \bigcup_{\varphi \in H^{T \times S}} \bigcap_{(t,s) \in T \times S} B_{t,s,\varphi(t,s)}$$
$$= \bigcup_{\varphi \in H^{T \times S}} \bigcap_{t \in T} \bigcap_{s \in S} B_{t,s,\varphi(t,s)} .$$

Any function $\varphi \in H^{T \times S}$ can be interpreted as an element of $(H^S)^T$ in a natural way, viz. as the mapping which associates to every $t \in T$ the function $\varepsilon_t = \varphi(t)$ from S into H, defined as follows:

$$\varepsilon_t(s) = \varphi(t, s) .$$

Adopting this interpretation, we can identify $H^{T \times S}$ with $(H^S)^T$. Moreover

$$\bigcap_{s \in S} B_{t,s,\varphi(t,s)} = \bigcap_{s \in S} B_{t,s,\varepsilon_t(s)} = A_{t,\varepsilon_t} = A_{t,\varphi(t)}$$

by (4). Hence it follows that

$$V = \bigcup_{\varphi \in (H^S)^T} \bigcap_{t \in T} \bigcap_{s \in S} B_{t,s,\varphi(t,s)} = \bigcup_{\varphi \in (H^S)^T} \bigcap_{t \in T} A_{t,\varphi(t)} .$$

Examples. J) There exists an $\mathfrak{m}$-field of sets $\mathfrak{F}$ which is not $(\mathfrak{m}, \mathfrak{m}^+)$-distributive[2], $\mathfrak{m}^+$ denoting the smallest cardinal greater than $\mathfrak{m}$.

---

[1] SMITH and TARSKI [1]. See also PIERCE [3].

[2] PIERCE [3].

Let T and S be sets of powers $\mathfrak{m}$ and $\mathfrak{m}^+$ respectively and let X be the set of functions $f \in T^S$ such that the set $T - f(S)$ is infinite. For any $t \in T$ and $s \in S$, let $A_{t,s}$ be the set of all $f \in X$ such that $f(s) = t$. It is easy to check that the class $\mathfrak{F}$ of all sets $A \subset X$ satisfying

(a') if $x \in A$, then $x \in \bigcap_{s \in S'} A_{t_s, s} \subset A$ for some $t_s \in T$ and some set $S' \subset S$, $\overline{\overline{S'}} \leq \mathfrak{m}$;

(a'') if $x \in -A$, then $x \in \bigcap_{s \in S''} A_{t_s, s} \subset -A$ for some $t_s \in T$ and some set $S'' \subset S$, $\overline{\overline{S''}} \leq \mathfrak{m}$;

is an $\mathfrak{m}$-field of subsets of X containing all the sets $A_{t,s}$. For every fixed $t \in T$, the set-theoretical intersection A_0 of all the sets $- A_{t,s}$ ($s \in S$) is composed of all functions $f \in X$ which do not assume the value t. Since A_0 does not contain any non-empty set of the form $\bigcap_{s \in S'} A_{t,s}$ ($\overline{\overline{S'}} \leq \mathfrak{m}$), by (a') the empty set is the only set $A \in \mathfrak{F}$ which is a subset of A_0, i.e. a subset of $- A_{t,s}$ for every $s \in S$. In other words, $\bigcap_{s \in S}^{\mathfrak{F}} - A_{t,s} = \wedge$, i.e.

$$\bigcup_{s \in S}^{\mathfrak{F}} A_{t,s} = V \quad \text{for every} \quad t \in T.$$

On the other hand, the set

$$\bigcap_{t \in T}^{\mathfrak{F}} A_{t, \varphi(t)}$$

is void for every $\varphi \in S^T$; for if $f \in A_{t, \varphi(t)}$ for every $t \in T$, then $f(\varphi(t)) = t$ for every $t \in T$ and consequently $f(S) = T$, i.e. $f \notin X$. Thus

$$\bigcap_{t \in T}^{\mathfrak{F}} \bigcup_{s \in S}^{\mathfrak{F}} A_{t,s} = V \neq \wedge = \bigcup_{\varphi \in S^T}^{\mathfrak{F}} \bigcap_{t \in T}^{\mathfrak{F}} A_{t, \varphi(t)}$$

which proves that $\mathfrak{F}$ is not $(\mathfrak{m}, \mathfrak{m}^+)$-distributive.

Note the following property of the field $\mathfrak{F}$ which will be used in § 35 N):

$$\bigcap_{t \in T} \bigcup_{s \in S} A_{t, s} = \wedge$$

where $\bigcap$ and $\bigcup$ denote the set-theoretical intersection and union respectively. In fact, $f \in \bigcup_{s \in S} A_{t, s}$ if and only if f assumes t as its value. Thus if $f \in \bigcap_{t \in T} \bigcup_{s \in S} A_{t, s}$, then f would map S onto the whole set T; however such functions do not belong to X.

K) For every infinite regular[1] cardinal $\mathfrak{m}$ there exists a complete Boolean algebra $\mathfrak{A}$ that is $(\mathfrak{m}', \mathfrak{n})$-distributive for every cardinal $\mathfrak{m}' < \mathfrak{m}$ and every cardinal $\mathfrak{n}$ but is not $\mathfrak{m}$-distributive[2].

Let $H = (-1, 1)$ and $\overline{\overline{T_0}} = \mathfrak{m}$. Thus $X = H^{T_0}$ is the set of all $x = \{x_t\}_{t \in T_0}$ where $x_t = \pm 1$. Similarly as on p. 43, let D_t be the set of all x such that $x_t = 1$. Consider X as a topological space, the class of all the intersections

$$\bigcap_{t \in T} \varepsilon_t \cdot D_t \quad \text{where} \quad \varepsilon_t = \pm 1, \, T \subset T_0, \, \overline{\overline{T}} < \mathfrak{m}$$

---

[1] A cardinal $\mathfrak{m}$ is said to be *regular* if it is not a sum of less than $\mathfrak{m}$ cardinals which are smaller than $\mathfrak{m}$.

[2] Scott [1].

being an open basis in X. The zero-dimensional topological space X has the following properties[1]: (a) the intersection of less than $\mathfrak{m}$ open (regular open) sets is open (regular open); (b) the union of less than $\mathfrak{m}$ nowhere dense sets is nowhere dense. Thus the intersection of less than $\mathfrak{m}$ dense open sets is a dense open set.

The complete Boolean algebra $\mathfrak{A}$ of all regular open sets in X has the required properties. For suppose that $\{A_{t,s}\}_{t \in T, s \in S}$ is an $(\mathfrak{m}', \mathfrak{n})$-indexed set of elements in $\mathfrak{A}$ $(\mathfrak{m}' < \mathfrak{m})$ such that

$$\bigcup_{s \in S}^{\mathfrak{A}} A_{t,s} = V \quad \text{for every} \quad t \in T .$$

Let $\cup$ and $\cap$ without superscripts denote set-theoretical operations. For every $t \in T$, the union $\bigcup_{s \in S} A_{t,s}$ is a dense open set. Since $\mathfrak{m}' < \mathfrak{m}$, the intersection

$$\bigcap_{t \in T} \bigcup_{s \in S} A_{t,s} = \bigcup_{\varphi \in S^T} \bigcap_{t \in T} A_{t,\varphi(t)}$$

is a dense open set. We have

$$\bigcap_{t \in T} A_{t,\varphi(t)} \subset \bigcap_{t \in T}^{\mathfrak{A}} A_{t,\varphi(t)}$$

since the set on the left side is regular open. Thus

$$\bigcup_{\varphi \in S^T}^{\mathfrak{A}} \bigcap_{t \in T}^{\mathfrak{A}} A_{t,\varphi(t)} = V .$$

By 19.2 (d_2) $(21')$ this proves that $\mathfrak{A}$ is $(\mathfrak{m}', \mathfrak{n})$-distributive. On the other hand, assuming $A_{t,s} = s \cdot D_t$ for $s \in S = H$ and $t \in T = T_0$, we have

$$V = \bigcap_{t \in T}^{\mathfrak{A}} \bigcup_{s \in S}^{\mathfrak{A}} A_{t,s} \neq \bigcup_{\varphi \in S^T}^{\mathfrak{A}} \bigcap_{t \in T}^{\mathfrak{A}} A_{t,\varphi(t)} = \Lambda$$

which proves that $\mathfrak{A}$ is not $\mathfrak{m}$-distributive.

A Boolean algebra $\mathfrak{A}$ is said to satisfy the $\mathfrak{m}$-*chain condition* provided every set of disjoint elements in $\mathfrak{A}$ has power $\leq \mathfrak{m}$.

Example. L) It is known from topology that every class of disjoint open subsets of a Cantor space $\mathscr{D}_\mathfrak{n}$ is at most enumerable[2]. This implies (see § 14) that every free Boolean algebra and, consequently, every sub-algebra of a free Boolean algebra satisfies the σ-chain condition. In particular, every Boolean algebra with property (a′) defined in § 15 satisfies the σ-chain condition.

The following simple criterion for the completeness of a Boolean algebra is very useful in practice.

20.5. *Every* $\mathfrak{m}$-*complete Boolean algebra* $\mathfrak{A}$ *satisfying the* $\mathfrak{m}$-*chain condition is a complete Boolean algebra*[3].

In fact, the join of any indexed set of disjoint elements exists. Thus, by 20.1, $\mathfrak{A}$ is complete.

---

[1] SIKORSKI [12].

[2] This follows from a more general theorem on separability of Cartesian products. See MARCZEWSKI [9], ŠANIN [1]

[3] TARSKI [3].

Example M) A real function m defined on a Boolean algebra $\mathfrak{A}$ is said to be an m-*measure* (or an m-*additive measure*) provided:

(5) $0 \leq m(A) \leq \infty$ for every $A \in \mathfrak{A}$; there exists an element $A_0 \in \mathfrak{A}$ such that $m(A_0) < \infty$;

(6) $$m\left(\bigcup_{t \in T} A_t\right) = \Sigma_{t \in T} m(A_t)$$

for every m-indexed set $\{A_t\}_{t \in T}$ of disjoint elements in $\mathfrak{A}$, such that $\bigcup_{t \in T} A_t$ exists.

Of course, the infinite sum $\Sigma_{t \in T}$ in (6) should be understood as follows: if $m(A_t) = 0$ for all t except an finite or enumerable sequence $t_n \in T$ ($t_i \neq t_j$ for $i \neq j$), and $\Sigma_n m(A_{t_n}) = c < \infty$, then $\Sigma_{t \in T} m(A_t) = c$; otherwise $\Sigma_{t \in T} m(A_t) = \infty$. If $\mathfrak{A}$ is an m-algebra, then the condition "such that $\bigcup_{t \in T} A_t$ exists" is superfluous in the above definition.

Obviously every m-measure is a measure in the sense defined in § 3 C). Consequently it has the properties § 3 (3), (4), (5).

If m is an m-measure on a Boolean m-algebra $\mathfrak{A}$, then

(7) $$m\left(\bigcup_{t \in T} A_t\right) \leq \Sigma_{t \in T} m(A_t)$$

for every m-indexed set $\{A_t\}_{t \in T}$ of elements in $\mathfrak{A}$.

In fact, it suffices to consider the case where T is the set of all ordinals of power $< \mathfrak{m}$. Let B_t be defined by (1). We have

$$m\left(\bigcup_{t \in T} A_t\right) = m\left(\bigcup_{t \in T} B_t\right) = \Sigma_{t \in T} m(B_t) \leq \Sigma_{t \in T} m(A_t)$$

since the elements B_t are disjoint, and $B_t \subset A_t$.

Notice that (7) is true also if m is a σ-measure on any Boolean algebra and $\bar{\bar{T}} \leq \aleph_0$ provided $\bigcup_{t \in T} A_t$ exists. The proof remains unchanged.

The case of σ-measures is the most important[1].

A σ-measure m is said to be σ-*finite* on a Boolean algebra $\mathfrak{A}$ if the unit element V is the join of an enumerable sequence $\{A_n\}$ of elements of finite measure. We may always suppose that the elements A_n are disjoint. Every finite measure is, of course, σ-finite. The Lebesgue measure is an example of σ-finite but non-finite σ-measure.

A measure m on a Boolean algebra $\mathfrak{A}$ is said to be *strictly positive* if $m(A) > 0$ for all $A \neq \wedge$ ($A \in \mathfrak{A}$).

It follows immediately from 20.5 that every Boolean σ algebra having a finite (or σ-additive and σ-finite) strictly positive measure is complete since it satisfies the σ-chain condition.

Note that the existence of strictly positive σ-finite σ-measure m on $\mathfrak{A}$ implies the existence of a strictly positive finite measure on $\mathfrak{A}$. For if $\{A_n\}$ is a sequence of disjoint elements such that $0 < m(A_n) < \infty$ and

---

[1] σ-measures on Boolean algebras have, roughly speaking, the same properties as σ-measures on fields of sets. We shall often use this fact without any reference. For details see e.g. AUMANN [3].

$\bigcup_{1 \leq n < \infty} A_n = V$, then the formula

$$m'(A) = \sum_{n=1}^{\infty} \frac{m(A \cap A_n)}{2^n m(A_n)} \quad \text{for} \quad A \in \mathfrak{A}$$

defines a strictly positive finite σ-measure m' on $\mathfrak{A}$.

§ 21. m-ideals and m-filters. Quotient algebras

An ideal $\varDelta$ of an m-complete Boolean algebra $\mathfrak{A}$ is said to be an m-*complete ideal* or an m-*ideal* provided

(a) if $A_t \in \varDelta$ for every $t \in T$, and $\overline{\overline{T}} \leq$ m, then $\bigcup_{t \in T} A_t \in \varDelta$.

It follows easily from 20.2 that an ideal $\varDelta$ is m-complete if the join of any m-indexed set of disjoint elements of $\varDelta$ belongs to $\varDelta$.

By duality, a filter V of an m-complete Boolean algebra $\mathfrak{A}$ is said to be an m-*complete filter* or an m-*filter* provided

(a') if $A_t \in V$ for every $t \in T$, and $\overline{\overline{T}} \leq$ m, then $\bigcap_{t \in T} A_t \in V$.

Examples. A) The class $\varDelta$ of all subsets A of a set X such that $\overline{\overline{A}} \leq$ m is an m-ideal of the Boolean algebra of all subsets of X. The set V of all sets $X - A$ where $\overline{\overline{A}} \leq$ m is an m-filter of this Boolean algebra.

B) Let $\mathfrak{A}$ be a complete Boolean algebra. An ideal (filter) of $\mathfrak{A}$ is m-complete for every m if and only if it is principal. Then it is generated by the join (meet) of all its elements.

21.1. *If $\varDelta$ is an* m-*ideal (if V is an* m-*filter) of an* m-*complete Boolean algebra $\mathfrak{A}$, then $\mathfrak{A}/\varDelta$ $(\mathfrak{A}/V)$ is an* m-*complete Boolean algebra, and, for every* m-*indexed set $A_t \in \mathfrak{A}$ $(t \in T)$,*

(1) $\bigcup_{t \in T} [A_t] = [\bigcup_{t \in T} A_t]$ and $\bigcap_{t \in T} [A_t] = [\bigcap_{t \in T} A_t]$.

It suffices to prove (1) for $\mathfrak{A}/\varDelta$.

Let $A = \bigcup_{t \in T} A_t$. We have $A_t - A = \varLambda \in \varDelta$. On the other hand, if $A_0 \in \mathfrak{A}$ be any element such that $A_t - A_0 \in \varDelta$ for every $t \in T$, then also $A - A_0 = \bigcup_{t \in T} (A_t - A_0) \in \varDelta$. The conditions § 18 (J_1^*), (J_2^*) being fulfilled, the first of equalities (1) holds. The proof of the second is analogous.

Nótice that the equalities (1) do not hold, in general, if $\overline{\overline{T}} >$ m. A counter-example can be easily obtained from the example in § 18 E) by replacing the ideal of all finite subsets by the ideal of all subsets whose cardinality is $\leq$ m.

The following important examples show that sometimes the degree of completeness of $\mathfrak{A}/\varDelta$ can be higher than that of $\mathfrak{A}$ and $\varDelta$ [1].

---

[1] We quote here [examples C), D), E)] only the simplest cases of the over-completeness of quotient algebras. For other interesting theorems of this kind, see SMITH and TARSKI [1].

Examples. C) Let $\mathfrak{A}$ be the σ field of all Borel subsets (or of all subsets having the Baire property) of a topological space X and let $\varDelta$ be the σ-ideal of all sets $A \in \mathfrak{A}$ of the first category. By 21.1, $\mathfrak{A}' = \mathfrak{A}/\varDelta$ is a σ-complete Boolean algebra. We shall prove that it is a complete Boolean algebra[1].

In fact, every set $A \in \mathfrak{A}$ is of the form

$$A = (G - A_1) \cup A_2$$

where G is open and $A_1, A_2 \in \varDelta$. Consequently $[A] = [G]$, i.e. every element of $\mathfrak{A}'$ is of the form $[G]$ where G is open in X.

Let $A'_t = [G_t]$ (G_t open) be any indexed set of elements of $\mathfrak{A}'$. Let G be the set-theoretical union of all G_t and let $A' = [G]$. The element A' is the join of all A'_t in $\mathfrak{A}'$. This follows from § 18 (J$_1^*$), (J$_2^*$). In fact, $G_t - G = \varLambda \in \varDelta$ for every t. Suppose now that $G_t - A_0 \in \varDelta$ for every t ($A_0 \in \mathfrak{A}$). The sets $G_t - A_0$ are of the first category and are relatively open in their union $G - A_0$ since $G_t - A_0 = (G - A_0) \cap G_t$. By a known topological theorem[2], their union $G - A_0$ is also of the first category, i.e. $G - A_0 \in \varDelta$.

Observe that we have just proved the commutativity (also for non-enumerable sets T):

(1')
$$\bigcup_{t \in T} [G_t] = \left[\bigcup_{t \in T} G_t \right] ,$$

but for open G_t only! In general, this non-enumerable commutativity does not hold for arbitrary sets in $\mathfrak{A}$ (for instance, for one-point subsets provided X is dense in itself). By passing to complements we also obtain the following (non-enumerable) commutativity law for closed sets F_t:

(1'')
$$\bigcap_{t \in T} [F_t] = \left[\bigcap_{t \in T} F_t \right] .$$

Notice that, in the case where X has an enumerable open basis, the completeness of $\mathfrak{A}/\varDelta$ can be deduced from 20.5 since the Boolean σ-algebra $\mathfrak{A}/\varDelta$ then satisfies the σ-chain condition.

The algebra $\mathfrak{A}/\varDelta$ is often called the *algebra of Borel subsets of X modulo sets of the first category*.

D) Suppose that m is an m-measure on a Boolean m-algebra. It follows immediately from § 20 (7) that the set of all elements of m measure zero is an m-ideal.

In particular, if m is a σ-measure on a Boolean σ-algebra $\mathfrak{A}$, then the set $\varDelta$ of all elements of m measure zero is a σ-ideal. By 21.1, $\mathfrak{A}/\varDelta$ is a Boolean σ-algebra. However, if the σ-measure m is finite (or, more generally, σ-finite), then $\mathfrak{A}/\varDelta$ is a complete Boolean algebra[3].

---

[1] This result is due to BIRKHOFF and ULAM. See BIRKHOFF [3]. See also VON NEUMANN [2].

[2] BANACH [3]. See also KURATOWSKI [3], p. 49.

[3] WECKEN [1]. See also BIRKHOFF [3].

This follows immediately from § 20 M) since the formula

$$m'([A]_\Delta) = m(A) \quad \text{for} \quad A \in \mathfrak{A}$$

defines a strictly positive σ-measure on $\mathfrak{A}/\Delta$, and m' is finite (σ-finite) if m is.

If $\mathfrak{A}$ is the σ-field of subsets (of a space X) measurable with respect to a σ-measure m, the algebra $\mathfrak{A}/\Delta$ is often called the *algebra of measurable subsets of X modulo sets of measure zero*.

E) The result obtained in Example D) can be reinforced as follows: If m is any finite measure on a σ-complete Boolean algebra $\mathfrak{A}$, and Δ is the ideal of elements of m measure zero, then $\mathfrak{A}/\Delta$ is a complete Boolean algebra[1].

This follows immediately from theorem 21.3 below which must be preceded by the following lemma.

21.2. *Let $\mathfrak{A}$ be a Boolean $\mathfrak{m}$-algebra, let Δ be an ideal in $\mathfrak{A}$, and let $\{A_t\}_{t \in T}$ be any $\mathfrak{m}$-indexed set of disjoint elements in $\mathfrak{A}/\Delta$. If Δ is $\mathfrak{m}'$-complete for every infinite cardinal $\mathfrak{m}' < \mathfrak{m}$, and the join*

$$(2) \qquad\qquad A = \bigcup_{t \in T}^{\mathfrak{A}/\Delta} A_t$$

exists, then the join

$$\bigcup_{t \in T'}^{\mathfrak{A}/\Delta} A_t$$

also exists for every set $T' \subset T$[2].

It is convenient to assume that T is the set of all ordinals of power $< \mathfrak{m}$. We have $A_t = [B_t]$ and $A = [B]$ for some elements $B_t, B \in \mathfrak{A}$. The elements

$$C_t = (B_t - \bigcup_{t' < t} B_{t'}) \cap B \in \mathfrak{A}$$

are disjoint and $[C_t] = A_t$.

Let $C = \bigcup_{t \in T'}^{\mathfrak{A}} C_t$. By definition,

$$[C] \subset A, \quad A_t \subset [C] \quad \text{for every} \quad t \in T', \quad \text{and}$$

$$[C] \cap A_t = \wedge \quad \text{for every} \quad t \in T - T'.$$

We have $[C] = \bigcup_{t \in T'}^{\mathfrak{A}/\Delta} A_t$. For suppose that this equality is false, i.e. there exists an element $D \neq \wedge$ $(D \in \mathfrak{A}/\Delta)$ such that

$$D \subset [C] \quad \text{and} \quad A_t \subset [C] - D \quad \text{for every} \quad t \in T'.$$

We have $A_t \subset A - D \neq A$ for every $t \in T$, which contradicts (2).

21.3. *If $\mathfrak{A}$ is a Boolean $\mathfrak{m}$-algebra, Δ is an ideal which is $\mathfrak{m}'$-complete for every infinite cardinal $\mathfrak{m}' < \mathfrak{m}$, and $\mathfrak{A}/\Delta$ satisfies the $\mathfrak{m}$-chain condition, then $\mathfrak{A}/\Delta$ is complete[3].*

Let $\{A_t\}_{t \in T'}$ be any indexed set of disjoint non-zero elements of $\mathfrak{A}/\Delta$.

---

[1] SMITH and TARSKI [1].
[2] SMITH and TARSKI [1].
[3] SMITH and TARSKI [1].

$\{A_t\}_{t \in T'}$ can be extended to a maximal indexed set $\{A_t\}_{t \in T}$ of disjoint non-zero elements in $\mathfrak{A}/\varDelta$. By maximality, we have $\bigcup_{t \in T} A_t = V$. Since $\mathfrak{A}/\varDelta$ satisfies the m-chain condition, we have $\overline{\overline{T}} \le \mathfrak{m}$. By theorem 21.2, the join $\bigcup_{t \in T'} A_t$ exists. Combining the last result with theorem 20.1 we infer that $\mathfrak{A}/\varDelta$ is complete.

Examples. F) Let $\mathfrak{A}$ be the σ-field of all Borel sets of real numbers, let $\varDelta_0$ be the ideal of all sets $A \in \mathfrak{A}$ of Lebesgue measure zero, and let $\varDelta_1$ be the ideal of all sets $A \in \mathfrak{A}$ of the first category. The complete Boolean algebras $\mathfrak{A}_0 = \mathfrak{A}/\varDelta_0$ and $\mathfrak{A}_1 = \mathfrak{A}/\varDelta_1$ are not isomorphic[1].

In fact, there exists a non-zero σ-finite measure m on $\mathfrak{A}_0$, viz. the measure m induced in the natural way by the Lebesgue measure μ:

$$(3) \qquad m([A]) = \mu(A) \quad \text{for} \quad A \in \mathfrak{A}.$$

On the other hand, every σ-finite σ-measure on $\mathfrak{A}_1$ vanishes identically. In fact, every σ-finite σ-measure m on $\mathfrak{A}_1$ determines uniquely a σ-finite σ-measure μ on $\mathfrak{A}$ [see (3)] and this measure μ vanishes on all sets $A \in \varDelta_1$. By a theorem in measure theory[2], the set X of all real numbers is the union of two disjoint Borel sets: $X = A_0 \cup A_1$ where A_0 has μ measure zero and $A_1 \in \varDelta_1$. Consequently

$$m([X]_{\varDelta_1}) = m([A_0]_{\varDelta_1}) + m([A_1]_{\varDelta_1}) = \mu(A_0) + m(\wedge) = 0.$$

This proves that m vanishes identically [see § 3 (1) and (4)][3].

Since the existence of a σ-finite non-zero σ-measure on a Boolean σ-algebra is invariant under isomorphism, we infer that $\mathfrak{A}_0$ and $\mathfrak{A}_1$ are not isomorphic.

G) Now let $\mathfrak{A}$ be the field of all sets of real numbers, let $\varDelta_0$ be the σ-ideal of all sets $A \in \mathfrak{A}$ of Lebesgue measure zero, and let $\varDelta_1$ be the σ-ideal of all sets $A \in \mathfrak{A}$ of the first category. The continuum hypothesis $2^{\aleph_0} = \aleph_1$ implies that the Boolean σ-algebras $\mathfrak{A}/\varDelta_0$ and $\mathfrak{A}/\varDelta_1$ are isomorphic.

In fact, every set of the first category is a subset of an F_σ-set[4] of the first category. The class of all F_σ-sets of the first category has power $2^{\aleph_0}$, therefore, by the continuum hypothesis, the class can be arranged in a transfinite sequence $\{B_t\}_{t < \Omega}$ where Ω is the least uncountable ordinal. With the help of this sequence we shall define, by induction, a transfinite sequence $\{A_t\}_{t < \Omega}$ of disjoint uncountable sets of the first category

---

[1] This remark is due to BIRKHOFF and ULAM. See BIRKHOFF [3].

[2] See e.g. MARCZEWSKI [1] or MARCZEWSKI and SIKORSKI [2].

[3] For another proof of the non-existence of any σ-measure on $\mathfrak{A}_1$, see HORN and TARSKI [1]. The non-existence of any strictly positive σ-measure on $\mathfrak{A}_1$ and the non-isomorphism of $\mathfrak{A}_0$ and $\mathfrak{A}_1$ follow also from § 29 C) and D).

[4] A subset of a topological space is said to be an F_σ-set provided it is the union of a sequence of closed sets.

such that

(4)   for every set $A \in \varDelta_1$ there exists an ordinal $t' < \varOmega$ such that
$A \subset \bigcup_{t<t'} A_t$.

Namely, let A_0 be any uncountable set in $\varDelta_1$. Suppose that sets $A_{t'}$ are
defined for all $t' < t$. The union $\bigcup_{t'<t} A_{t'}$ being of the first category, its
complement contains an uncountable set C_t of the first category. Let
$A_t = C_t \cup (B_t - \bigcup_{t'<t} A_{t'})$.

Using G_δ-sets[1] of measure zero instead of F_σ-sets of the first category,
we define a transfinite sequence $\{A_t'\}_{t<\varOmega}$ of disjoint uncountable sets of
measure zero, such that

(5)   for every set $A \in \varDelta_0$ there exists an ordinal $t' < \varOmega$ such that
$A \subset \bigcup_{t<t'} A_t'$.

It follows from (4) and (5) that

$$\bigcup_{t<\varOmega} A_t = \bigcup_{t<\varOmega} A_t' = \text{the set } X \text{ of all real numbers.}$$

It follows from the continuum hypothesis that the sets A_t and A_t'
have the same cardinality. Therefore there exists a one-to-one mapping
φ of X onto itself, such that

$$\varphi(A_t) = A_t' \quad \text{for every} \quad t < \varOmega.$$

This, together with (4) and (5), implies that

$$A \in \varDelta_1 \quad \text{if and only if} \quad \varphi(A) \in \varDelta_0.\text{[2]}$$

Hence it follows that the formula

$$h([A]_{\varDelta_1}) = [\varphi(A)]_{\varDelta_0} \quad \text{for} \quad A \in \mathfrak{A}$$

defines an isomorphism h of $\mathfrak{A}/\varDelta_1$ onto $\mathfrak{A}/\varDelta_0$.

The following theorem will be used in Example H).

21.4. *Let $\mathfrak{F}$ be the field of all subsets of a space X, and let $\varDelta$ be an ideal
(of $\mathfrak{F}$) containing all one-point subsets of X. Suppose that there exist two
$\mathfrak{m}$-indexed sets $\{A_t\}_{t \in T}$ and $\{B_t\}_{t \in T}$ such that*

(6) $A_t \cap A_{t'} = \varLambda$ *for* $t \neq t'$ *and* $A_t - B_t \notin \varDelta$ $(t, t' \in T)$;

(7)   *for every $B \in \varDelta$ there exists a $t \in T$ such that $B \subset B_t$.*

Then $\mathfrak{F}/\varDelta$ is not $\mathfrak{m}$-complete[3].

Namely the join $\bigcup_{t \in T} [A_t]$ does not exist in $\mathfrak{F}/\varDelta$ since, for every
$A \in \mathfrak{F}$, if

$$[A_t] \subset [A] \quad \text{for every} \quad t \in T,$$

---

[1] A subset of a topological space is said to be a G_δ-*set* provided it is the intersection of a sequence of open sets.

[2] The existence of a mapping φ with this property was proved by SIERPIŃSKI
[2, 6]. See also MARCZEWSKI [11], OXTOBY and ULAM [1].

[3] See SIKORSKI [10].

then there exists a $C \in \mathfrak{F}$ such that

$$[C] \subset [A] \neq [C] \quad \text{and} \quad [A_t] \subset [C] \quad \text{for every} \quad t \in T.$$

In fact, since $A_t - B_t \notin \Delta$ and $A_t - A \in \Delta$, the sets

$$C_t = (A_t - B_t) - (A_t - A)$$

are not empty. They are disjoint because they are subsets of disjoint sets A_t. Take one point x_t from each set C_t. Let B be the set of all the points x_t ($t \in T$) and let $C = A - B$. Since $A_t - C \subset (A_t - A) \cup (x_t) \in \Delta$, we have $[A_t] \subset [C] \subset [A]$. On the other hand, $B = A - C \notin \Delta$ on account of (7). In fact, B is not contained in any set B_t since $x_t \in B$ and $x_t \notin B_t$. This proves that $[C] \neq [A]$.

Examples. H) Let $\mathfrak{A}$, Δ_0 and Δ_1 have the same meaning as in example G). The Boolean σ-algebras $\mathfrak{A}/\Delta_0$ and $\mathfrak{A}/\Delta_1$ are not $2^{\aleph_0}$-complete.

This follows immediately from 21.4 and the fact that the set of all real numbers is the union of $2^{\aleph_0}$ disjoint non-measurable sets (or of $2^{\aleph_0}$ disjoint sets without the Baire property). The existence of such a decomposition follows e.g. from the following general set-theoretical lemma (the case where $\mathfrak{m} = 2^{\aleph_0}$ and $\mathfrak{L}$ is the class of all Borel sets of real numbers of power $2^{\aleph_0}$).

If $\overline{\overline{X}} = \mathfrak{m}$, $\mathfrak{L}$ is a class of subsets of X such that $\overline{\overline{\mathfrak{L}}} = \mathfrak{m}$ and $\overline{\overline{B}} = \mathfrak{m}$ for every $B \in \mathfrak{L}$, then there exists a class $\mathfrak{R}$ of disjoint subsets of X such that $\overline{\overline{\mathfrak{R}}} = \mathfrak{m}$, $\overline{\overline{A \cap B}} = \mathfrak{m}$ for every $A \in \mathfrak{R}$ and every $B \in \mathfrak{L}$, and X is the union of all sets $A \in \mathfrak{R}$[1].

In fact, let T be the set of all ordinals of power $< \mathfrak{m}$, let $\{x_t\}_{t \in T}$ be a transfinite sequence formed of all points in X, and let $\{B_t\}_{t \in T}$ be a transfinite sequence formed of all sets in $\mathfrak{L}$, such that every set $B \in \mathfrak{L}$ appears in $\{B_t\}_{t \in T}$ $\mathfrak{m}$ times. We define, by transfinite induction, some points $x_{t_1, t_2} \in X$, where $t_1 \leq t_2$ ($t_1, t_2 \in T$), as follows. Suppose that the points

$$(8) \qquad x_{t_1', t_2'} \quad \text{where either} \quad t_2' < t_2, \quad \text{or} \quad t_2' = t_2 \quad \text{and} \quad t_1' < t_1$$

are defined. Then x_{t_1, t_2} is the first point in the transfinite sequence $\{x_t\}_{t \in T}$ such that $x_{t_1, t_2} \in B_{t_1}$ and x_{t_1, t_2} differs from all the points (8). Let A_t be the set of all the points $x_{t, t'}$ ($t \leq t' \in T$). The class $\mathfrak{R}$ of all the sets A_t has the required properties.

I) Let $\mathfrak{F}$ be the field of all subsets of an uncountable space X. The problem of whether there exists a non-principal σ-ideal Δ such that $\mathfrak{F}/\Delta$ is complete is unsolved[2].

21.5. *Let Δ be an* m-*ideal of an* m-*distributive* $2^{\mathfrak{m}}$-*complete Boolean algebra* $\mathfrak{A}$. *Then* $\mathfrak{A}/\Delta$ *is* m-*distributive if and only if* Δ *is* $2^{\mathfrak{m}}$-*complete*[3].

---

[1] This lemma is a generalization of an argument in the proof of F. BERNSTEIN's theorem on the existence of totally imperfect sets. See e.g. KURATOWSKI [3], p. 422.

[2] See SIKORSKI [10].

[3] PIERCE [3]. A part of this theorem was proved independently by SMITH and TARSKI [1]. See also SIKORSKI [21].

Let $\{A_{t,s}\}_{t\in T,\,s\in S}$ be any $\mathfrak{m}$-indexed set of elements in $\mathfrak{A}/\Delta$. We have $A_{t,s} = [B_{t,s}]$ for some $B_{t,s} \in \mathfrak{A}$. The $\mathfrak{m}$-distributivity of $\mathfrak{A}$ implies

$$\bigcap_{t\in T} \bigcup_{s\in S} B_{t,s} = \bigcup_{\varphi\in S^T} \bigcap_{t\in T} B_{t,\varphi(t)} \, .$$

If Δ is $2^{\mathfrak{m}}$-complete, then, by 21.1, [ ] commutes with the above operations of join and meet. This proves that the analogous identity holds for $A_{t,s}$, i.e. $\mathfrak{A}/\Delta$ is $\mathfrak{m}$-distributive.

To prove the second part of 21.5, let T and S be sets of power $\mathfrak{m}$. It will be convenient to present $2^{\mathfrak{m}}$-indexed sets in the form $\{B_{\varphi}\}_{\varphi\in S^T}$.

Suppose that the $\mathfrak{m}$-ideal Δ is not a $2^{\mathfrak{m}}$-ideal. Thus there exists an indexed set $\{B_{\varphi}\}_{\varphi\in S^T}$ of disjoint elements in Δ, such that the join

$$A = \bigcup_{\varphi\in S^T} B_{\varphi}$$

does not belong to Δ. For any $t \in T$ and $s \in S$ let $A_{t,s}$ be the join of all elements B_{φ} such that $\varphi(t) = s$. The elements B_{φ} being disjoint, we have

$$B_{\varphi} = \bigcap_{t\in T} A_{t,\varphi(t)} \, .$$

since the element on the right side is contained in A, it contains B_{φ} and it is disjoint from all the $B_{\varphi'}$ where $\varphi' \neq \varphi$. Thus

$$A = \bigcup_{\varphi\in S^T} \bigcap_{t\in T} A_{t,\varphi(t)} = \bigcap_{t\in T} \bigcup_{s\in S} A_{t,s}$$

since $\mathfrak{A}$ is $\mathfrak{m}$-distributive. Consequently

$$\bigcap_{t\in T} \bigcup_{s\in S} [A_{t,s}] = [A] \neq \Lambda$$

since $A \notin \Delta$. On the other hand,

$$\bigcup_{\varphi\in S^T} \bigcap_{t\in T} [A_{t,\varphi(t)}] = \bigcup_{\varphi\in S^T} [B_{\varphi}] = \Lambda$$

since $B_{\varphi} \in \Delta$ for every $\varphi \in S^T$. This proves that $\mathfrak{A}/\Delta$ is not $\mathfrak{m}$-distributive.

We have defined the notion of $\mathfrak{m}$-ideals and $\mathfrak{m}$-filters only in the case of Boolean $\mathfrak{m}$-algebras $\mathfrak{A}$. If the hypothesis of the $\mathfrak{m}$-completeness of $\mathfrak{A}$ is omitted, the definition presents a difficulty, since the following three generalizations arise:

(D) an ideal Δ (a filter V) of $\mathfrak{A}$ is said to be an $\mathfrak{m}$-ideal ($\mathfrak{m}$-filter) provided that, for every $\mathfrak{m}$-indexed set $\{A_t\}_{t\in T}$ of elements in Δ (in V), there exists an element $A \in \Delta$ ($A \in V$) such that $A_t \subset A$ ($A \subset A_t$) for every $t \in T$;

(D') an ideal Δ (a filter V) of $\mathfrak{A}$ is said to be an $\mathfrak{m}$-ideal ($\mathfrak{m}$-filter) provided that, for every $\mathfrak{m}$-indexed set $\{A_t\}_{t\in T}$ of elements in Δ (in V), if the join $\bigcup_{t\in T}^{\mathfrak{A}} A_t$ (the meet $\bigcap_{t\in T}^{\mathfrak{A}} A_t$) exists, it belongs to Δ (to V);

(D'') an ideal Δ (a filter V) of $\mathfrak{A}$ is said to be an $\mathfrak{m}$-ideal ($\mathfrak{m}$-filter) provided that, for every $\mathfrak{m}$-indexed set $\{A_t\}_{t\in T}$ of elements in Δ (in V), the join $\bigcup_{t\in T}^{\mathfrak{A}} A_t$ (the meet $\bigcap_{t\in T}^{\mathfrak{A}} A_t$) exists and belongs to Δ (to V).

Unfortunately these definitions are not equivalent. For some problems definition (D) is most suitable, for other problems, definition (D'), etc.

For instance, definition (D') is more convenient for a representation problem discussed in § 24 (see p. 98), and definition (D) is more convenient for another representation problem discussed in § 29.

Almost everywhere in this book we shall examine only m-ideals and m-filters in Boolean m-algebras. In this case all the definitions (D), (D'), (D'') coincide with the definition given at the beginning of § 21. If the hypothesis of m-completeness of the Boolean algebra under consideration is omitted, we shall state explicitly which definition among (D), (D'), (D'') is actually assumed.

We are going to characterize m-ideals and m-filters by properties of the corresponding open and closed subsets in Stone spaces (see p. 27). For this purpose we must introduce the following definition.

An open (closed) subset B of a topological space is said to have *lower (upper) character* m if, for every m-indexed set $\{B_t\}_{t \in T}$ of open-closed subsets, $B_t \subset B$ for all $t \in T$ implies that B contains the closure of the union of the sets B_t $(B \subset B_t$ for all $t \in T$ implies that B is contained in the interior of the intersection of the sets $B_t)$.

In the next theorem definition (D) is assumed.

21.6. *An ideal Δ (a filter V) in a Boolean algebra $\mathfrak{A}$ is an* m-*ideal* (m-*filter*) *if and only if it corresponds to an open (closed) set of lower (upper) character* m.

The easy proof is left to the reader.

E x a m p l e. J) If $\mathfrak{A}$ is the Boolean algebra of all open-closed subsets of the space X of all ordinals $\leq \Omega$ (where Ω is the first uncountable ordinal) with the usual order topology, then the nowhere dense one-point set (Ω) has upper character $\aleph_0$. The dense open set $X - (\Omega)$ has lower character $\aleph_0$.

§ 22. m-homomorphisms. The interpretation in Stone spaces

Let h be a homomorphism of a Boolean algebra $\mathfrak{A}$ into a Boolean algebra $\mathfrak{A}'$.

Suppose that

(1) $$A = \bigcup_{t \in T}^{\mathfrak{A}} A_t ,$$

and that the join

(2) $$\bigcup_{t \in T}^{\mathfrak{A}'} h(A_t)$$

exists. Then

(3) $$h(A) \supset \bigcup_{t \in T}^{\mathfrak{A}'} h(A_t)$$

since (1) implies that $A_t \subset A$, and consequently $h(A_t) \subset h(A)$ for every $t \in T$. We shall say that the homomorphism h of $\mathfrak{A}$ into $\mathfrak{A}'$ *preserves* the join (1) if (2) exists and

(4) $$h(A) = \bigcup_{t \in T}^{\mathfrak{A}'} h(A_t) .$$

Suppose now that

(1') $A = \bigcap_{t \in T}^{\mathfrak{A}} A_t$

and that the meet

(2') $\bigcap_{t \in T}^{\mathfrak{A}'} h(A_t)$

exists. Then

(3') $h(A) \subset \bigcap_{t \in T}^{\mathfrak{A}'} h(A_t)$

since $A \subset A_t$ and consequently $h(A) \subset h(A_t)$ for every $t \in T$. We shall say that the homomorphism h of $\mathfrak{A}$ into $\mathfrak{A}'$ *preserves* the meet (1') if (2') exists and

(4') $h(A) = \bigcap_{t \in T}^{\mathfrak{A}'} h(A_t)$.

Let $\mathfrak{m}$ be a given infinite cardinal. The homomorphism h is said to be an $\mathfrak{m}$-*complete homomorphism* (or simply an $\mathfrak{m}$-*homomorphism*) of $\mathfrak{A}$ *into* $\mathfrak{A}'$ provided it preserves all joins (1) where $\overline{\overline{T}} \leq \mathfrak{m}$. It follows from the de Morgan laws [see § 19 (3)] that h is an $\mathfrak{m}$-homomorphism of $\mathfrak{A}$ into $\mathfrak{A}'$ if and only if h preserves all meets (1') where $\overline{\overline{T}} \leq \mathfrak{m}$. The homomorphism h is said to be a *complete homomorphism of $\mathfrak{A}$ into $\mathfrak{A}'$* provided it is an $\mathfrak{m}$-homomorphism of $\mathfrak{A}$ into $\mathfrak{A}'$ for every infinite cardinal $\mathfrak{m}$.

The same terminology is assumed for isomorphisms.

If there exists an $\mathfrak{m}$-homomorphism of $\mathfrak{A}$ onto $\mathfrak{A}'$, then $\mathfrak{A}'$ is said to be an $\mathfrak{m}$-*homomorphic image* of $\mathfrak{A}$.

Observe that a homomorphism h of $\mathfrak{A}$ into $\mathfrak{A}'$ is an $\mathfrak{m}$-homomorphism if and only if

(a) $\bigcap_{t \in T}^{\mathfrak{A}} A_t = \wedge$ (where $A_t \in \mathfrak{A}$, $\overline{\overline{T}} \leq \mathfrak{m}$) implies $\bigcap_{t \in T}^{\mathfrak{A}'} h(A_t) = \wedge$.

The necessity of the condition (a) is obvious. To prove the sufficiency suppose that (1) holds. Then $\bigcap_{t \in T}^{\mathfrak{A}}(A - A_t) = \wedge$ and, by (a), $\bigcap_{t \in T}^{\mathfrak{A}'}(h(A) - h(A_t)) = \wedge$. This implies (4) since $A_t \subset A$ and consequently $h(A_t) \subset h(A)$ for every $t \in T$.

Most often we shall use the notion of $\mathfrak{m}$-homomorphism and of $\mathfrak{m}$-isomorphism in the case where $\mathfrak{A}$ and $\mathfrak{A}'$ are $\mathfrak{m}$-complete Boolean algebras. Observe that, in that case, if h is an $\mathfrak{m}$-homomorphism, then the set $h^{-1}(\wedge)$ (i.e. the set of all $A \in \mathfrak{A}$ such that $h(A) = \wedge$) is an $\mathfrak{m}$-ideal of $\mathfrak{A}$, and the set $h^{-1}(\vee)$ (i.e. the set of all $A \in \mathfrak{A}$ such that $h(A) = \vee$) is an $\mathfrak{m}$-filter of $\mathfrak{A}$.

In the following theorem definition (D') of $\mathfrak{m}$-filters is assumed.

22.1. *Let X be a set of maximal $\mathfrak{m}$-complete filters of a Boolean algebra $\mathfrak{A}$. For every $A \in \mathfrak{A}$, let $h(A)$ be the set of all filters $V \in X$ such that $A \in V$. Then h is an $\mathfrak{m}$-homomorphism of $\mathfrak{A}$ into the field of all subsets of X.*

If $\mathfrak{A}$ is an $\mathfrak{m}$-complete Boolean algebra, then the class $\mathfrak{F}$ of all sets $h(A)$ where $A \in \mathfrak{A}$ is a reduced $\mathfrak{m}$-field of subsets of X.

If, for every $A \neq \wedge (A \in \mathfrak{A})$, there exists a filter $V \in X$ such that $A \in V$, then h is an isomorphism of $\mathfrak{A}$ onto $\mathfrak{F}$.

By 8.1 it suffices to prove that if (1') holds, then $h(A)$ is the set-theoretical intersection of the sets $h(A_t)$ $(t \in T, \overline{\overline{T}} \leq \mathfrak{m})$.

If V belongs to this intersection, then $V \in h(A_t)$ for every $t \in T$, i.e. $A_t \in V$ for every $t \in T$. Since V is an m-filter, we have $A \in V$, i.e. $V \in h(A)$. This proves that $\bigcap_{t \in T} h(A_t) \subset h(A)$. The converse inclusion is always satisfied [see (3')].

Examples. A) It follows from § 18 (6) that every isomorphism of $\mathfrak{A}$ onto $\mathfrak{A}'$ is a complete isomorphism of $\mathfrak{A}$ onto $\mathfrak{A}'$. This remark fails if "onto" is replaced by "into". For instance, if $\mathfrak{B}$ is the field (of subsets of the set X of non-negative integers) defined in § 18 A), then the identity isomorphism of $\mathfrak{B}$ into the field $\mathfrak{B}'$ of all subsets of X is not a σ-complete isomorphism of $\mathfrak{B}$ into $\mathfrak{B}'$ since an enumerable join in $\mathfrak{B}$ does not coincide with the set-theoretical union.

On the other hand, if h is an isomorphism of a Boolean algebra $\mathfrak{A}$ into a Boolean algebra $\mathfrak{A}'$, then (4) always implies (1) ((4') always implies (1')). This statement can be deduced directly from the definition of infinite joins and meets, or it can be obtained in the following way. If (4) holds, then also $h(A) = \bigcup_{t \in T}^{h(\mathfrak{A})} h(A_t)$ by § 18 (5). Since h is an isomorphism of $\mathfrak{A}$ onto the algebra $h(\mathfrak{A})$, the last equation is equivalent to (1). The proof for meets is by duality.

B) Let $\mathfrak{A}$ be a Boolean algebra and, for every $A \in \mathfrak{A}$, let $h(A)$ be the class of all atoms contained in A. Then h is a complete homomorphism of $\mathfrak{A}$ into the field of all subsets of the set X of all atoms of $\mathfrak{A}$.

This follows immediately from 22.1 since every atom can be identified with a maximal principal filter and every principal filter is m-complete for every infinite cardinal $\mathfrak{m}$.

Notice that if $\mathfrak{A}$ is m-complete, then, by 22.1, $h(\mathfrak{A})$ is an m-field of subsets of X. If $\mathfrak{A}$ is a complete Boolean algebra, then $h(\mathfrak{A})$ is a complete field (of subsets of X) containing all one-point sets (images of atoms!), thus $h(\mathfrak{A})$ is the field of all subsets of X.

C) It follows from the distributive laws § 19 (9) that the homomorphism h
$$h(A) = A \cap E \quad \text{for} \quad A \in \mathfrak{A}$$
is a complete homomorphism of $\mathfrak{A}$ onto $\mathfrak{A}|E$ (see § 10 A)).

D) Let $\mathfrak{A}$ and $\mathfrak{A}'$ be two Boolean m-algebras, $\overline{\overline{T}} \leq \mathfrak{m}$ and
$$\bigcup_{t \in T} A_t = V_{\mathfrak{A}}, \quad A_{t_1} \cap A_{t_2} = \wedge_{\mathfrak{A}} \quad \text{for} \quad t_1 \neq t_2,$$
$$\bigcup_{t \in T} A_t' = V_{\mathfrak{A}'}, \quad A_{t_1}' \cap A_{t_2}' = \wedge_{\mathfrak{A}'} \quad \text{for} \quad t_1 \neq t_2.$$
If, for every $t \in T$, h_t is an m-homomorphism from $\mathfrak{A}|A_t$ into (onto) $\mathfrak{A}'|A_t'$, then the formula
$$h(A) = \bigcup_{t \in T} h_t(A \cap A_t) \qquad\qquad (A \in \mathfrak{A})$$

6*

defines an $\mathfrak{m}$-homomorphism of $\mathfrak{A}$ into (onto) $\mathfrak{A}'$. If each h_t is an isomorphism, then h is an isomorphism.

E) A complete Boolean algebra $\mathfrak{A}$ is isomorphic to the direct union of complete Boolean algebras $\{\mathfrak{A}_t\}_{t\in T}$ if and only if there exist disjoint elements $A_t \in \mathfrak{A}$ $(t \in T)$ such that

$$\bigcup{}_{t\in T} A_t = V,$$

and $\mathfrak{A}|A_t$ is isomorphic to $\mathfrak{A}_t$ for every $t \in T$. Namely, if h_t is an isomorphism from $\mathfrak{A}|A_t$ onto $\mathfrak{A}_t$, then the formula

$$h(A) = \{h_t(A \cap A_t)\}_{t\in T} \qquad\qquad (A \in \mathfrak{A})$$

defines an isomorphism h from $\mathfrak{A}$ onto the direct union of $\{\mathfrak{A}_t\}_{t\in T}$.

The above remark is true under the hypothesis that all the algebras $\mathfrak{A}$, $\mathfrak{A}_t$ are $\mathfrak{m}$-complete, provided $\overline{\overline{T}} \leq \mathfrak{m}$.

F) It follows from § 21 (1) that, if $\varDelta$ is an $\mathfrak{m}$-complete ideal (if V is an $\mathfrak{m}$-complete filter) of an $\mathfrak{m}$-complete Boolean algebra $\mathfrak{A}$, then the natural homomorphism $h(A) = [A]$ is an $\mathfrak{m}$-complete homomorphism of $\mathfrak{A}$ onto $\mathfrak{A}/\varDelta$ (onto $\mathfrak{A}/V$).

G) Let $\mathfrak{A}_1$ be the Boolean algebra of all regular closed subsets of a topological space X (see § 1 B) and § 20 C)). Let $\mathfrak{A}$ be the field of all Borel subsets of X, and let $\varDelta$ be the ideal of all sets $A \in \mathfrak{A}$ of the first category (see § 21 C)). The reader familiar with topology can easily verify that the mapping

(5) $\qquad\qquad h(A) = [A]_\varDelta \quad$ for $\quad A \in \mathfrak{A}_1$

is a complete homomorphism of the complete Boolean algebra $\mathfrak{A}_1$ onto the complete Boolean algebra $\mathfrak{A}/\varDelta$. If no open non-empty subset of X is of the first category, h is an isomorphism[1].

Analogously, if $\mathfrak{A}_2$ is the Boolean algebra of all regular open subsets of X, then the mapping (5) (where $\mathfrak{A}_1$ is replaced by $\mathfrak{A}_2$) is a complete homomorphism of the complete Boolean algebra $\mathfrak{A}_2$ onto the complete Boolean algebra $\mathfrak{A}/\varDelta$.

H) If $\mathfrak{A}'_0$ is a subalgebra of $\mathfrak{A}'$, h is an $\mathfrak{m}$-homomorphism of a Boolean algebra $\mathfrak{A}$ into $\mathfrak{A}'$ and $h(\mathfrak{A})$ is a subset of $\mathfrak{A}'_0$, then h is also an $\mathfrak{m}$-homomorphism of $\mathfrak{A}$ into $\mathfrak{A}'_0$. This follows from § 18 (5).

I) Let h_0 be an isomorphism of a Boolean algebra $\mathfrak{A}$ onto the field $\mathfrak{F}_0$ of all open-closed subsets of the Stone space X of $\mathfrak{A}$. If there exists an essentially infinite join (1) in $\mathfrak{A}$ with $\overline{\overline{T}} \leq \mathfrak{m}$, then h_0 is not an $\mathfrak{m}$-isomorphism of $\mathfrak{A}$ into the field $\mathfrak{A}'$ of all subsets of X. In fact, (4) does not hold then. For suppose that (4) is true, i.e. that the closed set $h_0(A)$ is the union of the open sets $h_0(A_t)$. Since X is compact, there would

---

[1] See BIRKHOFF [2], p. 178.

exist a finite sequence $t_1, \ldots, t_n \in T$ such that $h_0(A) = h_0(A_{t_1}) \cup \cdots$ $\cup \, h_0(A_{t_n})$. Since h_0 is an isomorphism, we obtain $A = A_{t_1} \cup \cdots \cup A_{t_n}$, contradicting the hypothesis that the join (1) is essentially infinite.

Of course it is true that h_0 is an m-isomorphism of $\mathfrak{A}$ onto $\mathfrak{F}_0$ (see A)), but not of $\mathfrak{A}$ into $\mathfrak{A}'$. Therefore the designation of the Boolean algebra $\mathfrak{A}'$ in expressions of the form "h is an m-homomorphism (m-isomorphism) of $\mathfrak{A}$ into $\mathfrak{A}'$" is always necessary.

A subset S of a topological space X is said to be m-*closed* (m-*open*) provided it is the intersection (the union) of at most m open-closed subsets. Obviously, every m-closed (m-open) set is closed (open).

We recall that, for any subset S of a topological space, $\mathbf{C}S$ and $\mathbf{I}S$ denote the closure and interior of S, respectively.

We complete the remarks in I) by the following theorem.

22.2. *Let h_0 be an isomorphism of a Boolean algebra $\mathfrak{A}$ onto the field $\mathfrak{F}_0$ of all open-closed subsets of the Stone space X of $\mathfrak{A}$.*

For any elements A, A_t in $\mathfrak{A}$,

$$(6) \qquad A = \mathbf{U}_{t \in T} A_t \qquad\qquad (T \leq \mathfrak{m})$$

if and only if

$$(7) \qquad h_0(A) = \mathbf{C}\,\mathbf{U}_{t \in T}\, h_0(A_t) \, ;$$

i.e. if the m-closed set

$$(8) \qquad S = h_0(A) - \mathbf{U}_{t \in T}\, h_0(A_t)$$

is nowhere dense and $h_0(A_t) \subset h_0(A)$ for every $t \in T$.

Similarly,

$$(6') \qquad A = \bigcap_{t \in T} A_t \qquad\qquad (\overline{\overline{T}} \leq \mathfrak{m})$$

if and only if

$$(7') \qquad h_0(A) = \mathbf{I}\bigcap_{t \in T}\, h_0(A_t) \, ;$$

i.e. if the m-closed set

$$(8') \qquad S' = \bigcap_{t \in T}\, h_0(A_t) - h_0(A)$$

is nowhere dense and $h_0(A) \subset h_0(A_t)$ for every $t \in T$.

In (7), (8), (7'), (8'), $\mathbf{U}$ and $\bigcap$ denote the set-theoretical union and intersection, respectively.

The set (8) is m-closed since it is the intersection of the open-closed sets $h_0(A)$ and $h_0(-A_t)$, $t \in T$.

Let $S_0 = h_0(A) - \mathbf{C}\,\mathbf{U}_{t \in T}\, h_0(A_t)$. By definition, S_0 is open and

$$S_0 = \mathbf{I}S$$

where S is defined by (8).

If (6) holds, then $A_t \subset A$ and consequently $h_0(A_t) \subset h_0(A)$ for every $t \in T$. Thus $\mathbf{U}_{t \in T}\, h_0(A_t) \subset h_0(A)$ and

$$\mathbf{C}\,\mathbf{U}_{t \in T}\, h_0(A_t) \subset \mathbf{C}h_0(A) = h_0(A) \, .$$

The open set S_0 is empty. For if not, then there is an element $A_0 \in \mathfrak{A}$ such that $A_0 \neq \wedge$ and $h_0(A_0) \subset S$. Thus $h_0(A_0) \subset h_0(A)$ and $h_0(A_0) \cap \cap h_0(A_t) = \wedge$. Hence $A_0 \subset A$ and $A_0 \cap A_t = \wedge$, i.e. $A_t \subset A - A_0 \neq A$ for every $t \in T$ which contradicts (6).

If (7) holds, then $h_0(A_t) \subset h_0(A)$ for every $t \in T$, and $\mathbf{I}S = S_0 = \wedge$, i.e. S is nowhere dense.

If $h_0(A_t) \subset h_0(A)$ for every $t \in T$, then $A_t \subset A$ for every $t \in T$. If, in addition, S is nowhere dense, then (6) holds. For if not, then there is an element $B \subset A$, $B \neq A$, such that $A_t \subset B$ for every $t \in T$. Let $A_0 = A - B$. Since $A_0 \neq \wedge$ and $A_t \cap A_0 = \wedge$ for every $t \in T$, the non-empty open set $h_0(A_0)$ is a subset of S, i.e. S is not nowhere dense.

The first part of 22.2 is proved. The second part follows from the first by passage to complements.

If (6) holds, then (8) is called the *defect set* corresponding to the join (6). Similarly, if (6') holds, then (8') is the *defect set* corresponding to the meet (6'). By 22.2, defect sets are always nowhere dense.

It follows directly from 22.2 that

22.3. *The join* $\bigcup_{t \in T}^{\mathfrak{A}} A_t$ *exists if and only if the closed set*

$$B = \mathbf{C} \bigcup_{t \in T} h_0(A_t)$$

is open. Then $h_0^{-1}(B) = \bigcup_{t \in T}^{\mathfrak{A}} A_t$.

The meet $\bigcap_{t \in T}^{\mathfrak{A}} A_t$ *exists if and only if the open set*

$$B' = \mathbf{I} \bigcap_{t \in T} h_0(A_t)$$

is closed. Then $h_0^{-1}(B') = \bigcap_{t \in T}^{\mathfrak{A}} A_t$.

The following theorem is an immediate corollary of 22.3.

22.4. *A Boolean algebra* $\mathfrak{A}$ *is* $\mathfrak{m}$-*complete if and only if in the Stone space of* $\mathfrak{A}$ *the closure of every* $\mathfrak{m}$-*open set is open or, equivalently, if the interior of every* $\mathfrak{m}$-*closed set is closed.*

$\mathfrak{A}$ *is complete if and only if its Stone space is extremally disconnected.*

We recall that a totally disconnected topological space is said to be *extremally disconnected* if the closure of every open set is open, or equivalently (by passage to complements) if the interior of every closed set is closed[1].

To formulate the next theorem it is convenient to introduce the following definition.

A subset A of a topological space X is said to be $\mathfrak{m}$-*nowhere dense* provided it is a subset of a nowhere dense $\mathfrak{m}$-closed set. Any union of at most $\mathfrak{m}$ sets $\mathfrak{m}$-nowhere dense in X is called a *set of the* $\mathfrak{m}$-*category*.

It is easy to see that any continuous mapping φ of a topological space X' into another topological space X has the following property:

---

[1] For a detailed exposition of properties of extremally disconnected spaces, see GILLMAN and JERISON [1].

If A is an m-closed subset of X, then $\varphi^{-1}(A)$ is an m-closed subset of X'.

A mapping φ of X' into X is said to be m-*continuous* provided it is continuous and has the following property:

If A is an m-nowhere dense subset of X, then $\varphi^{-1}(A)$ is an m-nowhere dense subset of X'.

Of course, in order to prove that a continuous mapping φ is m-continuous, it suffices to show that $\varphi^{-1}(A)$ is nowhere dense for every m-closed nowhere dense set A.

Suppose now that h is a homomorphism of a Boolean algebra $\mathfrak{A}$ into a Boolean algebra $\mathfrak{A}'$. Let X and X' be the Stone spaces of $\mathfrak{A}$ and $\mathfrak{A}'$ respectively and let h_0 and h_0' be the isomorphisms of $\mathfrak{A}$ onto the field $\mathfrak{F}_0$ of all open-closed subsets of X, and of $\mathfrak{A}'$ onto the field $\mathfrak{F}_0'$ of all open-closed subsets of X', respectively. The homomorphism h determines, in a natural way, a corresponding homomorphism h' of $\mathfrak{F}_0$ into $\mathfrak{F}_0'$, viz. the homomorphism

$$(9) \qquad h'(B) = h_0'\big(h(h_0^{-1}(B))\big) \quad \text{for} \quad B \in \mathfrak{F}_0 .$$

By 11.1 (see also the remarks on p. 34), h' is induced by a continuous mapping φ of X' into X. By definition,

$$\varphi^{-1}(B) = h_0' h h_0^{-1}(B) \quad \text{for every} \quad B \in \mathfrak{F}_0 .$$

22.5. *In order that h preserve an infinite join* (6) *(an infinite meet* (6')*) it is necessary and sufficient that counterimage $\varphi^{-1}(S)$ of the corresponding defect set S be nowhere dense.*

Thus h is an m-*homomorphism if and only if φ is* m-*continuous.*

If (6) holds, then $A_t \subset A$ and consequently $h_0' h(A_t) \subset h_0' h(A)$ for every $t \in T$. By 22.2,

$$h(A) = \bigcup_{t \in T}^{\mathfrak{A}'} h(A_t)$$

if and only if the set

$$S' = h_0' h(A) - \bigcup_{t \in T} h_0' h(A_t)$$
$$= h_0' h h_0^{-1}(h_0(A)) - \bigcup_{t \in T} h_0' h h_0^{-1}(h_0(A_t))$$
$$= \varphi^{-1}(h_0(A_t)) - \bigcup_{t \in T} \varphi^{-1}(h_0(A_t)) = \varphi^{-1}(S)$$

is nowhere dense. By duality we get the similar statement for the meet (6').
The second part of 22.5 follows directly from the first.

It follows from 22.5 that if h preserves a join (6) or a meet (6'), then h also preserves many other joins and meets, viz. h preserves the joins and meets whose corresponding defect sets are subsets of the defect set of (6) or (6').

Sometimes we have to examine homomorphisms h from $\mathfrak{A}$ into $\mathfrak{A}'$ which preserve a given set of infinite joins and meets. For this purpose it is convenient to introduce the following terminology.

Let **J** be a class of non-void subsets of $\mathfrak{A}$ such that for every $\mathfrak{S} \in \mathbf{J}$ the join

(10) $$\bigcup_{A \in \mathfrak{S}} A$$

exists, and let **M** be a class of non-void subsets of $\mathfrak{A}$ such that for every $\mathfrak{S} \in \mathbf{M}$ the meet

(10') $$\bigcap_{A \in \mathfrak{S}} A$$

exists. A homomorphism h from $\mathfrak{A}$ into another Boolean algebra $\mathfrak{A}'$ is said to be a $(\mathbf{J}, \mathbf{M})$-*homomorphism* provided h preserves all the infinite joins (10) and meets (10').

Note that h preserves (10) if and only if h preserves the meet $\bigcap_{A \in \mathfrak{S}} -A$, by the de Morgan formula. Thus the preservation of a class of joins can always be reduced to the preservation of a given class of meets. Similarly the preservation of a class of meets can always be reduced to the preservation of a class of joins.

If **J** is the class of all non-void subsets $\mathfrak{S}$ of $\mathfrak{A}$ such that $\overline{\overline{\mathfrak{S}}} \leq \mathfrak{m}$ and (10) exists, and if **M** is the class of all non-void subsets $\mathfrak{S}$ of $\mathfrak{A}$ such that $\overline{\overline{\mathfrak{S}}} \leq \mathfrak{m}$ and (10') exists, then $(\mathbf{J}, \mathbf{M})$-homomorphisms coincide with $\mathfrak{m}$-homomorphisms.

Let $\mathfrak{N}$ be a class of nowhere dense subsets of a topological space X. A mapping φ from a topological space X' into X is said to be $\mathfrak{N}$-*continuous* if for every set $A \in \mathfrak{N}$ the set $\varphi^{-1}(A)$ is nowhere dense in X'.

Suppose that $\mathfrak{N}$ is the class of all the defect sets corresponding to infinite joins (10) and meets (10'). The second part of 22.5 can now be generalized as follows: h is a $(\mathbf{J}, \mathbf{M})$-homomorphism if and only if the mapping φ is $\mathfrak{N}$-continuous.

If $\mathfrak{N}$ is a class of nowhere dense subsets of the Stone space X of A, then by an $\mathfrak{N}$-*homomorphism* we shall understand a $(\mathbf{J}, \mathbf{M})$-homomorphism where **J** and **M** are respectively the classes of all non-void sets $\mathfrak{S} \subset \mathfrak{A}$ such that (10) or (10') exists and the corresponding defect sets of (10) or (10') belong to $\mathfrak{N}$.

If h is an isomorphism and a $(\mathbf{J}, \mathbf{M})$-homomorphism *($\mathfrak{N}$-homomorphism)*, then h is called a $(\mathbf{J}, \mathbf{M})$-*isomorphism ($\mathfrak{N}$-isomorphism)*.

Examples. J) Let $\{\{i_t\}_{t \in T}, \mathfrak{B}\}$ be a Boolean product of an indexed set $\{\mathfrak{A}_t\}_{t \in T}$ of non-degenerate Boolean algebras (see § 13). For every $t \in T$, the isomorphism i_t is a complete isomorphism from $\mathfrak{A}_t$ into $\mathfrak{B}$.

Since all Boolean products of $\{\mathfrak{A}_t\}_{t \in T}$ are isomorphic, it suffices to prove that the Boolean product

$$\{\{h_t^*\}_{t \in T}, \mathfrak{F}\}$$

defined in § 13 (7) has this property. We shall use below the notation from § 13. $\bigcup$ without superscripts denotes the set-theoretical union.

If $A = \bigcup_{u \in U}^{\mathfrak{A}_t} A_u$ $(A, A_u \in \mathfrak{A}_t)$, then the defect set $S = h_t(A) - \bigcup_{u \in U} h_t(A_t) \subset X_t$ is nowhere dense in X_t. By a known topological theorem, the set $S^* = h_t^*(A) - \bigcup_{u \in U} h_t^*(A_u)$ is nowhere dense in the Cartesian product X of all the spaces X_t. Since X is the Stone space of $\mathfrak{F}$, by 22.2 we have $h_t^*(A) = \bigcup_{u \in U}^{\mathfrak{F}} h_t^*(A_u)$. Similarly we prove that h_t^* preserves infinite meets in $\mathfrak{A}_t$.

K) A Boolean algebra $\mathfrak{A}$ satisfies the m-chain condition if and only if, in its Stone space X, every nowhere dense set is m-nowhere dense.

Let h_0 have the same meaning as in 22.2.

Suppose that N is a nowhere dense subset of X. Let $\{A_t\}_{t \in T}$ be a maximal class of non-zero disjoint elements in $\mathfrak{A}$ such that no set $h_0(A_t)$ intersects N. Since the class is maximal, the union G of all $h_0(A_t)$ is dense in X, i.e. its complement $N_0 = X - G$ is a nowhere dense set. We have $N \subset N_0$. If $\mathfrak{A}$ satisfies the m-chain condition, then $\overline{\overline{T}} \leq \mathfrak{m}$, and consequently the set $N_0 = \bigcap_{t \in T} h_0(-A_t)$ is m-closed. This proves that N is m-nowhere dense.

Suppose now that every nowhere dense subset of X is m-nowhere dense. We shall prove that, for every indexed set $\{A_t\}_{t \in T}$ of disjoint non-zero elements in $\mathfrak{A}$, we have $\overline{\overline{T}} \leq \mathfrak{m}$. It suffices to prove this in the case where $\{A_t\}_{t \in T}$ is a maximal class of disjoint elements, i.e. when the union G of all sets $h_0(A_t)$ is dense in X. The nowhere dense set $X - G$ is contained in a nowhere dense m-closed set N. Thus there exists an m-indexed set $\{B_s\}_{s \in S}$ of elements in $\mathfrak{A}$ such that the union G_0 of all sets $h_0(B_s)$ $(s \in S)$ satisfies

$$G_0 = X - N \subset G.$$

The set $h_0(B_s)$ is compact and $h_0(A_t)$ is open in the compact space X, so for every fixed s there exists only a finite number of indices t such that $h_0(A_t)$ intersects $h_0(B_s)$. Since G_0 is dense in X, every set $h_0(A_t)$ intersects at least one set $h_0(B_s)$. This proves that $\overline{\overline{T}} \leq \mathfrak{m}$.

L) The set of all cluster points of the Stone space X of a Boolean σ-algebra $\mathfrak{A}$ is dense in itself $\big($we recall that a point $x \in X$ is a cluster point of X if $x \in \mathbf{C}(X - (x))\big)$.[1]

Suppose the contrary, i.e. the set A of all cluster points in X has an isolated point x_0. Then there exists an open-closed set $B \subset X$ such that $A \cap B = (x_0)$. Let C be an enumerable subset of $B - (x_0)$ such that $B - C$ is infinite. The set C is σ-open since it is the sum of an enumerable sequence of one-point open-closed sets (x), $x \in C$ (points in C are isolated). The point x_0 belongs to the closures of the infinite disjoint sets C and $B - C - (x_0)$. This proves that the closure $C \cup (x_0)$ of the σ-open set C is not open. Contradiction with the hypothesis that $\mathfrak{A}$ is σ-complete.

---

[1] SEMADENI [1].

It follows from the statement just proved that no infinite superatomic Boolean algebra is σ-complete.

A maximal ideal $\varDelta$ (a maximal filter V) is said to *preserve the join* (1) provided the corresponding two-valued natural homomorphism of $\mathfrak{A}$ onto $\mathfrak{A}/\varDelta$ (onto $\mathfrak{A}/V$) preserves (1), i.e. the following condition is satisfied: if $A_t \in \varDelta$ (if $A_t \notin V$) for every $t \in T$, then $A \in \varDelta$ (then $A \notin V$).

A maximal ideal $\varDelta$ (a maximal filter V) is said to *preserve the meet* (1') provided the corresponding two-valued natural homomorphism of $\mathfrak{A}$ onto $\mathfrak{A}/\varDelta$ (onto $\mathfrak{A}/V$) preserves (1'), i.e. the following condition is satisfied: if $A_t \notin \varDelta$ (if $A_t \in V$) for every $t \in T$, then $A \notin \varDelta$ (then $A \in V$).

Every point of the Stone space X of a Boolean algebra $\mathfrak{A}$ can be interpreted as a maximal filter or as a maximal ideal (see § 8). With this interpretation, theorem 22.2 can be formulated as follows: the set of all maximal filters (maximal ideals) which do not preserve a given infinite join (6) or meet (6') is $\mathfrak{m}$-closed and nowhere dense in the Stone space of $\mathfrak{A}$. In fact, the set S in this interpretation is the set of all maximal filters (maximal ideals) which do not preserve (6), and S' is the set of all maximal filters (maximal ideals) which do not preserve (6') [see (8) and (8')].

Example. M) Consider (as above) points of the Stone space X of a Boolean algebra $\mathfrak{A}$ as maximal filters in $\mathfrak{A}$. The set-theoretical complement of the union of all $\mathfrak{m}$-nowhere dense subsets of X is the set of all maximal $\mathfrak{m}$-filters of $\mathfrak{A}$ (we assume here definition (D') of $\mathfrak{m}$-filters; see p. 80).

In fact, a maximal filter V is not $\mathfrak{m}$-complete if and only if it does not preserve an infinite join (1) with $\bar{\bar{T}} \leq \mathfrak{m}$, i.e. if it belongs to a set S of the form (8).

Observe that, if $\mathfrak{A}$ is a Boolean $\mathfrak{m}$-algebra, then in the natural one-to-one correspondence between maximal ideals, maximal filters, two-valued homomorphisms and two-valued measures (see § 6), the maximal $\mathfrak{m}$-ideals, maximal $\mathfrak{m}$-filters, two-valued $\mathfrak{m}$-homomorphisms and two-valued $\mathfrak{m}$-measures correspond with each other.

We finish this section with the following theorem.

22.6. *Let $\mathfrak{A}$ be a Boolean σ-algebra, $A_1, A_2 \in \mathfrak{A}$ and $A_1 \subset A_2$. If $\mathfrak{A}$ is isomorphic to $\mathfrak{A}|A_1$, then $\mathfrak{A}$ is also isomorphic to $\mathfrak{A}|A_2$* [1].

Let h be an isomorphism from $\mathfrak{A}$ onto $\mathfrak{A}|A_1$, let

$$B_1 = -A_2 \quad \text{and} \quad B_n = h(B_{n-1}) \quad \text{for} \quad n = 2, 3, \ldots,$$

---

[1] Sikorski [1] and Tarski [8]. The hypothesis that $\mathfrak{A}$ is σ-complete is essential. See Kinoshita [1] and Hanf [1]. The set-theoretical meaning of this theorem is explained on p. 193.

and let

$$B_0 = - \mathsf{U}_{1 \leq n < \infty} B_n \, .$$

Thus

(11) $$V = B_0 \cup \mathsf{U}_{1 \leq n < \infty} B_n \, .$$

By definition, $B_0 \cap B_n = \wedge$ for $n = 1, 2, \ldots$. Moreover $B_1 \cap B_{m+1} = \wedge$ for $m = 1, 2, \ldots$ since $B_{m+1} \subset A_1 \subseteq - B_1$. Hence $B_2 \cap B_{m+2} = h(B_1 \cap B_{m+1}) = \wedge$. By induction, $B_n \cap B_{n+m} = \wedge$. Thus all the elements

(12) $$B_0, B_1, B_2, \ldots$$

are disjoint. (11) and (12) imply that

(13) $$B_0 \cup \mathsf{U}_{2 \leq n < \infty} B_n = V_{\mathfrak{A} | A_2} = A_2 = - B_1 \, .$$

Let h_0 be the identity isomorphism from $\mathfrak{A} | B_0$ onto itself. The mapping $h_n = h | (\mathfrak{A} | B_n)$ is an isomorphism from $\mathfrak{A} | B_n$ onto $\mathfrak{A} | B_{n+1}$, $n = 1, 2, \ldots$ By D) (where $\mathfrak{A}' = \mathfrak{A} | A_2$) and (11), (12), (13), the formula

$$h'(A) = \mathsf{U}_{0 \leq n < \infty} h_n(A \cap B_n) = (A \cap B_0) \cap h(A - B_0) \qquad (A \in \mathfrak{A})$$

defines an isomorphism of $\mathfrak{A}$ onto $A | \mathfrak{A}_2$.

§ 23. m-subalgebras

Let $\mathfrak{B}$ be a subalgebra of a Boolean algebra $\mathfrak{A}$, and let $\mathfrak{m}$ be an infinite cardinal.

The subalgebra $\mathfrak{B}$ is said to be an $\mathfrak{m}$-*subalgebra of* $\mathfrak{A}$ (or an $\mathfrak{m}$-*complete subalgebra of* $\mathfrak{A}$) provided, for every $\mathfrak{m}$-indexed set $A_t \in \mathfrak{B}$, $t \in T$, the join $\mathsf{U}_{t \in T}^{\mathfrak{A}} A_t$ belongs to $\mathfrak{B}$ whenever it exists. By § 18 (5), then, this join is also the join of all the A_t in $\mathfrak{B}$:

(1) $$\mathsf{U}_{t \in T}^{\mathfrak{A}} A_t = \mathsf{U}_{t \in T}^{\mathfrak{B}} A_t \, .$$

It follows from the de Morgan laws [see § 19 (3)] that $\mathfrak{B}$ is an $\mathfrak{m}$-subalgebra of $\mathfrak{A}$ if and only if, for every $\mathfrak{m}$-indexed set $A_t \in \mathfrak{B}$, $t \in T$, the meet $\mathsf{\cap}_{t \in T}^{\mathfrak{A}} A_t$ belongs to $\mathfrak{B}$ whenever it exists. Then by § 18 (5'), this meet is also the meet of all the A_t in $\mathfrak{B}$:

(1') $$\mathsf{\cap}_{t \in T}^{\mathfrak{A}} A_t = \mathsf{\cap}_{t \in T}^{\mathfrak{B}} A_t \, .$$

Of course if $\mathfrak{A}$ is an $\mathfrak{m}$-algebra, then the words "whenever it exists" can be omitted in the above definition and condition.

If $\mathfrak{B}$ is an $\mathfrak{m}$-subalgebra of $\mathfrak{A}$ for every infinite cardinal $\mathfrak{m}$, then $\mathfrak{B}$ is said to be a *complete subalgebra of* $\mathfrak{A}$.

Observe that every $\mathfrak{m}$-subalgebra $\mathfrak{B}$ of a Boolean $\mathfrak{m}$-algebra $\mathfrak{A}$ is also a Boolean $\mathfrak{m}$-algebra. Every complete subalgebra of a complete Boolean algebra is also a complete Boolean algebra.

Examples. A) If h is an $\mathfrak{m}$-homomorphism of a Boolean $\mathfrak{m}$-algebra $\mathfrak{A}$ into a Boolean algebra $\mathfrak{A}'$, then the class $h(\mathfrak{A})$ of all elements $h(A)$, $A \in \mathfrak{A}$, is an $\mathfrak{m}$-subalgebra of $\mathfrak{A}'$ and a Boolean $\mathfrak{m}$-algebra.

B) A field $\mathfrak{F}$ of subsets of a space X is an $\mathfrak{m}$-field if and only if it is an $\mathfrak{m}$-subalgebra of the field of all subsets of X.

C) An indexed set $\{A_t\}_{t \in T}$ of elements of a Boolean algebra $\mathfrak{A}$ is said to be *monotonic* if, for any $t_1, t_2 \in T$, either $A_{t_1} \subset A_{t_2}$ or $A_{t_2} \subset A_{t_1}$ (i.e. if the set of all A_t is linearly ordered by the Boolean inclusion $\subset$).

If $\mathfrak{B}$ is a subalgebra of $\mathfrak{A}$ such that $\bigcap_{t \in T}^{\mathfrak{A}} B_t \in \mathfrak{B}$ for every monotonic $\mathfrak{m}$-indexed set $\{B_t\}_{t \in T}$ of elements of $\mathfrak{B}$, then $\mathfrak{B}$ is an $\mathfrak{m}$-subalgebra of $\mathfrak{A}$. Moreover, if h is a homomorphism of $\mathfrak{B}$ into a Boolean $\mathfrak{m}$-algebra $\mathfrak{C}$, such that $h(\bigcap_{t \in T}^{\mathfrak{B}} B_t) = \bigcap_{t \in T}^{\mathfrak{C}} h(B_t)$ for every monotonic $\mathfrak{m}$-indexed set $\{B_t\}_{t \in T}$ of elements of $\mathfrak{B}$, then h is an $\mathfrak{m}$-homomorphism from $\mathfrak{B}$ into $\mathfrak{C}$.

To prove the first part of the statement it suffices to show that if $\{A_t\}_{t \in T}$ is any $\mathfrak{m}$-indexed set of elements of $\mathfrak{B}$, then $\bigcap_{t \in T}^{\mathfrak{A}} A_t \in \mathfrak{B}$. We may suppose that T is the set of all ordinals less than an ordinal t_0 of cardinality $\leq \mathfrak{m}$. Suppose that $\bigcap_{t < t_1}^{\mathfrak{A}} A_t \notin \mathfrak{B}$ for an ordinal $t_1 \leq t_0$. Let t_1 be the smallest ordinal with this property. Thus for every $t < t_1$ the element $B_t = \bigcap_{t' < t}^{\mathfrak{A}} A_{t'}$ is in $\mathfrak{B}$. Since $\{B_t\}_{t < t_1}$ is a monotonic $\mathfrak{m}$-indexed set, we have $\bigcap_{t < t_1}^{\mathfrak{A}} B_t \in \mathfrak{B}$. On the other hand, it follows directly from the definition of the infinite meet that $\bigcap_{t < t_1}^{\mathfrak{A}} A_t = \bigcap_{t < t_1}^{\mathfrak{A}} B_t \in \mathfrak{B}$. Contradiction.

The proof of the second part of the statement is analogous.

Note that, by duality, the statement remains true if infinite meets are replaced by infinite joins.

A subalgebra $\mathfrak{B}$ of a Boolean algebra $\mathfrak{A}$ is said to be an $\mathfrak{m}$-*regular* subalgebra of $\mathfrak{A}$ provided that, for every $\mathfrak{m}$-indexed set $A_t \in \mathfrak{B}$, $t \in T$, whenever the join $\bigcup_{t \in T}^{\mathfrak{B}} A_t$ exists, it is also the join of all the A_t in $\mathfrak{A}$, i.e. the equality (1) holds. By the de Morgan laws [see § 19 (3)], the subalgebra $\mathfrak{B}$ of $\mathfrak{A}$ is an $\mathfrak{m}$-regular subalgebra of $\mathfrak{A}$ if and only if, for every $\mathfrak{m}$-indexed set $A_t \in \mathfrak{B}$, $t \in T$, the existence of the meet $\bigcap_{t \in T}^{\mathfrak{B}} A_t$ implies the existence of $\bigcap_{t \in T}^{\mathfrak{A}} A_t$ and the equality (1′) holds. The following condition is also necessary and sufficient for a subalgebra $\mathfrak{B}$ to be an $\mathfrak{m}$-regular subalgebra of $\mathfrak{A}$:

(2)   for every $\mathfrak{m}$-indexed set $A_t \in \mathfrak{B}$ $(t \in T)$, if $\bigcap_{t \in T}^{\mathfrak{B}} A_t = \wedge$, then also $\bigcap_{t \in T}^{\mathfrak{A}} A_t = \wedge$.

The necessity is obvious. The sufficiency follows from the fact that if $A \subset A_t$ for every $t \in T$ $(A \in \mathfrak{B}, A_t \in \mathfrak{B})$, then $A = \bigcap_{t \in T}^{\mathfrak{B}} A_t$ is equivalent to $\bigcap_{t \in T}^{\mathfrak{B}} (A_t - A) = \wedge$, and $A = \bigcap_{t \in T}^{\mathfrak{A}} A_t$ is equivalent to $\bigcap_{t \in T}^{\mathfrak{A}} (A_t - A) = \wedge$.

If a subalgebra $\mathfrak{B}$ of $\mathfrak{A}$ is m regular for every infinite cardinal m, then $\mathfrak{B}$ is said to be a *regular* subalgebra of $\mathfrak{A}$.

It follows directly from the definition that a subalgebra $\mathfrak{B}$ of $\mathfrak{A}$ is an m-regular (regular) subalgebra of $\mathfrak{A}$ if and only if the identity mapping of $\mathfrak{B}$ into $\mathfrak{A}$

$$h(A) = A \quad \text{for} \quad A \in \mathfrak{B}$$

is an m-complete (complete) homomorphism of $\mathfrak{B}$ into $\mathfrak{A}$.

More generally, if h is an isomorphism of $\mathfrak{A}$ into $\mathfrak{A}'$, then $h(\mathfrak{A})$ is an m-regular subalgebra of $\mathfrak{A}'$ if and only if h is an m-isomorphism of $\mathfrak{A}$ into $\mathfrak{A}'$.

Every m-subalgebra $\mathfrak{B}$ of a Boolean m-algebra $\mathfrak{A}$ is an m-regular subalgebra of $\mathfrak{A}$.

If $\mathfrak{C}$ is an m-regular (regular) subalgebra of $\mathfrak{B}$ and $\mathfrak{B}$ is an m-regular (regular) subalgebra of $\mathfrak{A}$, then $\mathfrak{C}$ is also an m-regular (regular) subalgebra of $\mathfrak{A}$.

We recall (see p. 37) that a set $\mathfrak{S}$ of elements of a Boolean algebra $\mathfrak{A}$ is said to be dense in $\mathfrak{A}$ provided that, for every element $A \in \mathfrak{A}$, $A \neq \wedge$, there exists an element $B \in \mathfrak{S}$ such that $\wedge \neq B \subset A$.

If $\mathfrak{S}$ is dense in $\mathfrak{A}$, then $\mathfrak{S} \cup (\wedge)$ and $\mathfrak{S} - (\wedge)$ are also dense in $\mathfrak{A}$.

A set $\mathfrak{S}$ containing $\wedge$ is dense in $\mathfrak{A}$ if and only if every element $A \in \mathfrak{A}$ is the join of all $B \in \mathfrak{S}$ such that $B \subset A$ (or, equivalently, if every $A \in \mathfrak{A}$ is the meet of all the elements $-B$ where $B \in \mathfrak{S}$ and $A \subset -B$).

Indeed, suppose that an element $A_1 \in \mathfrak{A}$ has the property that $B \subset A_1$ for every $B \in \mathfrak{B}$ such that $B \subset A$. The difference $A - A_1$ must be the zero element, for if not, it would contain a non-zero element $B_0 \in \mathfrak{B}$ and this element would be a subelement of A but not of A_1. However $A - A_1 = \wedge$ implies that $A \subset A_1$ which proves that A is the union of all $B \in \mathfrak{B}$ such that $B \subset A$. The proof of the analogous statement for meets can be obtained by passing to complements.

23.1. *If $\mathfrak{B}$ is a dense subalgebra of a Boolean algebra $\mathfrak{A}$, then $\mathfrak{B}$ is a regular subalgebra of $\mathfrak{A}$ and every element $A \in \mathfrak{A}$ is the join (the meet)of all elements $B \in \mathfrak{B}$ such that $B \subset A$ $(A \subset B)$*[1].

It suffices to prove only the first part of 23.1. Suppose $\bigcup_{t \in T}^{\mathfrak{B}} A_t = B \in \mathfrak{B}$ $(A_t \in \mathfrak{B}$ for every $t \in T)$. Let $A \in \mathfrak{A}$ be any element such that $A_t \subset A$ for every $t \in T$. If the element $B - A$ is not zero, then it contains a non-zero element $B_0 \in \mathfrak{B}$. Hence we have $A_t \subset B - B_0 \in \mathfrak{B}$ for every $t \in T$ and $B - B_0 \subset B \neq B - B_0$ which contradicts the fact that B is the join of all A_t in $\mathfrak{B}$. Thus if $A_t \subset A \in \mathfrak{A}$ for every $t \in T$, then $B - A = \wedge$, i.e. $B \subset A$. This proves that B is also the join of all A_t in $\mathfrak{A}$. Thus $\mathfrak{B}$ is a regular subalgebra of $\mathfrak{A}$.

---

[1] SIKORSKI [13].

23.2. *If $\mathfrak{B}'$ is a dense subalgebra of $\mathfrak{B}$, $\mathfrak{B}$ is a subalgebra of a Boolean algebra $\mathfrak{A}$, and $\mathfrak{B}'$ is a regular subalgebra of $\mathfrak{A}$, then $\mathfrak{B}$ is also a regular subalgebra of $\mathfrak{A}$* [1].

It suffices to prove (2). Let $A_t \in \mathfrak{B}$, $t \in T$. Suppose that $\bigcap_{t \in T}^{\mathfrak{B}} A_t = \wedge$. By 23.1, $A_t = \bigcap_{s \in S_t}^{\mathfrak{B}} A_{t,s}$ where $A_{t,s} \in \mathfrak{B}'$. Hence $\bigcap_{t \in T, s \in S_t}^{\mathfrak{B}} A_{t,s} = \wedge$ which implies that $\bigcap_{t \in T, s \in S_t}^{\mathfrak{B}'} A_{t,s} = \wedge$. Since $\mathfrak{B}'$ is a regular subalgebra of $\mathfrak{A}$, the last equality implies that

$$(3) \qquad\qquad \bigcap_{t \in T, s \in S_t}^{\mathfrak{A}} A_{t,s} = \wedge.$$

Suppose that $A \in \mathfrak{A}$ and $A \subset A_t$ for every $t \in T$. Then $A \subset A_{t,s}$ for every $t \in T$ and for every $s \in S_t$. Hence, by (3), $A = \wedge$. This proves that $\bigcap_{t \in T}^{\mathfrak{A}} A_t = \wedge$.

Examples. D) Let $\mathfrak{G}$ be an open basis of a topological space X, i.e. a class of open sets such that every open set is the union of sets in $\mathfrak{G}$. Let $\mathfrak{F}$ be a field of subsets of X such that $\mathfrak{G}$ is a subset of $\mathfrak{F}$ and every set $A \in \mathfrak{F}$ has the Baire property. Let $\varDelta$ be the ideal of all sets $A \in \mathfrak{F}$ of the first category in X. Then the class of all elements $[G]$ where $G \in \mathfrak{G}$ is a dense subset of the quotient algebra $\mathfrak{F}/\varDelta$.

In fact, if $A \in \mathfrak{F}$, then

$$A = (G_0 \cup B_1) - B_2$$

where G_0 is open and B_1, B_2 are of the first category. If $[A]_\varDelta \neq \wedge$, then G_0 is not of the first category. G_0 is the union of some sets belonging to $\mathfrak{G}$ and at least one of them, say G, is not of the first category (since any union of open sets of the first category is of the first category [2]). We have $\wedge \neq [G]_\varDelta \subset A$ which finishes the proof.

E) By an analogy with topological spaces, a Boolean algebra is said to be *separable* [3] provided it contains an at most enumerable dense subset $\mathfrak{G}$.

For instance, all enumerable Boolean algebras are separable. The field $\mathfrak{F}$ of all subsets of the set X of all positive integers is an example of an uncountable separable Boolean algebra. This algebra is atomic. There exist also uncountable complete atomless separable Boolean algebras. The algebra $\mathfrak{B}/\varDelta$, where $\mathfrak{B}$ is the field of all Borel sets of real numbers and $\varDelta$ is the ideal of all sets of the first category, is an example of an algebra of this kind (take as $\mathfrak{G}$ the set of all $[A]$ where A is any interval with rational endpoints!).

Any separable Boolean algebra $\mathfrak{A}$ is isomorphic to a subalgebra of $\mathfrak{F}$ [4]. In fact, suppose that the non-zero elements $A_1, A_2, \ldots$ form a dense subset

[1] SIKORSKI [13].

[2] BANACH [3]. See also KURATOWSKI [3], p. 49.

[3] For an investigation of separable Boolean algebras, see HORN and TARSKI [1].

[4] HORN and TARSKI [1].

of $\mathfrak{A}$. For every n, there exists a maximal filter V_n such that $A_n \in V_n$ (see 6.1). The mapping h given by

$$h(A) = \text{the set of all } n \text{ such that } A \in V_n$$

is the required isomorphism of $\mathfrak{A}$ into $\mathfrak{F}$ (see 8.1).

The converse statement is not true: $\mathfrak{F}$ contains non-separable subalgebras[1]. In fact, $\mathfrak{F}$ contains a subalgebra isomorphic to the free Boolean algebra $\mathfrak{F}_{0, 2^{\aleph_0}}$ which is not separable (see § 14 F)). On the other hand, a Boolean algebra is isomorphic to a subalgebra of $\mathfrak{F}$ if and only if its Stone space is a continuous image of the Stone space of $\mathfrak{F}$, i. e. of the Čech-Stone compactification $\beta(X)$ of the discrete space X (see § 8 F)). The reader familiar with the Čech-Stone compactification can easily verify that a Boolean algebra is isomorphic to a subalgebra of $\mathfrak{F}$ if and only if its Stone space contains an enumerable dense set. Thus the Stone space of the algebra $\mathfrak{B}/\varDelta$ (defined above) of Borel sets modulo sets of the first category contains an enumerable dense set. It can be proved that the Stone space of the algebra all Borel sets (of real numbers) modulo sets of the Lebesgue measure zero does not contain any enumerable dense subset[2] since this algebra is not separable (see the remark at the end of § 35 E)).

F) If $\mathfrak{S}$ is a dense subset of a Boolean algebra $\mathfrak{A}$ satisfying the n-chain condition, then every element of $\mathfrak{A}$ is the join of an n-indexed set $\{A_t\}_{t \in T}$ of disjoint elements in $\mathfrak{S}$. In fact, let $\{A_t\}_{t \in T}$ be a maximal set of disjoint elements $A_t \subset A$, $A_t \in \mathfrak{S}$. We have $A = \bigcup_{t \in T} A_t$ since $\mathfrak{S}$ is dense and $\{A_t\}_{t \in T}$ is maximal, and $\overline{\overline{T}} \leq n$ since A satisfies the n-chain condition.

A subset $\mathfrak{S}$ of a Boolean algebra $\mathfrak{A}$ is said to m-*generate* an m-subalgebra $\mathfrak{B}$ of $\mathfrak{A}$ provided $\mathfrak{B}$ is the smallest m-subalgebra of $\mathfrak{A}$ such that $\mathfrak{S}$ is a subset of $\mathfrak{B}$. The elements of $\mathfrak{S}$ are then called m-*generators* of $\mathfrak{B}$.

Thus $\mathfrak{S}$ m-generates the algebra $\mathfrak{A}$ (i.e. $\mathfrak{S}$ is a set of m-generators of the algebra $\mathfrak{A}$) if and only if the smallest m-subalgebra containing $\mathfrak{S}$ coincides with the whole algebra $\mathfrak{A}$. Observe that if a set $\mathfrak{S}$ of power $\leq n$ m-generates a Boolean algebra $\mathfrak{A}$, then

(4) $$\overline{\overline{\mathfrak{A}}} \leq n^m .$$

Suppose $\mathfrak{S}$ m-generates an m-subalgebra $\mathfrak{B}$ of $\mathfrak{A}$. Then if h_1 and h_2 are two m-homomorphisms of $\mathfrak{B}$ into a Boolean m-algebra $\mathfrak{A}'$ and $h_1(A) = h_2(A)$ for every $A \in \mathfrak{S}$, we have

(5) $$h_1(A) = h_2(A) \quad \text{for every} \quad A \in \mathfrak{B}$$

---

[1] HORN and TARSKI [1], TARSKI [6].
[2] See SEMADENI [2, 3].

since the class of all elements $A \in \mathfrak{A}$ at which h_1 and h_2 assume the same values is an $\mathfrak{m}$-subalgebra of $\mathfrak{A}$ and therefore contains $\mathfrak{B}$. Hence it follows that if a mapping f of $\mathfrak{S}$ into $\mathfrak{A}'$ can be extended to an $\mathfrak{m}$-homomorphism h of $\mathfrak{B}$ into $\mathfrak{A}'$, then this extension h is unique.

23.3. *Let f be a one-to-one mapping of a set $\mathfrak{S}$ $\mathfrak{m}$-generating a Boolean algebra $\mathfrak{A}$ onto a set $\mathfrak{S}'$ $\mathfrak{m}$-generating a Boolean algebra $\mathfrak{A}'$. If f can be extended to an $\mathfrak{m}$-homomorphism of $\mathfrak{A}$ into $\mathfrak{A}'$ and f^{-1} can be extended to an $\mathfrak{m}$-homomorphism of $\mathfrak{A}'$ into $\mathfrak{A}$, then f can be extended to an isomorphism of $\mathfrak{A}$ onto $\mathfrak{A}'$.*

The proof of this lemma is similar to that of the analogous lemma 12.1.

23.4. *Let $\mathfrak{A}_0$ be a subalgebra of a Boolean $\mathfrak{m}$-algebra $\mathfrak{A}$. If a set $\mathfrak{R}$ of elements of $\mathfrak{A}$ satisfies the following conditions:*

(a) *$\mathfrak{A}_0$ is a subset of $\mathfrak{R}$,*

(b) *$\bigcup_{t \in T}^{\mathfrak{A}} A_t \in \mathfrak{R}$ and $\bigcap_{t \in T}^{\mathfrak{A}} A_t \in \mathfrak{R}$ for every monotonic $\mathfrak{m}$-indexed set $\{A_t\}_{t \in T}$ of elements of $\mathfrak{R}$,*

then $\mathfrak{R}$ contains the $\mathfrak{m}$-subalgebra $\mathfrak{B}$ $\mathfrak{m}$-generated by $\mathfrak{A}_0$ (in other words, $\mathfrak{B}$ is the smallest class $\mathfrak{R}$ satisfying (a) and (b)).

If, moreover, h is a mapping from $\mathfrak{R}$ into a Boolean $\mathfrak{m}$-algebra $\mathfrak{A}'$, such that

(a') *$h|\mathfrak{A}_0$ is a homomorphism of $\mathfrak{A}_0$ into $\mathfrak{A}'$,*

(b') *$h(\bigcup_{t \in T}^{\mathfrak{A}} A_t) = \bigcup_{t \in T}^{\mathfrak{A}'} h(A_t)$ and $h(\bigcap_{t \in T}^{\mathfrak{A}} A_t) = \bigcap_{t \in T}^{\mathfrak{A}'} h(A_t)$ for every monotonic $\mathfrak{m}$-indexed set $\{A_t\}_{t \in T}$ of elements of $\mathfrak{R}$,*

then $h|\mathfrak{B}$ is an $\mathfrak{m}$-homomorphism of $\mathfrak{B}$ into $\mathfrak{A}'$.

Let $\mathfrak{R}_0$ denote the least class satisfying (a) and (b). For every element $B \in \mathfrak{A}$ let $\mathfrak{R}(B)$ be the class of all $A \in \mathfrak{A}$ such that simultaneously

(6) $A \cup B \in \mathfrak{R}_0, \quad A - B \in \mathfrak{R}_0 \quad \text{and} \quad B - A \in \mathfrak{R}_0 ,$

(6') $h(A \cup B) = h(A) \cup h(B), \quad h(A - B) = h(A) - h(B),$

$$\text{and} \quad h(B - A) = h(B) - h(A) .$$

It follows immediately from the symmetry of conditions (6) and (6') that

(7) $A \in \mathfrak{R}(B) \quad \text{if and only if} \quad B \in \mathfrak{R}(A) .$

For every fixed $B \in \mathfrak{A}$, the class $\mathfrak{R} = \mathfrak{R}(B)$ satisfies the condition (b). If $B \in \mathfrak{A}_0$, then (a) is also satisfied. Hence $\mathfrak{R}_0 \subset \mathfrak{R}(B)$, i.e.

if $B \in \mathfrak{A}_0$ and $A \in \mathfrak{R}_0$,   then $A \in \mathfrak{R}(B)$.

This implies, by (7), the following statement:

if $B \in \mathfrak{A}_0$ and $A \in \mathfrak{R}_0$,   then $B \in \mathfrak{R}(A)$.

Consequently, if $A \in \mathfrak{R}_0$, then the class $\mathfrak{R} = \mathfrak{R}(A)$ satisfies (a). Since it satisfies also (b), we have $\mathfrak{R}_0 \subset \mathfrak{R}(A)$. Hence, if $A, B \in \mathfrak{R}_0$, then $B \in \mathfrak{R}(A)$, i.e. the conditions (6), (6') are satisfied for arbitrary $A, B \in \mathfrak{R}_0$. This implies that $\mathfrak{R}_0$ is a subalgebra of $\mathfrak{A}$ and $h|\mathfrak{R}_0$ is a homomorphism.

By (b), (b') and C), $\Re_0$ is an m-subalgebra of $\mathfrak{A}$ and $h\,|\,\Re_0$ is an m-homomorphism. Hence $\mathfrak{B} \subset \Re_0$. On the other hand, $\Re_0 \subset \mathfrak{B}$ since the class $\Re = \mathfrak{B}$ satisfies (a) and (b). This proves that $\Re_0 = \mathfrak{B}$.[1]

To prove only the first part of 23.4 it suffices to avoid (6') everywhere in the above proof.

A subset $\mathfrak{S}$ of a Boolean algebra $\mathfrak{A}$ is said to *generate completely* a complete subalgebra $\mathfrak{B}$ of $\mathfrak{A}$ if $\mathfrak{B}$ is the smallest complete subalgebra of $\mathfrak{A}$ such that $\mathfrak{S}$ is a subset of $\mathfrak{B}$. The elements of $\mathfrak{S}$ are then called *complete generators of* $\mathfrak{B}$.

Thus $\mathfrak{S}$ completely generates the algebra $\mathfrak{A}$ (i.e., $\mathfrak{S}$ is a set of complete generators of $\mathfrak{A}$) if and only if the smallest complete subalgebra containing $\mathfrak{S}$ coincides with the whole algebra $\mathfrak{A}$. For instance, every dense subset of $\mathfrak{A}$ completely generates $\mathfrak{A}$ (see the remark before theorem 23.1).

§ 24. Representations by m-fields of sets

We have proved in § 8 that every Boolean algebra $\mathfrak{A}$ is isomorphic to a field of sets $\mathfrak{F}$. If $\mathfrak{A}$ is a Boolean m-algebra, then $\mathfrak{F}$ is also a Boolean m-algebra but, in general, $\mathfrak{F}$ is not an m-field of sets (see § 18 C)). Moreover, we can show that, for every infinite cardinal m, there are Boolean m-algebras that are not isomorphic to any m-field of sets.[2]

Example. A) Every m-field of sets is m-distributive. Consequently every Boolean m-algebra isomorphic to an m-field is m-distributive. If $\mathfrak{F}$ is the field of all Borel sets of real numbers, and $\varDelta$ is either the ideal of all sets of the first category or the ideal of all sets of measure zero, then for every m, $\mathfrak{F}/\varDelta$ is a Boolean m-algebra (see § 21 C) and D)) that is not isomorphic to any m-field of sets since $\mathfrak{F}/\varDelta$ is not m-distributive (see § 19 A)). Analogously, if $\mathfrak{F}$ has the same meaning and $\varDelta$ is the ideal of all at most enumerable sets, then $\mathfrak{F}/\varDelta$ is an example of a Boolean σ-algebra that is not isomorphic to any σ-field of sets since it is not σ-distributive.

By an easy analysis of the proof of 8.2 we obtain the following theorem[3].

24.1. *Let $\mathfrak{A}$ be a Boolean m-algebra. The following conditions are equivalent:*

(o) $\mathfrak{A}$ *is isomorphic to an m-field of sets;*

---

[1] This proof of theorem 23.4 is a slight modification of the proof of a similar theorem in HALMOS [1], p. 27.

[2] This fact was first observed by TARSKI [3]. See also MARCZEWSKI [4], SIKORSKI [4].

[3] PAUC [2], HORN and TARSKI [1], PIERCE [2], SIKORSKI [4], TARSKI [3].

(i) *for every element $A \neq \bigwedge (A \in \mathfrak{A})$ there exists a maximal $\mathfrak{m}$-filter ∇ such that $A \in \nabla$;*

(ii) *for every element $A \neq \bigvee (A \in \mathfrak{A})$ there exists a maximal $\mathfrak{m}$-ideal Δ such that $A \in \Delta$;*

(iii) *for every element $A \neq \bigwedge (A \in \mathfrak{A})$ there exists a two-valued $\mathfrak{m}$-measure m such that $m(A) = 1$;*

(iv) *for every element $A \neq \bigwedge (A \in \mathfrak{A})$ there exists a two-valued $\mathfrak{m}$-homomorphism h of $\mathfrak{A}$ such that $h(A) = \nabla$.*

(v) *in the Stone space of $\mathfrak{A}$, no open non-empty set is a union of $\mathfrak{m}$-nowhere dense sets.*

The equivalence of (i), (ii), (iii), (iv) follows from the natural one-to-one correspondence between maximal $\mathfrak{m}$-filters, maximal $\mathfrak{m}$-ideals, two-valued $\mathfrak{m}$-homomorphisms and two-valued $\mathfrak{m}$-measures (see the remark at the end of § 22). Condition (i) is equivalent to (v) on account of § 22, example M). Condition (i) implies (o) on account of 22.1. Conversely, (o) implies (i). In fact, since condition (i) is invariant under isomorphisms, it suffices to show that every $\mathfrak{m}$-field $\mathfrak{F}$ (of subsets of a space X) has property (i). If $A \in \mathfrak{F}$ is not empty, let x_0 be any element in A. The filter ∇ determined by the point x_0 satisfies (i).

Observe that, without the hypothesis of the $\mathfrak{m}$-completeness of $\mathfrak{A}$, all the conditions (i), (ii), (iv), (v) [where the definition (D') of $\mathfrak{m}$-ideals and $\mathfrak{m}$-filters is assumed; see p. 80] are equivalent to the following statement:

(o') $\mathfrak{A}$ is isomorphic to an $\mathfrak{m}$-regular subalgebra of an $\mathfrak{m}$-field of sets. The idea of the proof is the same.

An $\mathfrak{m}$-field $\mathfrak{F}$ of subsets of a space X is said to be $\mathfrak{m}$-*perfect* provided every maximal $\mathfrak{m}$-filter (or, what is the same, every maximal $\mathfrak{m}$-ideal) is determined by a point $x_0 \in X$.

E x a m p l e. B) Let X be the set of all maximal $\mathfrak{m}$-filters of a Boolean $\mathfrak{m}$-algebra $\mathfrak{A}$, and for every $A \in \mathfrak{A}$ let $h(A)$ be the set of all $\nabla \in X$ such that $A \in \nabla$. By 22.1, $\mathfrak{F} = h(\mathfrak{A})$ is a reduced $\mathfrak{m}$-field. The $\mathfrak{m}$-field $\mathfrak{F}$ is $\mathfrak{m}$-perfect. The exact proof of the last statement is similar to the corresponding part of the proof of 8.2.

By 22.1, if condition (i) is satisfied, then h is an isomorphism. Consequently,

24.2. *If one of the conditions* (o), (i), (ii), (iii), (iv), (v) *in 24.1 is satisfied, then the Boolean $\mathfrak{m}$-algebra $\mathfrak{A}$ is isomorphic to a reduced $\mathfrak{m}$-perfect $\mathfrak{m}$-field of sets $\mathfrak{F}$.*

In particular,

24.3. *Every $\mathfrak{m}$-field of sets is isomorphic to a reduced $\mathfrak{m}$-perfect $\mathfrak{m}$-field of sets.*

It follows from § 22 B) that

24.4. *Every atomic Boolean* m-*algebra* $\mathfrak{A}$ *is isomorphic to an* m-*field of sets. Viz. the mapping*

$$h(A) = \text{the set of all atoms } a \subset A$$

is a complete isomorphism of $\mathfrak{A}$ *into the field of all subsets of the set of all atoms of* $\mathfrak{A}$.

The atomicity is a sufficient but not a necessary condition for a Boolean m-algebra to be isomorphic to an m-field of sets. This follows from the fact that, for every infinite cardinal m, there exist non-atomic m-fields of sets.

Example. C) Let $\mathfrak{F}_{\mathfrak{m},\mathfrak{n}}$ denote the smallest m-field of subsets of the Cantor space $\mathscr{D}_{\mathfrak{n}} = H^{T_0}$ (where $\overline{\overline{T_0}} = \mathfrak{n}$ and $H = (-1, 1)$) such that $\mathfrak{F}_{0,\mathfrak{n}} \subset \mathfrak{F}_{\mathfrak{m},\mathfrak{n}}$ (see § 14, p. 43). The field $\mathfrak{F}_{\mathfrak{m},\mathfrak{n}}$ is reduced since $\mathfrak{F}_{0,\mathfrak{n}}$ is reduced. Every set $A \in \mathfrak{F}_{\mathfrak{m},\mathfrak{n}}$ can be represented in the form

$$(1) \qquad\qquad A = (\mathsf{P}_{t \in T_0 - T} H_t) \times B$$

where B is a subset of $\mathsf{P}_{t \in T} H_t$ and $\overline{\overline{T}} \leq \mathfrak{m}$. This follows from the fact that the class of all sets of the form (1) is an m-field of sets and contains $\mathfrak{F}_{0,\mathfrak{n}}$, therefore it contains $\mathfrak{F}_{\mathfrak{m},\mathfrak{n}}$.

Consequently every set $A \in \mathfrak{F}_{\mathfrak{m},\mathfrak{n}}$ is a union of sets D of the form

$$(2) \qquad\qquad D = \bigcap_{t \in T} \varepsilon(t) \cdot D_t$$

where $\overline{\overline{T}} \leq \mathfrak{m}$ and $\varepsilon(t) = \pm 1$, the sets D_t being defined as in § 14, p. 43.

If $\mathfrak{m} \geq \mathfrak{n}$, then every one-point subset of $\mathscr{D}_{\mathfrak{n}}$ belongs to $\mathfrak{F}_{\mathfrak{m},\mathfrak{n}}$ since it can be represented in the form (2) where $T = T_0$.

If $\sigma \leq \mathfrak{m} < \mathfrak{n}$, then the field $\mathfrak{F}_{\mathfrak{m},\mathfrak{n}}$ is atomless since it does not contain any one-point set (every atom of a reduced field is a one-point set!).

As we have observed, m-distributivity is a necessary condition for a Boolean m-algebra to be isomorphic to an m-field of sets. m-distributivity is not a sufficient condition (a counterexample will be given in § 28 D)). The atomicity is a sufficient, but, in general, not a necessary condition. m-distributivity and atomicity are both necessary and sufficient for Boolean algebras with m generators. More exactly:

24.5. *For any Boolean* m-*algebra* $\mathfrak{A}$ m-*generated by at most* m *generators, the following conditions are equivalent:*

    (i) $\mathfrak{A}$ *is isomorphic to an* m-*field of sets;*

    (ii) $\mathfrak{A}$ *is* m-*distributive;*

    (iii) $\mathfrak{A}$ *is atomic.*

Only the implication (ii) → (iii) need be proved. If an m-indexed set $\{A_t\}_{t \in T}$ m-generates $\mathfrak{A}$, then every element of the form

$$(3) \qquad\qquad \bigcap_{t \in T} \varepsilon(t) \cdot A_t$$

7*

where $\varepsilon(t) = \pm 1$ is either the zero or an atom since it satisfies condition
§ 9 (3). Since

$$\bigcap_{t \in T} (A_t \cup -A_t) = V,$$

it follows from 19.2 (d_2) that every element $A \neq \wedge$ contains at least one
atom (3). Thus $\mathfrak{A}$ is atomic.

24.6. *For every Boolean* $\mathfrak{m}$-*algebra* $\mathfrak{A}$, *the following conditions are
equivalent:*

(i) $\mathfrak{A}$ *is* $\mathfrak{m}$-*distributive;*

(ii) *if* $\mathfrak{S}$ *is the set of all atoms of an* $\mathfrak{m}$-*subalgebra* $\mathfrak{m}$-*generated by at
most* $\mathfrak{m}$ *elements, then* $\bigcup_{A \in \mathfrak{S}}^{\mathfrak{A}} A = V;$

(iii) *every* $\mathfrak{m}$-*subalgebra* $\mathfrak{m}$-*generated by at most* $\mathfrak{m}$ *elements is atomic;*

(iv) *every* $\mathfrak{m}$-*subalgebra* $\mathfrak{m}$-*generated by at most* $\mathfrak{m}$ *elements is isomorphic to an* $\mathfrak{m}$-*field of sets.*

The proof of the implication (i) → (ii) is the same as the proof of the
implication (ii) → (iii) in 24.5. (ii) implies (iii) by § 18 (5) and § 18 G).
(iii) implies (iv) by 24.4. (iv) implies (i) since $\mathfrak{A}$ is $\mathfrak{m}$-distributive if and
only if every one of its $\mathfrak{m}$-subalgebras $\mathfrak{m}$-generated by at most $\mathfrak{m}$ elements
is $\mathfrak{m}$-distributive.

Suppose that $\mathfrak{F}$ is a reduced $\mathfrak{m}$-perfect field of subsets of a space X.
We can introduce a topology in X in the same way as in § 7, i.e. by
defining $\mathfrak{F}$ to be an open basis of X. Every set $A \in \mathfrak{F}$ is then both open
and closed in X, but the converse statement is not true in general (see
§ 27 A)), and different $\mathfrak{m}$-fields of subsets of X can induce the same
topology in X. Therefore this topology is generally of no practical im-
portance.

The full analogy with the Stone space can be obtained only in the
case where the $\mathfrak{m}$-field $\mathfrak{F}$ under consideration satisfies the following
condition analogous to 6.1 (i):

($e_{\mathfrak{m}}$) every proper $\mathfrak{m}$-ideal is contained in a maximal $\mathfrak{m}$-ideal.

The formulation of conditions equivalent to ($e_{\mathfrak{m}}$) but analogous to
6.1 (ii), (iii), (iv) respectively is left to the reader.

If the reduced $\mathfrak{m}$-field $\mathfrak{F}$ has property ($e_{\mathfrak{m}}$), then the space X topologized as above satisfies the following condition, which is an infinite
analogue of compactness:

if $X = \bigcup_{t \in T} G_t$ where the G_t are open, then there exists a subset
$T' \subset T$ such that $\overline{\overline{T'}} \leq \mathfrak{m}$ and $X = \bigcup_{t \in T'} G_t$.

Property ($e_{\mathfrak{m}}$) implies also that $\mathfrak{F}$ is then the field of all open-closed
subsets of X.

The proof of the two last statements is analogous to the proof of the
analogous statements in 7.1[1].

---

[1] See SIKORSKI [12].

To give a criterion for existence of an extension of a proper m-ideal to a maximal m-ideal, we introduce the following definitions.

Let α be the least ordinal of power n. Denote by S (by S_0) the set of all sequences $s = \{\varepsilon_\gamma\}_{\gamma < \beta}$ where $0 < \beta \leq \alpha$ (where $\beta = \alpha$) and $\varepsilon_\gamma = \pm 1$ for every γ. The ordinal β is said to be the *length* of the transfinite sequence s. For any $s, s' \in S$, we write $s' < s$ if s' is a proper initial segment of s. If $s \in S - S_0$, then

$$s, \pm 1$$

denotes the sequence obtained from s by adding ± 1 at the end of s.

By an n-*dyadic system* we shall understand any indexed set $\{A_s\}_{s \in S}$ of elements of a Boolean algebra $\mathfrak{A}$ such that

$$A_{-1} \cup A_1 = V, \quad A_{-1} \cap A_1 = \wedge,$$

$$A_{s,-1} \cup A_{s,1} = A_s \quad \text{and} \quad A_{s,-1} \cap A_{s,1} = \wedge \quad \text{for every} \quad s \in S - S_0,$$

$$A_s = \bigcap_{s' < s} A_{s'} \quad \text{if the length of } s \in S \text{ is a limit ordinal.}$$

A Boolean algebra $\mathfrak{A}$ is said to be *dyadically n-distributive* provided

$$V = \bigcup_{s \in S_0} A_s$$

for every n-dyadic system $\{A_s\}_{s \in S}$ in $\mathfrak{A}$.

24.7. *Let Δ be a proper m-ideal of a Boolean m-algebra $\mathfrak{A}$. Suppose that there exists an infinite cardinal n such that $2^{n'} \leq m$ for every $n' < n$, $\mathfrak{A}$ is dyadically n-distributive, and $\mathfrak{A}/\Delta$ satisfies the n_0-chain condition for some $n_0 < n$. Then Δ is contained in a maximal m-ideal[1].*

It suffices to prove that there exists an element $E \notin \Delta$ such that

(4) the intersection $\Delta_0 = \Delta \cap (\mathfrak{A}|E)$ is a maximal ideal in $\mathfrak{A}|E$.

In fact, the set of all $A \in \mathfrak{A}$ such that $A \cap E \in \Delta_0$ is then a maximal m-ideal containing Δ.

Suppose that there is no element E with property (4), i.e. for every element $E \notin \Delta$ there exists an element, denoted by $f(E)$, such that $f(E) \subset E$, $f(E) \notin \Delta$ and $E - f(E) \notin \Delta$. We extend the mapping f over Δ by letting $f(E) = E$ for $E \in \Delta$.

Now we define an n-dyadic system as follows:

$$A_1 = f(V), \quad A_{-1} = -f(V),$$

$$A_{s,1} = f(A_s) \quad \text{and} \quad A_{s,-1} = A_s - f(A_s) \quad \text{for every} \quad s \in S - S_0,$$

$$A_s = \bigcap_{s' < s} A_{s'} \quad \text{if the length of } s \text{ is a limit ordinal.}$$

By dyadic n-distributivity,

$$V = \bigcup_{s \in S_0} A_s.$$

Let S' denote the set of all $s' \in S - S_0$ such that $A_{s'} \in \Delta$. For every $s \in S_0$ there exists an $s' \in S'$ such that $s' < s$. For if not, then the set of all

---

[1] SMITH and TARSKI [1].

elements $[A_{s'} - A_{s', \varepsilon_{s'}}]_{\Delta}$, where $s' < s$ and $\varepsilon_{s'} = \pm 1$ is chosen in such a way that $(s', \varepsilon_{s'}) < s$, would be a class of $\mathfrak{n}$ disjoint elements in $\mathfrak{A}/\Delta$. Contradiction.

Since $A_s \subset A_{s'}$ for $s' < s$, we have

$$V = \bigcup\nolimits_{s' \in S'} A_{s'} .$$

Since $S' \subset S - S_0$, we have $\overline{\overline{S'}} \leq \mathfrak{m}$. This implies that $V \in \Delta$, i.e. Δ is not proper. Contradiction.

Let $\mathfrak{A}$ be a Boolean algebra, and let $\mathbf{J}$ and $\mathbf{M}$ be two classes of non-void subsets of $\mathfrak{A}$ such that all the joins

(5) $$\bigcup\nolimits_{A \in \mathfrak{S}}^{\mathfrak{A}} A \qquad\qquad (\mathfrak{S} \in \mathbf{J})$$

and meets

(5') $$\bigcap\nolimits_{A \in \mathfrak{S}}^{\mathfrak{A}} A \qquad\qquad (\mathfrak{S} \in \mathbf{M})$$

exist. Sometimes it is important to know whether there exists a $(\mathbf{J}, \mathbf{M})$-homomorphism h of $\mathfrak{A}$ into the field $\mathfrak{F}$ of all subsets of a space X i.e. a homomorphism h of $\mathfrak{A}$ onto a field of sets that transforms the infinite joins (5) and meets (5') onto the corresponding set-theoretical unions and intersections, respectively.

An easy analysis of the proof of 22.1 shows that

24.8. *If X_0 is any set of maximal filters preserving all the joins* (5) *and meets* (6') *(see* § 22 *p.* 90), *then the mapping h defined by*

(6) $$h(A) = \text{the set of all } V \in X_0 \text{ such that } A \in V$$

is a $(\mathbf{J}, \mathbf{M})$-*homomorphism of $\mathfrak{A}$ into the field of all subsets of X_0.*

An analysis of the proof of the theorem 24.1 yields the following theorem.

24.9. *A necessary and sufficient condition for the existence of a* $(\mathbf{J}, \mathbf{M})$-*isomorphism h of $\mathfrak{A}$ into the field of all subsets of a space X is that, for every element $A \neq \Lambda$, there exists a maximal filter V preserving all the joins* (5) *and meets* (5') *and such that $A \in V$.*

Of course, the necessary and sufficient condition given in 24.9 is not always satisfied (for instance, it is not satisfied if $\mathfrak{A}$ is a Boolean σ-algebra that is not isomorphic to any σ-field of sets, and (5) and (5') are the sets of all enumerable joins and meets which hold in $\mathfrak{A}$). However it is always satisfied if the given set of joins and meets is at most enumerable:

24.10. *If $\overline{\overline{\mathbf{J}}} \leq \aleph_0$ and $\overline{\overline{\mathbf{M}}} \leq \aleph_0$, then there exists a* $(\mathbf{J}, \mathbf{M})$-*isomorphism h of $\mathfrak{A}$ into the field of all subsets of a space*[1].

In fact, let X be the Stone space of $\mathfrak{A}$, and let h_0 be the isomorphism of $\mathfrak{A}$ onto the field of all open-closed subsets of X. By 22.2, the union Z

---

[1] Theorem 24.10 was found by Sikorski. See Rasiowa and Sikorski [1] and Rasiowa [2]. Another algebraic proof of 24.10 was found by A. Tarski.

of all the defect sets corresponding to the joins (5) and meets (5') is a set of the first category in X. On the other hand, every set of the first category in a compact Hausdorff space is a boundary set, i.e. its complement is dense[1]. Hence, the set $X_0 = X - Z$ is dense in X and, consequently, the set

$$h(A) = X_0 \cap h_0(A)$$

is not empty for $A \neq \Lambda$. By 24.8, h is the required isomorphism.

By the de Morgan laws, a homomorphism preserves an infinite join (5) if and only if it preserves the infinite meet $\bigcap_{A \in \mathfrak{S}}^{\mathfrak{A}} - A$. Thus in the representation problem examined here it suffices to consider only the preservation of a set $\mathbf{M}$ of infinite meets. This remark will be used below to simplify the formulation of the next theorem.

Let $\mathbf{M}$ be a class of non-void subsets of a Boolean algebra $\mathfrak{A}$, such that all the meets (5') exist. $\mathfrak{A}$ is said to be $\mathbf{M}$-*distributive* if it has the following property (where $H = (-1, 1)$):

(7) $$B = \bigcup_{\varepsilon \in H^T} (B \cap \bigcap_{t \in T} \varepsilon(t) \cdot A_t)$$

for every element $B \in \mathfrak{A}$ and for every indexed set $\{A_t\}_{t \in T}$ of elements of $\mathfrak{A}$, such that for every $\varepsilon \in H^T$ the set composed of all the elements

(8) $$B, \quad \varepsilon(t) \cdot A_t \qquad\qquad (t \in T)$$

belongs to $\mathbf{M}$. In (7) and in the sequel $B \cap \bigcap_{t \in T} \varepsilon(t) \cdot A_t$ denotes the meet of all the elements (8) (this meet can exist even if $\bigcap_{t \in T} \varepsilon(t) \cdot A_t$ does not exist).

Let $\mathbf{M}$ be any class of non-void subsets of a Boolean algebra $\mathfrak{A}$, such that all the meets (5') exist. We shall denote by $\mathbf{M}'$ the smallest class of non-void subsets of $\mathfrak{A}$ such that

(a$_0$) $\mathbf{M}$ is a subclass of $\mathbf{M}'$;

(a$_1$) all one-element sets are in $\mathbf{M}'$;

(a$_2$) the union of two sets belonging to $\mathbf{M}'$ also belongs to $\mathbf{M}'$;

(a$_3$) if $\mathfrak{S} \in \mathbf{M}'$ and the meet (5') of all elements in $\mathfrak{S}$ is equal to Λ, then every set $\mathfrak{S}' \supset \mathfrak{S}$ belongs to $\mathbf{M}'$.

The smallest class $\mathbf{M}'$ always exists (viz. it can be defined as the intersection of all classes satisfying (a$_0$)—(a$_3$)). For any $\mathfrak{S} \in \mathbf{M}'$, the meet $\bigcap_{A \in \mathfrak{S}} A$ exists.

24.11. *In order that there exist an isomorphism h of $\mathfrak{A}$ into the field $\mathfrak{F}$ of all subsets of a set X preserving all the infinite meets (5') in $\mathbf{M}$, it is necessary and sufficient that $\mathfrak{A}$ be $\mathbf{M}'$-distributive*[2].

To prove the necessity, let us first observe that if h preserves all the infinite meets in $\mathbf{M}$, then h also preserves all the infinite meets in $\mathbf{M}'$.

---

[1] ČECH [1]. See also SIKORSKI [4].

[2] BRUNS [1].

For any elements $B, A_t \in \mathfrak{A}$ $(t \in T)$,

$$h(B) = \bigcup_{\varepsilon \in H^T}^{\mathfrak{F}} \left(h(B) \cap \bigcap_{t \in T}^{\mathfrak{F}} \varepsilon(t) \cdot h(A_t) \right)$$

because all the operations on the right-hand side are set-theoretical. If all the sets (8) are in $\mathbf{M}'$, then

$$h(B) \cap \bigcap_{t \in T}^{\mathfrak{F}} \varepsilon(t) \cdot h(A_t) = h(B \cap \bigcap_{t \in T} \varepsilon(t) \cdot A_t) .$$

Thus

$$h(B) = \bigcup_{\varepsilon \in H^T}^{\mathfrak{F}} \left(h(B \cap \bigcap_{t \in T} \varepsilon(t) \cdot A_t) \right)$$

which proves (7) (see the last remark in § 22 A), p. 83).

To prove the sufficiency, suppose that $\mathfrak{A}$ is $\mathbf{M}'$-distributive. Let $\{A_t\}_{t \in T}$ be an indexed set containing all elements in $\mathfrak{A}$. For any function $\varepsilon \in H^T$, let $\mathfrak{S}_\varepsilon$ be the set of all the elements $\varepsilon(t) \cdot A_t$. Thus, by (7)

(9) $$B = \bigcup_{\varepsilon \in H^T} (B \cap \bigcap_{A \in \mathfrak{S}_\varepsilon} A)$$

provided that for every $\varepsilon \in H^T$, the set composed of B and all the elements in $\mathfrak{S}_\varepsilon$ belongs to $\mathbf{M}'$.

For every $\varepsilon \in H^T$, let V_ε be the set of all the elements $C \in \mathfrak{A}$ such that

(10) $$C \supset \bigcap_{A \in \mathfrak{S}} A \quad \text{for a set } \mathfrak{S} \in \mathbf{M}', \mathfrak{S} \subset \mathfrak{S}_\varepsilon .$$

It follows from (a_2) that V_ε is a filter in $\mathfrak{A}$. It follows from (a_1) that for every $t \in T$, either A_t or $- A_t$ is in V_ε. Thus the filter V_ε is either maximal or not proper. It follows immediately from definition (10) (see also p. 90) that if V_ε is maximal then V_ε preserves all the infinite meets in $\mathbf{M}'$, in particular all the infinite meets in $\mathbf{M}$ (see (a_0)).

By 24.9, it suffices to prove that for every non-zero element $B \in \mathfrak{A}$ there is an $\varepsilon \in H^T$ such that $B \in V_\varepsilon$ and V_ε is maximal. Suppose the contrary, i.e. that there is a $B \neq \wedge$ such that for every $\varepsilon \in H^T$, either V_ε is not proper or $B \notin V_\varepsilon$. In other words, the filter generated by B and V_ε is not proper, i.e., by (10), there exists a set $\mathfrak{S} \in \mathbf{M}'$, $\mathfrak{S} \subset \mathfrak{S}_\varepsilon$ such that

$$B \cap \bigcap_{A \in \mathfrak{S}} A = \wedge .$$

By (a_1) and (a_2) the set composed of B and all the elements of $\mathfrak{S}$ belongs to $\mathbf{M}'$. By (a_3), the set composed of B and all the elements of $\mathfrak{S}_\varepsilon$ belongs to $\mathbf{M}'$. Since $B \cap \bigcap_{A \in \mathfrak{S}_\varepsilon} A = \wedge$, we obtain $B = \wedge$ from (9).

Example. D) Suppose $\mathbf{M}$ is the class of all subsets of a Boolean $\mathfrak{m}$-algebra $\mathfrak{A}$ which are of power $\leq \mathfrak{m}$. Theorem 24.11 yields then a necessary and sufficient condition for $\mathfrak{A}$ to be isomorphic to an $\mathfrak{m}$-field of sets.

Note that then the algebra $\mathfrak{A}$ is $\mathbf{M}$-distributive if and only if it is $\mathfrak{m}$-distributive, on account of 19.1 and 19.2. $\mathbf{M}'$-distributivity is a stronger condition than $\mathfrak{m}$-distributivity.

§ 25. Complete Boolean algebras

If $\mathfrak{A}$ is a complete Boolean algebra, then the converse to theorem 24.4 is also true. More precisely:

25.1. *A complete Boolean algebra $\mathfrak{A}$ is isomorphic to a complete field of sets if and only if it is atomic. In this case $\mathfrak{A}$ is isomorphic to the field of all subsets of the set of all atoms of $\mathfrak{A}$* [1].

It suffices to prove that if $\mathfrak{A}$ is isomorphic to a complete field of sets, then $\mathfrak{A}$ is atomic. The remaining part of 25.1 follows from 24.4.

Therefore it suffices to show that every complete field $\mathfrak{F}$ of subsets of a space X is atomic. Let A be a non-empty set in $\mathfrak{F}$ and let $x_0 \in A$. The intersection A_0 of all sets $B \in \mathfrak{F}$ such that $x_0 \in B$ is an atom of $\mathfrak{F}$ and $A_0 \subset A$.

It follows from 25.1 that two complete atomic Boolean algebras are isomorphic if and only if the cardinals of their sets of atoms are equal (see the partial results in § 8 B) and § 9 D) for finite Boolean algebras which are a particular case of complete Boolean algebras).

Another criterion is given by the following theorem.

25.2. *A complete Boolean algebra $\mathfrak{A}$ is isomorphic to a complete field of sets if and only if $\mathfrak{A}$ is completely distributive.*

The necessity of complete distributivity follows from the fact that every complete field of sets is completely distributive.

To prove the sufficiency, suppose that $\{A_t\}_{t \in T}$ contains all elements of a completely distributive complete Boolean algebra $\mathfrak{A}$. Let S be the set composed of numbers -1 and 1 only, and let $A_{t,s} = s \cdot A_t$ for $t \in T$ and $s \in S$. For every $\varphi \in S^T$, the element

$$(1) \qquad\qquad \bigcap_{t \in T} A_{t, \varphi(t)}$$

is either equal to $\wedge$ or it is an atom [since it satisfies condition § 9 (3)]. By distributivity [see 19.2 (d$_2$)] every element $A \neq \wedge$ ($A \in \mathfrak{A}$) contains an atom (1). This proves that $\mathfrak{A}$ is atomic.

A Boolean algebra $\mathfrak{A}$ is said to be *homogeneous* if, for every $A \in \mathfrak{A}$, $A \neq \wedge$, the algebra $\mathfrak{A} | A$ is isomorphic to $\mathfrak{A}$. We give below two important examples of complete homogeneous algebras.

Examples. A) Let $\mathfrak{F}$ be the field of all Borel subsets of the space X of all real numbers, and let Δ be the ideal of all Borel sets of Lebesgue measure zero. The complete Boolean algebra $\mathfrak{A} = \mathfrak{F}/\Delta$ is homogeneous.

Indeed, if $A \in \mathfrak{F}$, $[A] \neq \wedge$, then there exists a one-to-one mapping φ from A onto X such that, for every $B \subset A$, $\varphi(B)$ is a Borel set if and

---
[1] Theorem 25.1 is due to A. LINDENBAUM and A. TARSKI (see TARSKI [2]), theorem 25.2 is due to TARSKI [2]. See also HORN and TARSKI [1].

only if B is a Borel set, $\varphi(B)$ has a positive measure if and only if B has a positive measure[1].

Hence it follows that φ induces an isomorphism of $\mathfrak{A} = \mathfrak{F}/\varDelta$ onto $\mathfrak{F}_A/\varDelta_A$ where $\mathfrak{F}_A = \mathfrak{F} \,|\, A$ and $\varDelta_A = \varDelta \cap \mathfrak{F}_A$. By § 10 C), $\mathfrak{F}_A/\varDelta_A$ is isomorphic to $\mathfrak{A} \,|\, [A]$. Consequently $\mathfrak{A}$ is isomorphic to $\mathfrak{A} \,|\, [A]$.

B) Let X be a topological space such that

(h) every non-empty open set contains an open subset homeomorphic to X.

Let $\mathfrak{F}$ be the field of all Borel subsets of X, and let $\varDelta$ be the ideal of all Borel sets of the first category. The complete Boolean algebra $\mathfrak{A} = \mathfrak{F}/\varDelta$ is homogeneous.

In fact, every element $A \in \mathfrak{A}$ can be represented in the form $A = [G]$ where G is open. If $A \neq \varLambda$, then G is not empty and contains an open subset G_0 homeomorphic to X. The field $\mathfrak{F}_0$ of all Borel subsets of the topological space G_0 and the ideal $\varDelta_0$ of all Borel subsets of the first category in the space G_0 satisfy the identities

$$\mathfrak{F}_0 = \mathfrak{F} \,|\, G_0 , \quad \varDelta_0 = \varDelta \cap \mathfrak{F}_0 .$$

Thus, by § 10 C), $\mathfrak{A} \,|\, A_0$ where $A_0 = [G_0] \in \mathfrak{A}$ is isomorphic to $\mathfrak{F}_0/\varDelta_0$. On the other hand, the homeomorphism of G_0 onto X induces an isomorphism from $\mathfrak{F}/\varDelta$ onto $\mathfrak{F}_0/\varDelta_0$. This proves that $\mathfrak{A} \,|\, A_0$ is isomorphic to $\mathfrak{A}$. Since $A_0 \subset A$, the algebra $\mathfrak{A} \,|\, A$ is also isomorphic to $\mathfrak{A}$ by 22.6.

The space of all real numbers is an example of a space X with property (h). It is easy to see that in this case the algebra $\mathfrak{F}/\varDelta$ has power $2^{\aleph_0}$.

Another important example[2] of a topological space with property (h) is given by the Cantor space $\mathscr{D}_\mathfrak{n} = H^{T_0} (\mathfrak{n} \geq \aleph_0)$ defined on p. 43. In fact, every non-empty open set $G \subset \mathscr{D}_\mathfrak{n}$ is the union of sets

$$(2) \qquad\qquad \varepsilon_1 \cdot D_{t_1} \cap \cdots \cap \varepsilon_n \cdot D_{t_n} \qquad (t_i \in T_0, \varepsilon_i = \pm 1)$$

where the notation is that of p. 43. It is easy to verify that every set (2) is homeomorphic to $\mathscr{D}_\mathfrak{n}$.

Suppose $\mathfrak{n}$ is an infinite cardinal. The class of all sets (2) has power $\mathfrak{n}$. Since every class of disjoint sets (2) is finite or enumerable (see § 20, example L)), every open set in $\mathscr{D}_\mathfrak{n}$ is the union of a finite or countable sequence of sets (2) and a nowhere dense set. This implies that the number of elements in $\mathfrak{F}/\varDelta$ is $\leq \mathfrak{n}^{\aleph_0}$. This implies also that every element in $\mathfrak{F}/\varDelta$ can be represented in the form

$$(3) \qquad\qquad\qquad [G \times H^{T_0 - T'}]$$

where T' is finite or enumerable, and G is an open subset of the Cantor space $H^{T'}$. Moreover, for every enumerable set $T \subset T_0$ we can find a set

---

[1] See e.g. Marczewski [15].
[2] Pierce [4].

$G_T \subset H^T$ such that if G is another open set such that $G \subset H^{T'}$ and

$$[G_T \times H^{T_0-T}] = [G \times H^{T_0-T'}]$$

then $T \subset T'$ (to construct G_T it is convenient to think of H^T for an enumerable set T as the Cantor set of real numbers defined on p. 28; see footnote on p. 43). This proves that $[G_{T_1}] \neq [G_{T_2}]$ for distinct enumerable sets T_1, $T_2 \subset T_0$. Since the class of all enumerable subsets of T_0 has power $\mathfrak{n}^{\aleph_0}$, we infer that $\mathfrak{F}/\varDelta$ contains at least $\mathfrak{n}^{\aleph_0}$ elements. Hence it follows that $\mathfrak{F}/\varDelta$ has exactly $\mathfrak{n}^{\aleph_0}$ elements.

Thus for every infinite cardinal $\mathfrak{n}$ there exists a complete homogeneous Boolean algebra of power $\mathfrak{n}^{\aleph_0}$.[1]

All known examples of complete Boolean algebras are isomorphic to direct unions of complete homogeneous Boolean algebras. It is not known whether every complete Boolean algebra has this property. The answer to this problem is affirmative if the homogeneity is replaced by a similar property called weak homogeneity. This follows from theorem 25.3 below.

For any non-zero element A of a fixed Boolean algebra $\mathfrak{A}$ we shall denote by card A the cardinal of the Boolean algebra $\mathfrak{A} | A$. By definition card A has the following monotonicity property:

$$\text{if} \quad A \subset B \ (A, B \in \mathfrak{A}), \quad \text{then} \quad \text{card} A \leq \text{card} B .$$

A Boolean algebra $\mathfrak{A}$ is said to be *weakly homogeneous* if card A = card V (i.e. card $A = \overline{\overline{\mathfrak{A}}}$) for every $A \in \mathfrak{A}$, $A \neq \wedge$.

25.3.[2] *For every non-degenerate complete Boolean algebra $\mathfrak{A}$, there exists a unique decomposition*

(4) $$V = \mathsf{U}_{t \in T} A_t$$

with non-zero disjoint elements A_t, and for every $t \in T$ there exists a (non-unique, in general) decomposition

(5) $$A_t = \mathsf{U}_{s \in S_t} A_{t,s}$$

with non-zero disjoint elements $A_{t,s}$, such that all the algebras $\mathfrak{A} | A_{t,s}$ are weakly homogeneous and

(6) $$\text{card} A_{t,s} = \text{card} A_{t',s'} \quad \text{if and only if} \quad t = t' .$$

---

[1] PIERCE [4]. Earlier GINSBURG [2] proved that for every infinite cardinal $\mathfrak{n}$ there is a complete homogeneous algebra of cardinality $2^{\mathfrak{n}}$.

[2] PIERCE [4] where the theorem is formulated more generally; card A is supposed to be any cardinal-valued mapping which has the monotonicity property. The proof is the same. See also PIERCE [9].

Let T_0 be the set of all cardinals $\leq \operatorname{card} V$. For every $t \in T_0$, let

(7) $$B_t = \bigcup_{\operatorname{card} A \leq t} A$$

and let

(8) $$A_t = B_t - \bigcup_{t' < t} B_{t'}.$$

The elements A_t are disjoint and have the following properties:

(9) if $\wedge \neq B \subset A_t$, then there is an A such $A \cap B \neq \wedge$ and $\operatorname{card} A = t$;

(10)   if $\wedge \neq C \subset A_t$, then $\operatorname{card} C \geq t$,

for the inequality $t' = \operatorname{card} C < t$ would imply $C \subset B_{t'}$ and $C \cap A_t = \wedge$.

The unit element is the union of all the elements A_t, $t \in T_0$, since the union of all A_t is equal to the union of all B_t and the last of the sets B_t is equal to V.

Let T be the set of all $t \in T_0$ such that $A_t \neq \wedge$. Clearly (4) holds.

For every $t \in T$, let $\{A_{t,s}\}_{s \in S_t}$ be a maximal set of disjoint non-zero subelements of A_t, such that

(11) $$\operatorname{card} A_{t,s} = t.$$

By definition, (6) holds. We shall prove (5).

Suppose (5) does not hold for some $t \in T$, i.e. there is a non-zero element $B \subset A_t$ such that $B \cap A_{t,s} = \wedge$ for every $s \in S_t$. Let A be an element satisfying (9). By (10), the element $C = A \cap B$ satisfies the inequality $\operatorname{card} C \geq t$. By the monotonicity, $\operatorname{card} C \leq \operatorname{card} A = t$. Thus $\operatorname{card} C = t$. This contradicts the maximality of $\{A_{t,s}\}_{s \in S_t}$ since C is disjoint from all $A_{t,s}$, $s \in S_t$.

The algebra $\mathfrak{A}|A_{t,s}$ is weakly homogeneous. For suppose $\wedge \neq C \subset A_{t,s}$. Then $\operatorname{card} C \leq t$ by the monotonicity, and $\operatorname{card} C \geq t$ by (10). Thus $\operatorname{card} C = t = \operatorname{card} A_{t,s}$, and $A_{t,s}$ is the unit element of $\mathfrak{A}|A_{t,s}$.

It remains to prove that (7) is unique. Suppose that $\{A_t\}_{t \in T}$ and $\{A_{t,s}\}_{s \in S_t}$ ($t \in T$) are two indexed sets satisfying the conditions mentioned in 25.3. By (6) we may assume that the index t coincides with the cardinal $\operatorname{card} A_{t,s}$. Then $A_{t,s} \subset B_t - \bigcup_{t' < t} B_t$ where B_t is defined by (7). Hence it follows by (4) and (5) that the elements A_t coincide with the elements (8). This completes the proof of 25.3.

The weak homogeneity of the $\mathfrak{A}|A_{t,s}$ implies that for every cardinal $t \in T$ either $t = 2$ or t is infinite, and consequently $t \geq 2^{\aleph_0}$ by § 20 E). Denote by T' the set of all infinite cardinals in T and by $\mathfrak{n}$ the cardinal of $\mathfrak{A}$. It follows from (4), (5) and (6) that

$$\mathfrak{n} = 2^{\bar{\bar{S}}_2} \cdot \Pi_{t \in T'} \, t^{\bar{\bar{S}}_t}$$

(see § 16 (1) and § 22 E)). If $2 \notin T$, we have $\bar{\bar{S}}_2 = 0$.

Now observe that if a Boolean algebra $\mathfrak{A}$ is infinite and weakly homogeneous, then the cardinal t of $\mathfrak{A}$ has the property

(12) $$t^{\aleph_0} = t.$$

In fact, then there exists an infinite sequence C_n of disjoint non zero elements such that $V = \bigcup_{1 \leq n < \infty} C_n$. Consequently

$$t = \overline{\overline{\mathfrak{A}}} = \prod_{n=1}^{\infty} \overline{\overline{\mathfrak{A}|C_n}} = t^{\aleph_0} .$$

By (12) we get the following identity for the cardinal $\mathfrak{n}$ of $\mathfrak{A}$:

$$\mathfrak{n}^{\aleph_0} = 2^{\overline{\overline{S}_2} \cdot \aleph_0} \cdot \prod_{t \in T} t^{\overline{\overline{S}_t}} .$$

Suppose now that the algebra $\mathfrak{A}$ is infinite. Then either S_2 is infinite, or T' is not empty.

It S_2 is infinite, then $\overline{\overline{S}}_2 \cdot \aleph_0 = \overline{\overline{S}}_2$ and consequently

(13) $\mathfrak{n}^{\aleph_0} = \mathfrak{n} .$

If S_2 is finite, then T' contains at least one cardinal t_0 and $t_0 \geq 2^{\aleph_0}$. Consequently (see § 16 (1) and § 22 E))

$$2^{\overline{\overline{S}_2} \cdot \aleph_0} \cdot t_0^{\overline{\overline{S}}_{t_0}} = t_0^{\overline{\overline{S}}_{t_0}} = 2^{\overline{\overline{S}}_2} \cdot t_0^{\overline{\overline{S}}_{t_0}} ,$$

and identity (13) also holds. Thus we have proved that the cardinal $\mathfrak{n}$ of any infinite complete Boolean algebra $\mathfrak{A}$ satisfies (13). On the other hand, if a cardinal $\mathfrak{n}$ satisfies (13), then there exists a complete Boolean algebra whose cardinal is $\mathfrak{n}$. Indeed, it follows from the remark at the end of B) that there exists a homogeneous complete algebra of power $\mathfrak{n}^{\aleph_0}$, i.e. of power $\mathfrak{n}$ by (13). Thus we have shown that

25.4. *In order that an infinite cardinal* $\mathfrak{n}$ *be the power of a complete Boolean algebra it is necessary and sufficient that* $\mathfrak{n}^{\aleph_0} = \mathfrak{n}$.[1]

We recall that it follows from the generalized Cantor continuum hypothesis $2^{\aleph_\alpha} = \aleph_{\alpha+1}$ that (13) holds for all infinite cardinals except those which are the sum of an enumerable sequence of smaller cardinals.

Now we shall give a topological characterization of complete Boolean algebras.

Suppose X is a topological space and $N \subset X$. If A is a set that is open-closed in X, then $B = A - N$ is open-closed in the topological space $X - N$. We shall say that a set N *separates* the space X if the class of all sets

(14) $B = A - N$ where A is both open and closed in X

is a proper subclass of the class of all open-closed subsets of the topological space $X - N$ (i.e. if $X - N$ contains at least one subset B which is open and closed in $X - N$ but is not of the form (14)).

25.5. *A Boolean algebra* $\mathfrak{A}$ *is complete if and only if no nowhere dense closed subset separates the Stone space of* $\mathfrak{A}$.[2]

---

[1] Pierce [4]
[2] Dwinger [2].

Since $\mathfrak{A}$ is complete if and only if the Stone space of $\mathfrak{A}$ is extremally disconnected it suffices to prove that a topological space X is extremally disconnected if and only if no nowhere dense closed subset N separates X. For any set $S \subset X$, $\mathbf{C}S$ will denote the closure of S in X.

Let B be a subset of the space $X - N$ that is open-closed. Since N is closed, the disjoint sets B and $B' = (X - N) - B$ are open in X. Consequently $\mathbf{C}B \cap B' = \wedge$. Thus $B = A - N$ where $A = \mathbf{C}B$. If X is extremally disconnected, then A is open-closed in X, i.e. N does not separate X.

Conversely, suppose that no nowhere dense closed set separates X. Let B be any open subset of X. The set $N = \mathbf{C}B - B$ is nowhere dense and closed, and B is both open and closed in $X - N$. Thus there exists a set A that is open and closed in X, such that $B = A - N$. Hence $B \subset A$ and consequently $\mathbf{C}B \subset \mathbf{C}A = A$. Since $B = A - (\mathbf{C}B - B)$, we infer that $\mathbf{C}B = A$. Thus the closure $\mathbf{C}B$ of any open set B is open, i.e. X is extremally disconnected.

§ 26. The field of all subsets of a set

According to the definition on p. 98, the complete field $\mathfrak{F}$ of all subsets of a set X is $\mathfrak{m}$-perfect if and only if every two-valued $\mathfrak{m}$-measure m on $\mathfrak{F}$ is determined by a point $x_0 \in X$. Notice that m is determined by x_0 if and only if $m((x_0)) = 1$.

If $\overline{\overline{X}} = \mathfrak{n} = \overline{\overline{Y}}$ and the field of all subsets of X is $\mathfrak{m}$-perfect, then the field of all subsets of Y is also $\mathfrak{m}$-perfect. Thus the property "to be $\mathfrak{m}$-perfect" depends only on the cardinal $\mathfrak{n}$ of X. Consequently we shall say that a cardinal $\mathfrak{n}$ is $\mathfrak{m}$-*perfect* if $\overline{\overline{X}} = \mathfrak{n}$ implies that the field of all subsets of X is $\mathfrak{m}$-perfect.

If $\mathfrak{n} \leqq \mathfrak{m}$, then $\mathfrak{n}$ is $\mathfrak{m}$-perfect since for every two-valued $\mathfrak{m}$-measure m on the field of all subsets of a set X of power $\mathfrak{n}$

(1) $\Sigma_{x \in X} m((x)) = m(X) = 1$

and therefore $m((x_0)) = 1$ for an $x_0 \in X$. Thus it is only interesting to examine the case $\mathfrak{n} \geqq \mathfrak{m}$. This inequality is assumed in the following theorem.

26.1. *The following conditions are equivalent:*

(i) $\mathfrak{n}$ *is* $\mathfrak{m}$-*perfect;*

(ii) *for every Boolean* $\mathfrak{n}$-*algebra* $\mathfrak{A}$, *every two-valued* $\mathfrak{m}$-*measure (every two-valued* $\mathfrak{m}$-*homomorphism) on* $\mathfrak{A}$ *is an* $\mathfrak{n}$-*measure (an* $\mathfrak{n}$-*homomorphism);*

(iii) *for every Boolean* $\mathfrak{n}$-*algebra* $\mathfrak{A}$, *every maximal* $\mathfrak{m}$-*filter (maximal* $\mathfrak{m}$-*ideal) in* $\mathfrak{A}$ *is an* $\mathfrak{n}$-*filter (* $\mathfrak{n}$-*ideal);*

(iv) *for every Boolean* $\mathfrak{n}$-*algebra* $\mathfrak{A}$, *every* $\mathfrak{m}$-*homomorphism* h *of* $\mathfrak{A}$ *into any complete atomic Boolean algebra* $\mathfrak{A}'$ *is an* $\mathfrak{n}$-*homomorphism.*

By the natural one-to-one correspondence between two-valued measures, two-valued homomorphisms, maximal filters and maximal ideals (see p. 18 and p. 90), the two versions of (ii) and the two versions of (iii) are equivalent.

(i) implies (ii). Let m be a two-valued $\mathfrak{m}$-measure on a Boolean $\mathfrak{n}$-algebra $\mathfrak{A}$. We have to prove that

(2) $$m\left(\bigcup_{x \in X} A_x \right) = \Sigma_{x \in X} m(A_x)$$

for every indexed set $\{A_x\}_{x \in X}$ of disjoint elements of $\mathfrak{A}$ with $\overline{\overline{X}} = \mathfrak{n}$. This equation holds if $m\left(\bigcup_{x \in X} A_x \right) = 0$ (see § 3 (4)). Suppose $m\left(\bigcup_{x \in X} A_x \right) = 1$. Then the formula

$$m'(B) = m\left(\bigcup_{x \in B} A_x \right) \quad \text{for} \quad B \subset X$$

defines a two valued $\mathfrak{m}$-measure m' on the field $\mathfrak{F}$ of all subsets B of X. By (i), m' is determined by a point $x_0 \in X$, i.e. $m'((x_0)) = 1$ and consequently $m(A_{x_0}) = 1$. Since m is two-valued and $A_x \cap A_{x_0} = \wedge$ for $x \neq x_0$, we have $m(A_x) = 0$ for $x \neq x_0$. This proves (2).

(ii) implies (i). Let m be a two-valued $\mathfrak{m}$-measure on the Boolean algebra $\mathfrak{A} = \mathfrak{F} =$ the field of all subsets of a set X, $\overline{\overline{X}} = \mathfrak{n}$. By (ii), m is an $\mathfrak{n}$-measure and consequently

$$\Sigma_{x \in X} m((x)) = m(X) = 1$$

which implies that $m((x_0)) = 1$ for a point $x_0 \in X$. The point x_0 determines the measure m.

(ii) implies (iv). By 25.1 it suffices to consider only the case where $\mathfrak{A}'$ is the field $\mathfrak{F}$ of all subsets of a space X. By § 16, we can consider $\mathfrak{A}'$ as the direct union of the two-element Boolean algebras $\{\mathfrak{B}_x\}_{x \in X}$ where $\mathfrak{B}_x$ is composed of the empty set and the one-point set (x) only. Then the homomorphism h can be represented in the form $h(A) = \{h_x(A)\}_{x \in X}$ where $h_x(A) = h(A) \cap (x)$ is an $\mathfrak{m}$-homomorphism of $\mathfrak{A}$ into $\mathfrak{B}_x$. By (ii), h_x is $\mathfrak{n}$-complete. This proves that h is $\mathfrak{n}$-complete.

(iv) implies (ii) (in the formulation for homomorphisms). It suffices to assume that $\mathfrak{A}'$ is a two-element Boolean algebra.

26.2. *Suppose* $\mathfrak{n}$ *is* $\mathfrak{m}$-*perfect*, $\mathfrak{n} \geqq \mathfrak{m}$. *If a Boolean* $\mathfrak{n}$-*algebra is isomorphic to an* $\mathfrak{m}$-*field of sets, then it is isomorphic to an* $\mathfrak{n}$-*field of sets*[1].

This follows immediately from 24.1 and 26.1.

26.3. *Suppose that* $\mathfrak{n}$ *is* $\mathfrak{m}$-*perfect*, $\mathfrak{n} \geqq \mathfrak{m}$, $\mathfrak{A}$ *is a Boolean* $\mathfrak{n}$-*algebra and* $\varDelta_0$ *is an* $\mathfrak{m}$-*ideal in* $\mathfrak{A}$. *If the Boolean algebra* $\mathfrak{A}/\varDelta_0$ *is isomorphic to an* $\mathfrak{m}$-*field of sets, then* $\varDelta_0$ *is an* $\mathfrak{n}$-*ideal*[2].

$\mathfrak{A}/\varDelta_0$ is a Boolean $\mathfrak{m}$-algebra. For every maximal $\mathfrak{m}$-ideal $\varDelta$ in $\mathfrak{A}/\varDelta_0$, the set $\varDelta'$ of all $A \in \mathfrak{A}$ such that $[A] \in \varDelta$ is a maximal $\mathfrak{m}$-ideal in $\mathfrak{A}$. By 26.1, $\varDelta'$ is an $\mathfrak{n}$-ideal. By 24.1 (ii) the hypothesis that $\mathfrak{A}/\varDelta_0$ is isomorphic to an $\mathfrak{m}$-field of sets implies that $\varDelta_0$ is the intersection of all ideals $\varDelta'$ just defined. Thus $\varDelta_0$ is an $\mathfrak{n}$-ideal.

---

[1] See SIKORSKI [4], SMITH and TARSKI [1].

[2] TARSKI [3]. See also SMITH and TARSKI [1]. They prove other interesting theorems stating a higher completeness of ideals.

26.4. *The class $\mathcal{R}$ of all* $\mathfrak{m}$*-perfect cardinals* $\mathfrak{n}$ *has the following properties:*

(a) $\mathfrak{m}$ *belongs to* $\mathcal{R}$*;*

(b) *if* $\mathfrak{a}$ *belongs to* $\mathcal{R}$*, and* $\mathfrak{b} \leq \mathfrak{a}$*, then* $\mathfrak{b}$ *belongs to* $\mathcal{R}$*;*

(c) *if, for every* $t \in T$*,* $\mathfrak{a}_t$ *is a cardinal in* $\mathcal{R}$ *and if* $\bar{\bar{T}}$ *is in* $\mathcal{R}$*, then the cardinal* $\Sigma_{t \in T} \mathfrak{a}_t$ *belongs to* $\mathcal{R}$*;*

(d) *if* $\mathfrak{a}$ *belongs to* $\mathcal{R}$*, then* $2^{\mathfrak{a}}$ *belongs to* $\mathcal{R}$.[1]

(a) was proved [see (1)].

Proof of (b). Suppose that $\mathfrak{a} \leq \mathfrak{b}$ and $\mathfrak{b}$ belongs to $\mathcal{R}$. Let $\bar{\bar{X}} = \mathfrak{a}$ and let X' be a set such that $X \subset X'$ and $\bar{\bar{X'}} = \mathfrak{b}$. Every two-valued $\mathfrak{m}$-measure m on the field of all subsets of X can be extended to a two-valued $\mathfrak{m}$-measure m' on the field of all subsets of X' by the formula

$$m'(A) = m(A \cap X) \quad \text{for} \quad A \subset X'.$$

The measure m' is determined by a point $x_0 \in X'$. Since $m'(X' - X) = 0$, we infer that $x_0 \in X$. The point x_0 determines the measure m.

Proof of (c). Suppose that $\bar{\bar{X}} = \mathfrak{n} = \Sigma_{t \in T} \mathfrak{a}_t$ where $\bar{\bar{T}}$ and all the cardinals $\mathfrak{a}_t$ belong to $\mathcal{R}$. The set X is the union of disjoint subsets X_t ($t \in T$) of cardinality $\mathfrak{a}_t$ respectively. Let m be a two-valued $\mathfrak{m}$-measure on the field of all subsets of the set X. The formula

$$m'(B) = m\left(\bigcup_{t \in B} X_t\right) \quad \text{for} \quad B \subset T$$

defines a two-valued $\mathfrak{m}$-measure m' on the field of all subsets B of the set T. Thus there exists a point $t_0 \in T$ which determines the measure m'. Since $m'((t_0)) = 1$, we have $m(X_{t_0}) = 1$. The measure m considered only on the field of all subsets of X_{t_0} is a two-valued $\mathfrak{m}$-measure. Hence it follows that there exists a point $x_0 \in X_{t_0}$ such that $m((x_0)) = 1$. The point x_0 determines the whole measure m.

Proof of (d). Let $\mathfrak{a}$ be in $\mathcal{R}$. Since $\bar{\bar{\mathscr{D}}}_{\mathfrak{a}} = 2^{\mathfrak{a}}$ it suffices to prove that every $\mathfrak{m}$-measure m defined on the field of all subsets of the Cautor space $\mathscr{D}_{\mathfrak{a}} = H^{T_0}$ ($\bar{\bar{T}}_0 = \mathfrak{a}$) is determined by a point x_0. For each $t \in T_0$ let D_t be the subset of $\mathscr{D}_{\mathfrak{a}}$ defined on p. 43. Observe that for every function $\varphi(t) = \pm 1$ ($t \in T_0$), the intersection

(2) $$\bigcap_{t \in T_0} \varphi(t) \cdot D_t$$

contains exactly one point.

---

[1] This theorem is due to ULAM [1]. Under a more restrictive hypothesis on the cardinality of X, it is a particular case of a general theorem stating that every σ-finite σ-measure on the field of all subsets of X is concentrated on an enumerable set. See BANACH and KURATOWSKI [1], BANACH [1], ULAM [1]. For the case $\bar{\bar{X}} = 2^{\aleph_0}$ see also MARCZEWSKI [1], SIERPIŃSKI [1]. Further generalizations were given by MARCZEWSKI and SIKORSKI [1], MAZUR [1], SIERPIŃSKI [2] Chapter V, TARSKI [3].

One of the sets D_t and $-D_t$ has measure equal to 1. Let $\varphi(t) = 1$ if $m(D_t) = 1$, and let $\varphi(t) = -1$ otherwise. By definition

$$m(\varphi(t) \cdot D_t) = 1 \quad \text{and} \quad m(-\varphi(t) \cdot D_t) = 0 .$$

By 26.1 the measure m is $\mathfrak{a}$-additive. Therefore the set (2) (for the function φ just defined) has measure 1 since its complement has measure zero [it is the union of $\mathfrak{a}$ sets $-\varphi(t) \cdot D_t$ of measure zero]. The set (2) is composed of only one point x_0. Since $m((x_0)) = 1$, the point x_0 determines the measure m.

Theorem 26.4 is proved.

The intersection $\mathfrak{K}_m$ of all classes $\mathfrak{K}$ of cardinals satisfying conditions (a), (b), (c), (d) also satisfies these conditions. A cardinal $\mathfrak{n}$ is said to be m-*attainable* if it belongs to $\mathfrak{K}_m$. If $m = \sigma = \aleph_0$, then we shall say simply *attainable* instead of σ-*attainable*. For instance, the cardinals $1, 2, \ldots, \aleph_0$, $2^{\aleph_0}, 2^{2^{\aleph_0}}, \ldots \aleph_1, \aleph_2, \ldots, \aleph_\omega, \aleph_{\omega+1}, \ldots$ are attainable and, consequently, m-attainable for every m.

The existence of non-m-attainable (in particular, of non-attainable) cardinals depends on the assumed set of axioms for the set theory. In fact, on the one hand, for all of the usually accepted sets of axioms for set theory, the class of sets whose cardinals are m-attainable satisfies all the axioms again, and in this restricted model of axiomatic set theory all cardinals are m-attainable. On the other hand, it seems to be very probable that by adding (to the assumed set of axioms of set theory) a new axiom stating the existence of non-m-attainable cardinals, we also obtain a consistent set of axioms.

It follows from 26.4 that every m-attainable cardinal is m-perfect. In particular, every attainable cardinal is σ-perfect. Obviously, every σ-perfect cardinal is m-perfect for every m.

The problem of whether non-attainable cardinals can be σ-perfect was open for a long time. Recently this problem was solved affirmatively[1]. The results just obtained show that the class of all σ-perfect cardinals is very large; it contains, roughly speaking, all cardinals usually appearing in mathematical reasonings. However, at present it is not known if all cardinals are σ-perfect, although everything suggests that this conjecture is true[2].

---

[1] The solution is due to TARSKI [18] who obtained it as a corollary to a metamathematical theorem of HANF [2, 3]. The first proof of the Tarski theorem was metamathematical. A mathematical proof was given by KEISLER [1]. For a full exposition see KEISLER and TARSKI [1]. The papers quoted concern also other related problems in the theory of Boolean algebras. See also ERDÖS and TARSKI [2] and HANF [4].

[2] SCOTT [4] proved that the existence of cardinals that are non-σ-perfect implies the existence of sets that are not constructive in the sense of GÖDEL [1]. This result also shows that practically it would be impossible to define a non-σ-perfect cardinal.

§ 27. The field of all Borel subsets of a metric space

Let X be a metric space with distance function ϱ. If $x \in X$ and the set $A \subset X$ is not empty, then $\varrho(x, A)$ denotes the greatest lower bound of the set of the numbers $\varrho(x, y)$ where $y \in A$. If the set A is empty, we let $\varrho(x, A) = 0$.

27.1. *The σ-field $\mathfrak{F}$ of all Borel subsets of the metric space X is σ-perfect if and only if the field $\mathfrak{F}'$ of all subsets of X is σ-perfect.*

In particular, if the cardinal $\overline{\overline{X}}$ is attainable, then $\mathfrak{F}$ is σ-perfect[1].

If $\mathfrak{F}'$ is not σ-perfect, then there exists a two-valued σ-measure m on $\mathfrak{F}'$ such that $m((x)) = 0$ for every $x \in X$. The measure m restricted to the field $\mathfrak{F}$ is an example of a two-valued σ-measure which is not determined by any point in X. Thus $\mathfrak{F}$ is not σ-perfect.

Suppose now that $\mathfrak{F}'$ is σ-perfect, i.e. that the cardinal $\mathfrak{n} = \overline{\overline{X}}$ is σ-perfect. Let m be a two-valued σ-measure on $\mathfrak{F}$, and let $\{G_t\}_{t < \alpha}$ be a transfinite sequence formed of all (different) open subsets of X of measure zero; i.e.

(1) $m(G_t) = 0$ for $0 \leq t < \alpha$.

Let T be the set of all ordinals t such that $0 \leq t < \alpha$, let H_0 be the empty set, and let H_t be the union

$$H_t = \bigcup_{t' < t} G_{t'}   (0 < t \leq \alpha) .$$

For every set $B \subset T$, the union

$$S_B = \bigcup_{t \in B} (H_{t+1} - H_t)$$

is a Borel subset[2] of X. In fact,

$$S_B = \bigcup_{1 \leq n < \infty} F_n$$

where F_n is the union

$$F_n = \bigcup_{t \in B} F_{n,t} ,$$

and $F_{n,t}$ is the set of all points $x \in X - H_t$ such that

$$\varrho(x, X - H_{t+1}) \geq 1/n .$$

The sets $F_{n,t}$ are closed. Moreover, if $x \in F_{n,t}$, $y \in F_{n,t'}$ and $t \neq t'$, then $\varrho(x, y) \geq 1/n$. Consequently the set F_n is also closed. This proves that S_B is a Borel set.

For every set $B \subset T$, let

$$m'(B) = m(S_B) .$$

m' is a σ-measure on the field of all subsets of T, and assumes at most the values 0 and 1. By (1), $m'((t)) = 0$ for every $t \in T$. The cardinal $\overline{\overline{T}}$ satisfies

---

[1] MARCZEWSKI and SIKORSKI [1]. Under a hypothesis on the cardinal $\overline{\overline{X}}$, it is a particular case of a general theorem stating that every finite σ-measure on the field of all Borel subsets of X is concentrated on a separable subset. See MARCZEWSKI and SIKORSKI [1]. For a generalization to non-metrizable spaces, see KATĚTOV [2].

[2] This statement is a particular case of a general theorem due to MONTGOMERY [1]. See also KURATOWSKI [2].

the inequality $\overline{\overline{T}} \leq 2^n$, therefore it is also σ-perfect by 20.4. This implies that the measure m' vanishes identically. In particular, $m'(T) = 0$, i.e. $m(H_\alpha) = 0$ since $H_\alpha = S_T$. The set H_α is the union of all open sets of m measure zero, i.e. it is the greatest open set of m measure zero. This implies that its complement has measure equal to 1 and is composed only of one point x_0. Since $m((x_0)) = 1$, the point x_0 determines the measure m.

It follows from the general theorem 27.1 that every two-valued σ-measure on the Borel field of a separable metric space is determined by a point since the cardinal of the space is then $\leq 2^{\aleph_0}$ and therefore it is σ-perfect. However this particular case of 27.1 can be proved much more simply. Let $\{G_n\}$ be a countable open basis of the space X, and let G be the union of all sets G_n such that $m(G_n) = 0$. The set G is the greatest open set with $m(G) = 0$, therefore its complement is composed only of one point which determines the measure m.

Example. A) If the metric space X is uncountable and σ-perfect, then the field $\mathfrak{F}$ of all Borel subsets of X and the field $\mathfrak{F}'$ of all subsets of X do not have property $(e_{\aleph_0})$ of § 24 (consider e.g. the σ-ideal of all at most enumerable subsets of X). The topology induced in X by $\mathfrak{F}$ or by $\mathfrak{F}'$ by the method described in § 24, p. 100, is the same, viz. it is the discrete topology.

§ 28. Representation of quotient algebras as fields of sets

Various examples of Boolean $\mathfrak{m}$-algebras are given in the form $\mathfrak{F}/\varDelta$ where $\mathfrak{F}$ is an $\mathfrak{m}$-field of subsets of a space X and $\varDelta$ is an $\mathfrak{m}$-ideal of $\mathfrak{F}$. Now we shall discuss the problem of when a Boolean $\mathfrak{m}$-algebra $\mathfrak{F}/\varDelta$ of this form is isomorphic to an $\mathfrak{m}$-field of sets. By 24.3 we can always assume, without any essential restriction, that $\mathfrak{F}$ is $\mathfrak{m}$-perfect.

28.1. *Let $\mathfrak{F}$ be an $\mathfrak{m}$-field of subsets of a space X, and let $\varDelta$ be an $\mathfrak{m}$-ideal of $\mathfrak{F}$. The following condition is sufficient and, if $\mathfrak{F}$ is $\mathfrak{m}$-perfect, also necessary for the quotient algebra $\mathfrak{F}/\varDelta$ to be isomorphic with an $\mathfrak{m}$-field of sets:*

(a) *there exists a set $X' \subset X$ (X' may belong to $\mathfrak{F}$ or not) such that, for every set $A \in \mathfrak{F}$,*

(1) $A \in \varDelta$ *if and only if* $A \cap X' = \varLambda$

(i.e. $\varDelta$ is the class of all sets $A \in \mathfrak{F}$ which are subsets of $X - X'$).

If condition (a) *is satisfied, then $\mathfrak{F}/\varDelta$ is isomorphic to the $\mathfrak{m}$-field $\mathfrak{F}' = \mathfrak{F}|X'$ of all sets $A \cap X'$ where $A \in \mathfrak{F}$.*[1]

If condition (a) is satisfied, then $\mathfrak{F}/\varDelta$ and $\mathfrak{F}|X'$ are isomorphic by § 10 B).

Suppose now that $\mathfrak{F}$ is $\mathfrak{m}$-perfect and that $\mathfrak{F}/\varDelta$ is isomorphic with an $\mathfrak{m}$-field of sets, i.e. $\mathfrak{F}/\varDelta$ satisfies the condition 24.1 (i). If V is a maximal

---

[1] Theorems 28.1—28.3 were proved by SIKORSKI [4].

m-filter of $\mathfrak{F}/\Delta$, then the set of all $A \in \mathfrak{F}$ such that $[A] \in V$ is a maximal m-filter of $\mathfrak{F}$ determined by a point x_V in X. By definition, for every $A \in \mathfrak{F}$,

(2) $\qquad\qquad\qquad [A] \in V$ if and only if $x_V \in A$.

The set X' of all points x_V has the required property (a). In fact, if $A \in \Delta$, i.e. $[A] = \wedge$, then $[A] \notin V$ and consequently, by (2), $x_V \notin A$ for every V. This proves that $A \cap X' = \wedge$. On the other hand, if $A \notin \Delta$, i.e. $[A] \neq \wedge$, then, by 24.1 (i), there is a maximal m-filter V such that $[A] \in V$. Consequently, by (2), $x_V \in A$, i.e. $A \cap X' \neq \wedge$.

The following two theorems are direct consequences of 28.1, 26.1 and 27.1.

28.2. *If the cardinal $\overline{\overline{X}}$ of the set X is m-perfect, $\mathfrak{F}$ is the field of all subsets of X and Δ is an m-ideal of $\mathfrak{F}$, then $\mathfrak{F}/\Delta$ is isomorphic to an m-field of sets if and only if the ideal Δ is principal.*

28.3. *If X is a metric space, the cardinal $\overline{\overline{X}}$ is σ-perfect, $\mathfrak{F}$ is the σ-field of all Borel subsets of X and Δ is a σ-ideal of $\mathfrak{F}$, then $\mathfrak{F}/\Delta$ is isomorphic to a σ-field of sets if and only if there exists a subset $X' \subset X$ such that*

$A \in \Delta$ *if and only if the intersection* $A \cap X'$ *is empty.*

Examples. A) If $\overline{\overline{X}}$ is σ-perfect, $\mathfrak{F}$ is the field of all subsets of X and Δ is a proper σ-ideal containing all one-point subsets, then $\mathfrak{F}/\Delta$ is not isomorphic to any σ-field of sets.

B) If X is a metric space, $\overline{\overline{X}}$ is σ-perfect, $\mathfrak{F}$ is the σ-field of all Borel subsets of X and Δ is a proper σ-ideal (of $\mathfrak{F}$) containing all one-point subsets, then $\mathfrak{F}/\Delta$ is not isomorphic to any σ-field of sets.

This result is a generalization of that in § 24 A).

C) Suppose that $\mathfrak{n}$ is an m-attainable cardinal ($\mathfrak{n} \geq \mathfrak{m}$), $\mathfrak{F}$ is an $\mathfrak{n}$-field of sets and Δ is an m-ideal of $\mathfrak{F}$. If Δ is not $\mathfrak{n}$-complete, then $\mathfrak{F}/\Delta$ is not isomorphic to any m-field of sets[1].

This follows immediately from 26.3.

D) As we have observed in § 24 A), m-distributivity is a necessary condition for a Boolean algebra to be isomorphic to an m-field of sets. m-distributivity is not, however, a sufficient condition.

In fact, let X be any space of power $2^{2^{\mathfrak{m}}}$, let $\mathfrak{F}$ be the field of all subsets of X and let Δ be the ideal of all subsets of power $\leq 2^{\mathfrak{m}}$. By 21.5, the $2^{\mathfrak{m}}$-algebra $\mathfrak{F}/\Delta$ is m-distributive. It follows from 28.2 that $\mathfrak{F}/\Delta$ is not isomorphic to any m-field of sets[2].

---

[1] TARSKI [3].

[2] The same construction is used in SIKORSKI [21] for the problem of axiomatization of the notion of σ-fields of sets. See also HORN and TARSKI [1], SIKORSKI [4], SMITH and TARSKI [1].

We recall that in § 24 (see theorem 24.11) another stronger con-dition of distributivity was formulated which is equivalent to the existence of an isomorphism with an m-field of sets.

§ 29. A fundamental representation theorem for Boolean σ-algebras. m-representability

The considerations in § 28 suggest the following question:

Is every Boolean m-algebra isomorphic to a quotient algebra $\mathfrak{F}/\Delta$ where $\mathfrak{F}$ is an m-field of sets and Δ is an m-ideal of $\mathfrak{F}$?

The answer to this question is negative for $\mathfrak{m} \geq 2^{\aleph_0}$. In fact, if a Boolean m-algebra $\mathfrak{A}$ is isomorphic to $\mathfrak{F}/\Delta$ where $\mathfrak{F}$ is an m-field and Δ is an m-ideal and $\mathfrak{m} \geq 2^{\aleph_0}$, then $\mathfrak{F}/\Delta$ is σ-distributive on account of 21.5. On the other hand, there exist complete Boolean algebras which are not σ-distributive (see § 19 A) and § 21 C) and D)).

Now we shall prove that the answer to our question is affirmative if $\mathfrak{m} = \aleph_0$.

29.1. *For every Boolean σ-algebra $\mathfrak{A}$ there exist a σ-field of sets $\mathfrak{F}$ and a σ-ideal Δ of $\mathfrak{F}$ such that $\mathfrak{A}$ is isomorphic to $\mathfrak{F}/\Delta$.*[1]

More precisely:

Let X be the Stone space of $\mathfrak{A}$, let $\mathfrak{F}$ be the least σ-field (of subsets of X) containing all open-closed subsets of X, and let Δ be the σ-ideal of all sets $B \in \mathfrak{F}$ of the first category in X. Then $\mathfrak{A}$ is isomorphic to $\mathfrak{F}/\Delta$. Namely, if h_0 is an isomorphism of $\mathfrak{A}$ onto the field $\mathfrak{F}_0$ of all open-closed subsets of X, then

$$h(A) = [h_0(A)]_\Delta \quad for \quad A \in \mathfrak{A}$$

is an isomorphism of $\mathfrak{A}$ onto $\mathfrak{F}/\Delta$.

The homomorphism h is an isomorphism. This follows from the topological theorem that no open non-empty subset of a compact Hausdorff space is of the first category[2]. Hence, if $A \neq \wedge$ ($A \in \mathfrak{A}$), then the open-closed set $h_0(A)$ does not belong to Δ and consequently $h(A)$ is not the zero element of $\mathfrak{F}/\Delta$.

To prove that h maps $\mathfrak{A}$ onto $\mathfrak{F}/\Delta$ it suffices to show that the class $\mathfrak{F}'$ of all sets $B \in \mathfrak{F}$ such that $[B] = h(A)$ for an element $A \in \mathfrak{A}$ is a σ-field of sets (and therefore it coincides with $\mathfrak{F}$ since $\mathfrak{F}_0 \subset \mathfrak{F}' \subset \mathfrak{F}$). If $B \in \mathfrak{F}'$,

---

[1] This fundamental representation theorem for Boolean σ-algebras was found independently by LOOMIS [1] and SIKORSKI [4]. It was published by LOOMIS in August 1947 and it was presented (with proof) by SIKORSKI at the Congress of the Polish Mathematical Society in Kraków on May 1947, but published one year later because of the printing difficulties in Poland after the second world war.

The proof presented here is that of SIKORSKI [4]. A similar proof was found independently by P. R. HALMOS. See also AUMANN [1]. A new proof based on meta-mathematical ideas was given by TARSKI [12].

[2] ČECH [1]. See also SIKORSKI [4].

i.e. $[B] = h(A)$ for an $A \in \mathfrak{A}$, then $[-B] = h(-A)$ and therefore
$-B \in \mathfrak{F}'$ also. If $B_n \in \mathfrak{F}'$ for $n = 1, 2, \ldots$, i.e. $[B_n] = h(A_n) = [h_0(A_n)]$
for some elements $A_n \in \mathfrak{A}$, then

$$[\bigcup_{1 \leq n < \infty} B_n] = [\bigcup_{1 \leq n < \infty} h_0(A_n)]$$
$$= [h_0(\bigcup_{1 \leq n < \infty}^{\mathfrak{A}} A_n)] = h(\bigcup_{1 \leq n < \infty}^{\mathfrak{A}} A_n)$$

by 22.2. Consequently the set-theoretical union $\bigcup_{1 \leq n < \infty} B_n$ also belongs to $\mathfrak{F}'$ which proves that $\mathfrak{F}'$ is a σ-field of sets.

A Boolean algebra is said to be $\mathfrak{m}$-*representable* provided it is iso-
morphic to an $\mathfrak{m}$-regular subalgebra of a quotient algebra $\mathfrak{F}/\varDelta$ where $\mathfrak{F}$
is an $\mathfrak{m}$-field of sets and $\varDelta$ is an $\mathfrak{m}$-ideal of $\mathfrak{F}$ (in other words, if there
exists an $\mathfrak{m}$-isomorphism of $\mathfrak{A}$ into $\mathfrak{F}/\varDelta$ where $\mathfrak{F}$ and $\varDelta$ have the properties
mentioned).

Thus a Boolean $\mathfrak{m}$-algebra is called $\mathfrak{m}$-*representable* if and only if it is
isomorphic to a quotient algebra $\mathfrak{F}/\varDelta$ where $\mathfrak{F}$ is an $\mathfrak{m}$-field of sets, and $\varDelta$
is an $\mathfrak{m}$-ideal of $\mathfrak{F}$.

A Boolean $\mathfrak{m}$-algebra is $\mathfrak{m}$-representable if and only if it is an $\mathfrak{m}$-homo-
morphic image of an $\mathfrak{m}$-field of sets, i.e. there exist an $\mathfrak{m}$-field of sets $\mathfrak{F}$
and an $\mathfrak{m}$-homomorphism h of $\mathfrak{F}$ onto $\mathfrak{A}$.

Indeed, if $\mathfrak{F}$ and h with the above properties exist, then the identity

$$h'([A]_\varDelta) = h(A) \quad \text{for} \quad A \in \mathfrak{F},$$

where $\varDelta$ is the $\mathfrak{m}$-ideal of all $B \in \mathfrak{F}$ with $h(B) = \varLambda$, defines an isomorphism of $\mathfrak{F}/\varDelta$ onto $\mathfrak{A}$. Conversely, if h' is an isomorphism of $\mathfrak{F}/\varDelta$
onto $\mathfrak{A}$ where $\mathfrak{F}$ is an $\mathfrak{m}$-field of sets and $\varDelta$ is an $\mathfrak{m}$-ideal in $\mathfrak{F}$, then the
same identity defines an $\mathfrak{m}$-homomorphism h of the $\mathfrak{m}$-field $\mathfrak{F}$ onto $\mathfrak{A}$.

It follows directly from the above characterization that any $\mathfrak{m}$-homo-
morphic image of an $\mathfrak{m}$-representable $\mathfrak{m}$-algebra is also $\mathfrak{m}$-representable
since it is also an $\mathfrak{m}$-homomorphic image of an $\mathfrak{m}$-field of sets [another
proof can be easily deduced from § 10 (10)].

In the terminology just introduced the results obtained can be
formulated as follows. Every Boolean σ-algebra is σ-representable. For
every $\mathfrak{m} \geq 2^{\aleph_0}$ there exist Boolean $\mathfrak{m}$-algebras which are not $\mathfrak{m}$-re-
presentable[1].

We are going to examine in more detail which Boolean algebras are
$\mathfrak{m}$-representable.

Let $\mathfrak{A}$ be a Boolean algebra, let h_0 be the isomorphism of $\mathfrak{A}$ onto the
field $\mathfrak{F}_0$ of all open-closed subsets of the Stone space X of $\mathfrak{A}$.

---

[1] Recently KARP [1] proved that for every $\mathfrak{m} > \aleph_0$ there is a non-$\mathfrak{m}$-represent-
able $\mathfrak{m}$-algebra. More precisely, for every $\mathfrak{m} \geq \aleph_0$ there exists a complete $\mathfrak{m}$-repre-
sentable algebra which is not $\mathfrak{m}^+$-representable, $\mathfrak{m}^+$ being the smallest cardinal
greater than $\mathfrak{m}$.

We recall (see § 22, p. 86) that a set $B \subset X$ is called $\mathfrak{m}$-nowhere dense if it is a subset of a nowhere dense $\mathfrak{m}$-closed set, i.e. a subset of the set-theoretical intersection

$$\bigcap_{t \in T} h_0(A_t) \quad \text{where} \quad \bigcap_{i \in T}^{\mathfrak{A}} A_t = \wedge \quad \text{and} \quad \overline{\overline{T}} \leq \mathfrak{m}.$$

A set B is said to be of the $\mathfrak{m}$-category provided it is the union of an $\mathfrak{m}$-indexed set of $\mathfrak{m}$-nowhere dense sets.

Let $\mathfrak{F}_{\mathfrak{m}}$ be the least $\mathfrak{m}$-field containing $\mathfrak{F}_0$, and let $\varDelta$ be the $\mathfrak{m}$-ideal of all sets $A \in \mathfrak{F}_{\mathfrak{m}}$ which are of the $\mathfrak{m}$-category.

The class $\mathfrak{F}$ of all sets of the form

$$(B_0 \cup B_1) - B_2 \quad \text{where} \quad B_0 \in \mathfrak{F}_0 \quad \text{and} \quad B_1, B_2 \in \varDelta$$

is a subfield of $\mathfrak{F}_{\mathfrak{m}}$, and $\mathfrak{F}/\varDelta$ is a subalgebra of $\mathfrak{F}_{\mathfrak{m}}/\varDelta$.

The quotient algebra $\mathfrak{F}/\varDelta$ is called the *canonical $\mathfrak{m}$-representation for* $\mathfrak{A}$. The homomorphism h of $\mathfrak{A}$ onto $\mathfrak{F}/\varDelta$ given by

$$h(A) = [h_0(A)]_\varDelta \quad \text{for} \quad A \in \mathfrak{A}$$

is called the $\mathfrak{m}$-*canonical homomorphism*. If h is one-to-one, then it is called the $\mathfrak{m}$-*canonical isomorphism*.

Using the above terminology and notation we shall prove the following two theorems.

29.2. *The $\mathfrak{m}$-canonical homomorphism h is an $\mathfrak{m}$-homomorphism of $\mathfrak{A}$ into $\mathfrak{F}_{\mathfrak{m}}/\varDelta$. Consequently, if h is an isomorphism, then $\mathfrak{F}/\varDelta$ is an $\mathfrak{m}$-regular subalgebra of $\mathfrak{F}_{\mathfrak{m}}/\varDelta$.*

If $\mathfrak{A}$ is an $\mathfrak{m}$-algebra, then $\mathfrak{F} = \mathfrak{F}_{\mathfrak{m}}$ and consequently $\mathfrak{F}/\varDelta = \mathfrak{F}_{\mathfrak{m}}/\varDelta$.

If $\bigcap_{t \in T}^{\mathfrak{A}} A_t = \wedge$ ($\overline{\overline{T}} \leq \mathfrak{m}$), then the intersection of all the sets $h_0(A_t)$ belongs to $\varDelta$ and consequently, by § 21 (1),

$$\bigcap_{t \in T}^{\mathfrak{F}_{\mathfrak{m}}/\varDelta} h(A_t) = [\bigcap_{t \in T} h_0(A_t)]_\varDelta = \wedge.$$

This proves the first part of the theorem [see § 22 (a)].

Suppose now that $\mathfrak{A}$ is $\mathfrak{m}$-complete. To prove that $\mathfrak{F} = \mathfrak{F}_{\mathfrak{m}}$ it suffices to show that $\mathfrak{F}$ is an $\mathfrak{m}$-field.

Observe that, by definition of $\mathfrak{F}$, $B \in \mathfrak{F}$ if and only if there exists an element $A \in \mathfrak{A}$ such that

(1) $$h_0(A) - B \in \varDelta \quad \text{and} \quad B - h_0(A) \in \varDelta.$$

Suppose that $B_t \in \mathfrak{F}$ for every $t \in \overline{T}$ ($\overline{\overline{T}} \leq \mathfrak{m}$), i.e.

$$h_0(A_t) - B_t \in \varDelta \quad \text{and} \quad B_t - h_0(A_t) \in \varDelta$$

for an $A_t \in \mathfrak{A}$. Let $A = \bigcup_{t \in T}^{\mathfrak{A}} A_t$ and let B the set-theoretical union of all B_t. We have

$$h_0(A) - B \subset (h_0(A) - \bigcup_{t \in T} h_0(A_t)) \cup \left(\bigcup_{t \in T} (h_0(A_t) - B_t) \right)$$

and

$$B - h_0(A) \subset \bigcup_{t \in T} (B_t - h_0(A_t))$$

where $\bigcup_{t\in T}$ denotes the set-theoretical union. This proves (see 22.2) that A and B satisfy (1). Hence it follows that $B \in \mathfrak{F}$. This completes the proof of the second part of 29.2.

In the next theorem we assume definition (D) (see § 21, p. 80) of $\mathfrak{m}$-filters in arbitrary Boolean algebras.

29.3. *For every Boolean algebra $\mathfrak{A}$, each of the following conditions is both necessary and sufficient for $\mathfrak{A}$ to be $\mathfrak{m}$-representable*[1]:

(r) *the $\mathfrak{m}$-canonical homomorphism is an isomorphism;*

(r_1) *for every $\mathfrak{m}$-indicated set $\{A_{t,s}\}_{t\in T, s\in S}$ (of elements in $\mathfrak{A}$) such that*

$$(2) \qquad \text{all the joins } \bigcup_{s\in S} A_{t,s} \text{ and the meet } \bigcap_{t\in T} \bigcup_{s\in S} A_{t,s}, \text{ exist,}$$

if

$$(3) \qquad \bigcap_{t\in T} \bigcup_{s\in S} A_{t,s} \neq \wedge,$$

then there exists a $\varphi \in S^T$ such that

$$(4) \qquad \bigcap_{t\in T'} A_{t,\varphi(t)} \neq \wedge \text{ for every finite set } T' \subset T;$$

(r_2) *for every $\mathfrak{m}$-indexed set $\{A_{t,s}\}_{t\in T, s\in S}$ such that (2) holds, if*

$$(5) \qquad \bigcap_{t\in T} \bigcup_{s\in S} A_{t,s} = V,$$

then for every element $A \neq \wedge$ $(A \in \mathfrak{A})$ there exists a $\varphi \in S^T$ such that

$$(6) \qquad A \cap \bigcap_{t\in T'} A_{t,\varphi(t)} \neq \wedge \text{ for every finite set } T' \subset T;$$

(r_2') *for every $\mathfrak{m}$-indexed set $\{A_{t,s}\}_{t\in T, s\in S}$ satisfying (2), if (5) holds, then for every proper $\mathfrak{m}$-filter V of $\mathfrak{A}$ there exists a $\varphi \in S^T$ such that*

$$(7) \quad A \cap \bigcap_{t\in T'} A_{t,\varphi(t)} \neq \wedge \text{ for every } A \in V \text{ and every finite set } T' \subset T;$$

(r_3) *in the Stone space of $\mathfrak{A}$, any intersection of at most $\mathfrak{m}$ dense $\mathfrak{m}$-open sets is dense, i.e. the intersection of at most $\mathfrak{m}$ dense $\mathfrak{m}$-open sets intersects every non-empty open set;*

(r_3') *in the Stone space of $\mathfrak{A}$, any intersection of at most $\mathfrak{m}$ dense $\mathfrak{m}$-open sets intersects any non-empty closed set of the upper character*[2] *$\mathfrak{m}$;*

(r_4) *in the Stone space of $\mathfrak{A}$, every set of the $\mathfrak{m}$-category is a boundary set, i.e. no open non-empty set is of the $\mathfrak{m}$-category;*

(r_4') *in the Stone space of $\mathfrak{A}$, no closed non-empty set of upper character $\mathfrak{m}$ is of the $\mathfrak{m}$-category;*

(r_5) *for every set of infinite joins and meets in $\mathfrak{A}$:*

$$(8) \qquad \begin{aligned} \bigcup_{s\in S_t'} A_{t,s} &= A_t \text{ where } \overline{\overline{S_t'}} \leq \mathfrak{m},\ t \in T',\ \overline{\overline{T'}} \leq \mathfrak{m}, \\ \bigcap_{s\in S_t''} B_{t,s} &= B_t \text{ where } \overline{\overline{S_t''}} \leq \mathfrak{m},\ t \in T'',\ \overline{\overline{T''}} \leq \mathfrak{m}, \end{aligned}$$

---

[1] Conditions (r_5) and (r_5') are due to CHANG [1]. See also SCOTT [3]. Condition (r_4) is due to PIERCE [2]. Condition (r_1) is a modification (SIKORSKI [25]) of a sufficient condition given by SMITH [1]. The whole theorem 29.3 was published by SIKORSKI [25].

[2] See the definition on p. 81 and theorem 21.6.

and for every $A \neq \wedge (A \in \mathfrak{A})$ there exists a maximal filter containing A and preserving all the joins and meets (8);

(r_5') *for every set* (8) *of infinite joins and meets in* $\mathfrak{A}$ *and for every proper* $\mathfrak{m}$*-filter* V *there exists a maximal filter containing* V *and preserving all the joins and meets* (8).

If $\mathfrak{A}$ *is a Boolean* $\mathfrak{m}$*-algebra, then each of the above conditions is necessary and sufficient for* $\mathfrak{A}$ *to be isomorphic to a quotient algebra* $\mathfrak{F}'/\varDelta'$ *where* $\mathfrak{F}'$ *is an* $\mathfrak{m}$*-field of sets and* $\varDelta'$ *is an* $\mathfrak{m}$*-ideal of* $\mathfrak{F}'$.

(r_1) implies (r_2). The proof is similar to the proof of the implication $(d_1) \rightarrow (d_2)$ in 19.2.

(r_2) implies (r_3). In fact, suppose that, for every $t \in T$ $(\overline{\overline{T}} \leqq \mathfrak{m})$, G_t is a dense $\mathfrak{m}$-open subset of X, i.e.

$$G_t = \mathsf{U}_{s \in S} h_0(A_{t,s}) \quad \text{where} \quad \mathsf{U}_{s \in S}^{\mathfrak{A}} A_{t,s} = V \ (\overline{\overline{S}} \leqq \mathfrak{m}).$$

Let G be any non-empty open subset of X. Since $\mathfrak{F}_0$ is an open basis for X, there is an element $A \neq \wedge (A \in \mathfrak{A})$ such that $h_0(A) \subset G$. By (r_2), there exists a mapping $\varphi \in S^T$ such that (6) holds, i.e.

$$h_0(A) \cap \mathsf{\cap}_{t \in T'} h_0(A_{t,\varphi(t)}) \neq \wedge$$

for every finite set $T' \subset T$. Since all the sets $h_0(A)$, $h_0(A_{t,s})$ are closed in the compact space X, we obtain

$$\wedge \neq h_0(A) \cap \mathsf{\cap}_{t \in T} h_0(A_{t,\varphi(t)}) \subset G \cap \mathsf{\cap}_{t \in T} G_t.$$

(r_3) implies (r_4) by passing to complements.

To prove the next two implications it will be convenient to identify maximal filters with points in X, i.e. to assume that X is the set of all maximal filters in $\mathfrak{A}$ and

$$h_0(A) = \text{the set of all } V \in X \text{ such that } A \in V$$

(see the proof of 8.2).

(r_4) implies (r_5). In fact, the set of all maximal filters which do not preserve a join or meet in (8) is of the $\mathfrak{m}$-category (see 22.2 and remarks on p. 90). By (r_4) there exists a point in $h_0(A)$ which does not belong to this set of the $\mathfrak{m}$-category. This point is a maximal filter preserving all the joins and meets (8), and $A \in V$.

(r_5) implies (r_2). Suppose (5) holds, and apply (r_5) to the joins

(9) $$\mathsf{U}_{s \in S} A_{t,s} = V \qquad (t \in T).$$

Let V_0 be a maximal filter preserving all the joins (9) and containing A. By definition, $V_0 \in h_0(A)$ and $V_0 \in \mathsf{U}_{s \in S} h_0(A_{t,s})$ for every $t \in T$. Thus there exists an $s = \varphi(t)$ such that $V_0 \in h_0(A_{t,\varphi(t)})$. Consequently

$$h_0(A) \cap \mathsf{\cap}_{t \in T'} h_0(A_{t,\varphi(t)}) \neq \wedge \quad \text{for every finite} \quad T' \subset T,$$

i.e. (6) holds.

(r_2') implies (r_3'), (r_3') implies (r_4'), (r_4') implies (r_5'), (r_5') implies (r_2'). The proof of these implications is similar to the proof of the implications $(r_2) \to (r_3)$, $(r_3) \to (r_4)$, $(r_4) \to (r_5)$, $(r_5) \to (r_2)$ respectively.

(r_2) implies (r_2')[1]. For suppose that (5) holds but (7) does not hold, i.e. for every $\varphi \in S^T$ there exist a finite set $T_\varphi \subset T$ and, for this set T_φ, an element $A_{T_\varphi} \in V$ such that

$$A_{T_\varphi} \cap \bigcap_{t \in T_\varphi} A_{t, \varphi(t)} = \wedge.$$

The set of all A_{T_φ} has power $\leq \mathfrak{m}$ (since the class of all finite subsets of T has cardinality $\leq \mathfrak{m}$). V being an $\mathfrak{m}$-filter, there exists an element $A \in V$ such that $A \subset A_{T_\varphi}$ for every $\varphi \in S^T$. We have $A \neq \wedge$ since V is proper. Thus

$$A \cap \bigcap_{t \in T_\varphi} A_{t, \varphi(t)} = \wedge$$

for every $\varphi \in S^T$, i.e. (r_2) does not hold.

(r_2') implies (r_2) (take as V the principal filter generated by A !).

(r_4) implies (r). In fact, if $A \neq \wedge$ $(A \in \mathfrak{A})$, then $h_0(A)$ is open and non-empty. By (r_4), $h_0(A) \not\subset \varDelta$, i.e. $h(A) \neq \wedge$. This proves that h is an isomorphism.

If (r) holds, then $\mathfrak{A}$ is $\mathfrak{m}$-representable by theorem 29.2.

If $\mathfrak{A}$ is $\mathfrak{m}$-representable, then (r_1) holds. It suffices to prove this implication in the case where $\mathfrak{A}$ is an $\mathfrak{m}$-regular subalgebra of $\mathfrak{F}'/\varDelta'$ where $\mathfrak{F}'$ is an $\mathfrak{m}$-field of sets and $\varDelta'$ is an $\mathfrak{m}$-ideal of $\mathfrak{F}'$.

Suppose that (3) holds in $\mathfrak{A}$. Since $\mathfrak{A}$ is an $\mathfrak{m}$-regular subalgebra of $\mathfrak{F}'/\varDelta'$, all joins and meets in (3) can be considered as joins and meets in $\mathfrak{F}'/\varDelta'$. We have

$$A_{t, s} = [B_{t, s}] \quad \text{for some sets} \quad B_{t, s} \in \mathfrak{F}'.$$

Let B be the union of all finite intersections

$$B_{t_1, s_1} \cap \cdots \cap B_{t_n, s_n}$$

which belong to $\varDelta'$, and let

$$C_{t, s} = B_{t, s} - B.$$

We have

$$A_{t, s} = [C_{t, s}]$$

since $B \in \varDelta'$. Moreover, for any finite intersection,

(10)   if $C_{t_1, s_1} \cap \cdots \cap C_{t_n, s_n} \neq \wedge$, then $C_{t_1, s_1} \cap \cdots \cap C_{t_n, s_n} \not\subset \varDelta'$, i.e.

$$A_{t_1, s_1} \cap \cdots \cap A_{t_n, s_n} \neq \wedge.$$

---

[1] The proof of this implication is a slight modification of A. BIAŁYNICKI-BIRULA's proof (not published) of CHANG's [1] representation theorem.

By (3), $[\bigcap_{t \in T} \bigcup_{s \in S} C_{t,s}] \neq \Lambda$. Thus the set $\bigcap_{t \in T} \bigcup_{s \in S} C_{t,s}$ contains a point x. Consequently, for every $t \in T$ there exists an $s = \varphi(t)$ such that $x \in C_{t, \varphi(t)}$. Therefore

$$C_{t_1, \varphi(t_1)} \cap \cdots \cap C_{t_n, \varphi(t_n)} \neq \Lambda.$$

This implies (4) on account of (10).

The final part follows directly from the part just proved and the definition of m-representable m-algebras.

In the case $\mathfrak{m} = \aleph_0$ theorem 29.3 yields the following generalization of 29.1:

29.4. *Every Boolean algebra is σ-representable.*

This follows immediately from 29.3 (r_4) since every set of the σ-category is of the first category and no open non-empty subset of a compact Hausdorff space is of the first category.

In the case where $\mathfrak{A}$ is a σ-algebra, the quotient algebra $\mathfrak{F}/\Delta$ defined in 29.1 is identical with the canonical σ-representation of $\mathfrak{A}$. In fact, in both cases $\mathfrak{F}$ is the least σ-field containing all open-closed subsets of the Stone space. Though the ideals Δ mentioned in 29.1 and in the definition of the canonical σ-representation are defined differently, they are equal. Indeed, it follows immediately from the definitions that the ideal Δ mentioned in the definition of the canonical σ-representation is a subset of the ideal Δ mentioned in 29.1. On the other hand, if $A \in \mathfrak{F}$, then $A = (B_0 \cap B_1) - B_2$ where B_0 is both open and closed, and B_1, B_2 belong to the ideal Δ mentioned in the definition of the canonical σ-representation. If A is of the first category, i.e. A belongs to the ideal Δ in 29.1, then B_0 is empty and consequently the set $A = B_1 - B_2$ belongs to the ideal Δ in the definition of the canonical σ-representation. Thus the two ideals coincide.

29.5. *A Boolean m-algebra is m-representable if and only if each of its m-subalgebras m-generated by at most* m *elements is m-representable.*

This follows directly from 29.3 since $\mathfrak{A}$ satisfies (r_1) if and only if each of its m-subalgebras m-generated by at most m elements satisfies (r_1).

Examples. A) Let $\mathfrak{A}$ be an m-representable Boolean algebra, let X be the Stone space of $\mathfrak{A}$, and let $\mathfrak{F}/\Delta$ be the canonical m-representation of $\mathfrak{A}$. Then there exists an isomorphism g of $\mathfrak{F}/\Delta$ into $\mathfrak{F}$ such that

(11) $$[g(A)]_\Delta = A$$

for every $A \in \mathfrak{F}/\Delta$. In fact, X is then the Stone space of $\mathfrak{F}/\Delta$, and the natural isomorphism g of $\mathfrak{F}/\Delta$ onto the field of all open-closed subsets of X has property (11).

The question arises whether, for every field $\mathfrak{F}'$ of subsets of a space X' and for every ideal Δ' of $\mathfrak{F}'$, there exists an isomorphism g of $\mathfrak{F}'/\Delta'$ into $\mathfrak{F}'$

such that (11) holds. A negative answer to this question is given by the next example[1].

B) If $\mathfrak{m} < \mathfrak{n} < \mathfrak{n}'$, $\mathfrak{F}'$ is the field of all subsets of a space with cardinal $\mathfrak{n}'$ and $\varDelta'$ is the ideal of all subsets of cardinals $\leq \mathfrak{m}$, then there exists no isomorphism g of $\mathfrak{F}'/\varDelta'$ into $\mathfrak{F}'$ such that (11) holds.

To simplify the proof, let us assume that $\mathfrak{F}'$ is the field of all subsets of the Cartesian product $X \times X'$, where X, X' are sets with cardinals $\mathfrak{n}$ and $\mathfrak{n}'$ respectively. For every $x \in X$ let $A_x = (x) \times X'$ and, similarly, for every $x' \in X'$ let $A^{x'} = X \times (x')$. Suppose that there exists a homomorphism g of $\mathfrak{F}'/\varDelta'$ into $\mathfrak{F}'$ such that (11) holds. Let U (let U') be the set of all points (x, x') which do not belong to $g([A_x])$ (to $g([A^{x'}])$). We have

(12) $U \cup U' = X \times X'$

since if $(x, x') \notin U \cup U'$, then $(x, x') \in g([A_x]) \cap g([A^{x'}]) = g([[(x, x')]]) = \varLambda$ which is impossible.

For every $x' \in X'$,

(13)   the intersection $A^{x'} \cap U'$ has power $< \mathfrak{n}$

since $A^{x'} \cap U' \subset A^{x'} - g([A^{x'}]) \in \varDelta'$ on account of (11). Similarly we prove that for every $x \in X$, the intersection $A_x \cap U$ has power $< \mathfrak{n}$. Hence it follows that the projection of U onto the axis X' has power $\leq \mathfrak{n}$. Consequently there exists a point x'_0 which does not belong to this projection, i.e. such that $A^{x'_0} \subset U'$ [see (12)]. Thus $A^{x'_0} \cap U'$ has power $\mathfrak{n}$, contradicting (13).

Observe that the hypothesis of the statement just proved can be weakened. For instance, it suffices to suppose that $\mathfrak{n} < \mathfrak{n}'$ and that $\varDelta$ is the ideal of all subsets of power $< \mathfrak{n}$ (in this formulation the case of the ideal of all finite subsets of an uncountable space is included).

The representation problem discussed in this section can be generalized as follows.

Let $\mathfrak{A}$ be a Boolean algebra. Let $\mathbf{J}$ be a class of non-void subsets $\mathfrak{S}$ of $\mathfrak{A}$ such that $\overline{\overline{\mathfrak{S}}} \leq \mathfrak{m}$ and the join

(14) $\mathsf{U}_{A \in \mathfrak{S}} A = A_{\mathfrak{S}}$

exists, for every $\mathfrak{S} \in \mathbf{J}$. Let $\mathbf{M}$ be a class of non-void subsets $\mathfrak{S}$ of $\mathfrak{A}$ such that $\overline{\overline{\mathfrak{S}}} \leq \mathfrak{m}$ and the meet

(14') $\mathsf{\cap}_{A \in \mathfrak{S}} A = B_{\mathfrak{S}}$

exists, for every $\mathfrak{S} \in \mathbf{M}$. Under what conditions do an $\mathfrak{m}$-field of sets $\mathfrak{F}$, an $\mathfrak{m}$-ideal $\varDelta$ of $\mathfrak{F}$ and a $(\mathbf{J}, \mathbf{M})$-isomorphism h from $\mathfrak{A}$ into $\mathfrak{F}/\varDelta$ exist? We

---

recall (see p. 88) that by a $(\mathbf{J}, \mathbf{M})$-isomorphism $((\mathbf{J}, \mathbf{M})$-homomorphism) we understand an isomorphism (a homomorphism) preserving all the joins (14) and meets (14′). If the isomorphism h having the required properties exists, $\mathfrak{A}$ is said to be $(\mathbf{J}, \mathbf{M}, \mathfrak{m})$-*representable*.

To simplify the formulation of the answer to this question, let us observe that a homomorphism h defined on $\mathfrak{A}$ preserves all the joins (14) and meets (14′) if and only if it preserves all the joins

$$
\begin{aligned}
&\bigcup_{A \in \mathfrak{S}} (A \cup - A_{\mathfrak{S}}) = V && (\mathfrak{S} \in \mathbf{J}), \\
&\bigcup_{A \in \mathfrak{S}} (-A \cup B_{\mathfrak{S}}) = V. && (\mathfrak{S} \in \mathbf{M}).
\end{aligned}
$$
(15)

Denote by $\mathbf{J}_0$ the class of all the sets of elements

$$A \cup - A_{\mathfrak{S}} \qquad\qquad (A \in \mathfrak{S})$$

and of all the sets of elements

$$-A \cup B_{\mathfrak{S}} \qquad\qquad (A \in \mathfrak{S})$$

where $\mathfrak{S} \in \mathbf{J}$ or $\mathfrak{S} \in \mathbf{M}$ respectively. It will be convenient to represent $\mathbf{J}_0$ as an indexed set of indexed sets $\{\{A_{t,s}\}_{s \in S}\}_{t \in T_0}$ where

$$\overline{\overline{S}} = \mathfrak{m}.$$

By definition,

(15′) $\qquad\qquad \bigcup_{s \in S}^{\mathfrak{A}} A_{t,s} = V \quad$ for every $\quad t \in T_0$.

The class of all the joins (15′) coincides with the class of all the joins (15). A homomorphism h defined on A is a $(\mathbf{J}, \mathbf{M})$-homomorphism if and only if h preserves all the joins (15′).

Let h_0 be an isomorphism of $\mathfrak{A}$ onto the field $\mathfrak{F}_0$ of all open-closed subsets of the Stone space X of $\mathfrak{A}$. The defect sets corresponding to the infinite joins in $\mathbf{J}_0$:

(16) $\qquad\qquad X - \bigcup_{s \in S} h_0(A_{t,s}) \qquad\qquad (t \in T_0)$

(where $\bigcup$ denotes the set-theoretical union) are nowhere dense $\mathfrak{m}$-closed sets by 22.2. The sets (16) are exactly the defect sets corresponding to the joins (14) and meets (14′). Every subset of a set (16) will be called $(\mathbf{J}, \mathbf{M})$-*nowhere dense*. Any union of at most $\mathfrak{m}$ $(\mathbf{J}, \mathbf{M})$-nowhere dense sets will be called a set of the $(\mathbf{J}, \mathbf{M}, \mathfrak{m})$-*category*. The sets

$$\bigcup_{s \in S} h_0(A_{t,s}) \qquad\qquad (t \in T_0)$$

are dense $\mathfrak{m}$-open subsets of X. We shall call them *dense* $(\mathbf{J}, \mathbf{M})$-*open sets*.

Let $\mathfrak{F}_\mathfrak{m}$ denote, as previously, the least $\mathfrak{m}$-field containing $\mathfrak{F}_0$, and let $\varDelta$ denote now the $\mathfrak{m}$-ideal of all sets $A \in \mathfrak{F}_\mathfrak{m}$ of the $(\mathbf{J}, \mathbf{M}, \mathfrak{m})$-category. The homomorphism h of $\mathfrak{A}$ into $\mathfrak{F}_\mathfrak{m}/\varDelta$ given by

$$h(A) = [h_0(A)]_\varDelta \quad \text{for} \quad A \in \mathfrak{A}$$

is called the $(\mathbf{J}, \mathbf{M}, \mathfrak{m})$-*canonical homomorphism*.

The $(\mathbf{J}, \mathbf{M}, \mathfrak{m})$-canonical homomorphism h (of $\mathfrak{A}$ into $\mathfrak{F}_{\mathfrak{m}}/\varDelta$) preserves all the joins (15'). The proof is the same as that of the corresponding part of 29.2. Consequently h preserves all the joins (14) and meets (14'), i.e. it is a $(\mathbf{J}, \mathbf{M})$-homomorphism from $\mathfrak{A}$ into $\mathfrak{F}_{\mathfrak{m}}/\varDelta$.

An answer to our question is given by the following theorem, the formulation of which is analogous to that of 29.3 thanks to the terminology introduced above.

29.6. *The following conditions are equivalent:*

(R) *the* $(\mathbf{J}, \mathbf{M}, \mathfrak{m})$-*canonical homomorphism is an isomorphism;*

(R_1) $\mathfrak{A}$ *is* $(\mathbf{J}, \mathbf{M}, \mathfrak{m})$-*representable;*

(R_2) *for every set* $T \subset T_0$ *of power* $\leq \mathfrak{m}$ *and for every element* $A \neq \wedge$ $(A \in \mathfrak{A})$, *there exists a* $\varphi \in S^T$ *such that*

$$A \cap \bigcap_{t \in T'} A_{t, \varphi(t)} \neq \wedge \quad \text{for every finite set} \quad T' \subset T;$$

(R_2') *for every set* $T \subset T_0$ *of power* $\leq \mathfrak{m}$ *and for every proper* $\mathfrak{m}$-*filter* V *of* $\mathfrak{A}$, *there exists a* $\varphi \in S^T$ *such that*

$A \cap \bigcap_{t \in T'} A_{t, \varphi(t)} \neq \wedge$ *for every* $A \in V$ *and every finite set* $T' \subset T;$

(R_3) *in the Stone space of* $\mathfrak{A}$, *any intersection of at most* $\mathfrak{m}$ *dense* $(\mathbf{J}, \mathbf{M})$-*open sets is dense, i.e. any intersection of at most* $\mathfrak{m}$ *dense* $(\mathbf{J}, \mathbf{M})$-*open sets intersects every non-empty open set;*

(R_3') *in the Stone space of* $\mathfrak{A}$, *any intersection of at most* $\mathfrak{m}$ *dense* $(\mathbf{J}, \mathbf{M})$-*open sets intersects any non-empty closed set of upper character* $\mathfrak{m};$

(R_4) *in the Stone space of* $\mathfrak{A}$, *every set of the* $(\mathbf{J}, \mathbf{M}, \mathfrak{m})$-*category is a boundary set, i.e. no open non-empty set is of the* $(\mathbf{J}, \mathbf{M}, \mathfrak{m})$-*category;*

(R_4') *in the Stone space of* $\mathfrak{A}$, *no closed non-empty set of upper character* $\mathfrak{m}$ *is of the* $(\mathbf{J}, \mathbf{M})$-*category;*

(R_5) *for every set* $T \subset T_0$ *of power* $\leq \mathfrak{m}$ *and for every* $A \neq \wedge$ $(A \in \mathfrak{A})$ *there exists a maximal filter containing* A *and preserving all the joins* $\bigcup_{s \in S} A_{t, s} = V$ *where* $t \in T;$

(R_5') *for every set* $T \subset T_0$ *of power* $\leq \mathfrak{m}$ *and for every proper* $\mathfrak{m}$-*filter* V *of* $\mathfrak{A}$, *there exists a maximal filter containing* V *and preserving all the joins* $\bigcup_{s \in S} A_{t, s} = V$ *where* $t \in T.$[1]

Of course, conditions (R), (R_2), (R_2'), (R_3), (R_3'), (R_4), (R_4'), (R_5), (R_5') are analogues of conditions (r), (r_2), (r_2'), (r_3), (r_3'), (r_4), (r_4'), (r_5), (r_5') in 29.3 respectively.

The proof of 29.6 is an easy modification of that of 29.3; one proves the implications $(\mathrm{R}_2) \to (\mathrm{R}_3) \to (\mathrm{R}_4) \to (\mathrm{R}_5) \to (\mathrm{R}_2)$, $(\mathrm{R}_2') \to (\mathrm{R}_3') \to$ $\to (\mathrm{R}_5') \to (\mathrm{R}_2')$, $(\mathrm{R}_2) \to (\mathrm{R}_2') \to (\mathrm{R}_2)$, $(\mathrm{R}_4) \to (\mathrm{R}) \to (\mathrm{R}_1) \to (\mathrm{R}_2)$.

Observe that if $\mathbf{J}$ contains all sets $\mathfrak{S}$ of power $\leq \mathfrak{m}$ such that (14) exists, or if $\mathbf{M}$ contains all the sets $\mathfrak{S}$ of power $\leq \mathfrak{m}$ such that (14') exists, then $(\mathbf{J}, \mathbf{M})$-nowhere dense sets coincide with $\mathfrak{m}$-nowhere dense sets, sets of $(\mathbf{J}, \mathbf{M}, \mathfrak{m})$-category coincide with sets of $\mathfrak{m}$-category, and the $(\mathbf{J}, \mathbf{M}, \mathfrak{m})$-

---

[1] Similarly as in 29.3, we use in 29.6 definition (D) (see p. 80) of $\mathfrak{m}$-filters.

canonical homomorphism coincides with the m-canonical homomorphism from p. 119. Then $\mathfrak{A}$ is $(\mathbf{J}, \mathbf{M}, \mathfrak{m})$-representable if and only if $\mathfrak{A}$ is $\mathfrak{m}$-representable and theorem 29.6 becomes a part of 29.3.

§ 30. Weak (m, n)-distributivity

The conditions (r_1) and (r_2) in the representation theorem 29.3 are similar to conditions (d_1), (d_2) in 19.2 (for $\mathfrak{n} = \mathfrak{m}$), i.e. they express a distributivity property weaker than the $\mathfrak{m}$-distributivity defined in § 19. We shall now introduce a new distributivity property, called weak $\mathfrak{m}$-distributivity, that is weaker than $\mathfrak{m}$-distributivity but stronger than (r_1) and (r_2), i.e. stronger than $\mathfrak{m}$-representability[1]. For this purpose we assume the following notation.

T and S will denote non-empty sets. $\mathcal{S}$ will denote the class of all finite non-void subsets of S. According to the convention assumed on p. 2, $\mathcal{S}^T$ is the set of all functions Φ defined on T with values in $\mathcal{S}$, i.e. such that, for every $t \in T$, $\Phi(t)$ is a finite non-void subset of S. If $\{A_{t,s}\}_{t \in T, s \in S}$ is any indexed set of elements of a Boolean algebra, and $\Phi \in \mathcal{S}^T$, then

$$A_{t, \Phi(t)} = \bigcup_{s \in \Phi(t)} A_{t, s}.$$

A Boolean algebra $\mathfrak{A}$ is said to be *weakly* (m, n)-*distributive* provided

$$\bigcap_{t \in T} \bigcup_{s \in S} A_{t,s} = \bigcup_{\Phi \in \mathcal{S}^T} \bigcap_{t \in T} A_{t, \Phi(t)}$$

for every (m, n)-indexed set $\{A_{t,s}\}_{t \in T, s \in S}$ of elements in $\mathfrak{A}$, such that all the meets $\bigcap_{t \in T} A_{t, \Phi(t)}$ exist and

(1)   all the joins $\bigcup_{s \in S} A_{t,s}$ and the meet $\bigcap_{t \in T} \bigcup_{s \in S} A_{t,s}$ exist.

$\mathfrak{A}$ is said to be *weakly* $\mathfrak{m}$-*distributive* if it is weakly (m, m)-distributive.

By definition, every Boolean algebra is weakly (m, n)-distributive if one of the cardinals $\mathfrak{m}$, $\mathfrak{n}$ is finite. To exclude this trivial case, we shall assume in the sequel that $\mathfrak{m}$ and $\mathfrak{n}$ are infinite.

30.1. *The following conditions are equivalent for any Boolean algebra* $\mathfrak{A}$[2]:
(w) $\mathfrak{A}$ *is weakly* (m, n)-*distributive;*
(w_1) *for every* (m, n)-*indexed set* $\{A_{t,s}\}_{t \in T, s \in S}$ *satisfying* (1), *if*

(2) $$\bigcap_{t \in T} \bigcup_{s \in S} A_{t,s} \neq \wedge,$$

then there exists a $\Phi \in \mathcal{S}^T$ *such that*[3]

(3) $$\bigcap_{t \in T} A_{t, \Phi(t)} \neq \wedge;$$

---

[1] For a discussion of the connection between $\mathfrak{m}$-distributivity, weak $\mathfrak{m}$-distributivity and $\mathfrak{m}$-representability, see SIKORSKI [27].

[2] See SIKORSKI [25] where the theorem is proved for $\mathfrak{m} = \mathfrak{n}$.

[3] The inequality (3) should be read: either the meet (3) does not exist or it exists but it is not equal to $\wedge$. An analogous remark holds for (5).

(w_2) *for every* $(\mathfrak{m}, \mathfrak{n})$-*indexed set* $\{A_{t,s}\}_{t \in T, s \in S}$ *satisfying* (1), *if*

(4) $$\bigcap_{t \in T} \bigcup_{s \in S} A_{t,s} = V ,$$

then for every $A \neq \wedge$ ($A \in \mathfrak{A}$) *there exists a* $\Phi \in S^T$ *such that*

(5) $$A \cap \bigcap_{t \in T} A_{t, \Phi(t)} \neq \wedge ;$$

(w_3) *in the Stone space of* $\mathfrak{A}$, *the interior of any intersection of at most* $\mathfrak{m}$ *dense* $\mathfrak{n}$-*open subsets is dense;*

(w_4) *in the Stone space of* $\mathfrak{A}$, *every union of at most* $\mathfrak{m}$ $\mathfrak{n}$-*nowhere dense sets is nowhere dense;*

(w_5) *for every set of infinite joins and meets in* $\mathfrak{A}$:

(6)
$$\bigcup_{s \in S'_t} A_{t,s} = A_t \quad where \quad \overline{\overline{S'_t}} \leq \mathfrak{n}, \quad t \in T', \overline{\overline{T'}} \leq \mathfrak{m} ,$$
$$\bigcap_{s \in S''_t} B_{t,s} = B_t \quad where \quad \overline{\overline{S''_t}} \leq \mathfrak{n}, \quad t \in T'', \overline{\overline{T''}} \leq \mathfrak{m} ,$$

each non-zero element $A \in \mathfrak{A}$ *contains a non-zero element* B *such that every maximal filter containing* B *preserves all the joins and meets* (6).

Observe that in the case when $\mathfrak{m} = \mathfrak{n}$, condition ($w_4$) can be formulated as follows:

(w'_4) *in the Stone space of* $\mathfrak{A}$, *every set of the* $\mathfrak{m}$-*category is nowhere dense.*

(w) implies (w_1), (w_1) implies (w_2), (w_2) implies (w). The proof of these implications is similar to the proof of the implications (d) → (d_1), (d_1) → (d_2), (d_2) → (d) in 19.2.

In the proof of the following implications, h_0 denotes the isomorphism of $\mathfrak{A}$ onto the field of all open-closed subsets of the Stone space X of $\mathfrak{A}$.

(w_2) implies (w_3). For suppose that, for every $t \in T$ ($\overline{\overline{T}} \leq \mathfrak{m}$), G_t is an $\mathfrak{n}$-open dense subset of X, i.e.

$$G_t = \bigcup_{s \in S} h_0(A_{t,s}) \quad where \quad \bigcup^{\mathfrak{A}}_{s \in S} A_{t,s} = V$$

($\overline{\overline{S}} \leq \mathfrak{n}$). Let G_0 be the interior of the intersection of the sets G_t, $t \in T$, and let $G \subset X$ be any non-empty open subset of X. There exists a non-zero element A in $\mathfrak{A}$ such that $h_0(A) \subset G$. By (w_2) there exists a $\Phi \in S^T$ such that the interior C of the intersection

$$h_0(A) \cap \bigcap_{t \in T} h_0(A_{t, \Phi(t)})$$

is not empty. Since C is open and $C \subset h_0(A) \cap \bigcap_{t \in T} G_t$, we infer that $C \subset G \cap G_0$. The intersection of G_0 with any non-empty open set G being non-empty, the set G_0 is dense.

(w_3) implies (w_4) by passing to complements.

In the proof of the next two implications we identify points in X with maximal filters, i.e. we assume that X is the set of all maximal filters in $\mathfrak{A}$, and

$$h_0(A) = \text{the set of all } V \in X \text{ such that } A \in V$$

(see the proof of 8.2).

(w_4) implies (w_5). In fact, the set N of all maximal filters which do not preserve any of the joins or meets (6) is the union of at most $\mathfrak{m}$ $\mathfrak{n}$-nowhere dense sets, viz. the defect sets corresponding to the joins and meets (6) (see 22.2 and the remark on p. 90), and consequently N is nowhere dense on account of (w_4). Thus the set $h_0(A) - N$ has a non-empty interior, i.e. there exists an element $B \neq \wedge$ $(B \subset A)$ such that $h_0(B) \subset$ $\subset h_0(A) - N$. The element B has all the required properties.

(w_5) implies (w_2). Suppose that (4) holds and apply (w_5) to the joins

$$(7) \qquad\qquad \bigcup_{s \in S} A_{t,s} = V \qquad\qquad (t \in T) .$$

Since all maximal filters containing B (i.e. belonging to $h_0(B)$) preserve all the joins (7), we have

$$h_0(B) \subset \bigcup_{s \in S} h_0(A_{t,s}) \quad \text{for every} \quad t \in T .$$

Since $h_0(B)$ is closed and the $h_0(A_{t,s})$ are open, there exists a finite set $\Phi(t) \subset S$ such that

$$h_0(B) \subset \bigcup_{s \in \Phi(t)} h_0(A_{t,s}) = h_0(A_{t,\Phi(t)}) .$$

This implies (together with the inclusion $B \subset A$) that

$$\wedge \neq B \subset A \cap A_{t,\Phi(t)} \quad \text{for every} \quad t \in T .$$

Thus (5) holds.

30.2. *Every* $(\mathfrak{m}, \mathfrak{n})$-*distributive Boolean algebra is weakly* $(\mathfrak{m}, \mathfrak{n})$-*distributive.*

This follows directly from the definition.

30.3. *Every weakly* $\mathfrak{m}$-*distributive Boolean algebra is* $\mathfrak{m}$-*representable.*

This follows immediately from 29.3 and 30.1 since (w_3) implies (r_3).

Examples. A) Every Boolean σ-algebra with a strictly positive finite (or, more generally, σ-finite) σ-measure m is weakly σ-distributive[1].

We shall prove only the case of finite measures because the case of σ-finite measures can be reduced to the case of finite measures (see the remark at the end of § 20).

Suppose that

$$\bigcap_{1 \leq n < \infty} \bigcup_{1 \leq j < \infty} A_{n,j} = V$$

and let $A \neq \wedge$. The measure m being finite and σ-additive, there exists a finite set $\Phi(n)$ of positive integers such that

$$m\left(\bigcup_{j \in \Phi(n)} A_{n,j} \right) \geq m(V) - \frac{m(A)}{2^{n+1}} .$$

Hence

$$m\left(\bigcap_{1 \leq n < \infty} A_{n,\Phi(n)} \right) \geq m(V) - \frac{m(A)}{2} .$$

This proves that $A \cap \bigcap_{1 \leq n < \infty} A_{n,\Phi(n)} \neq \wedge$, i.e. (w_2) holds.

---

[1] This theorem was explicitly formulated and proved by HORN and TARSKI [1] but it was used earlier by BANACH and KURATOWSKI [1]. See also VON NEUMANN [2].

B) Let $\mathfrak{F}$ be the field of all Borel sets of real numbers, and let Δ be the ideal of all sets of the first category. The complete Boolean algebra $\mathfrak{F}/\Delta$ is not weakly σ-distributive[1].

Let W be the set of all irrational numbers in the unit interval $0 < x < 1$, and let $A_{n,j}$ be the set of all $x \in W$ whose development into a continued fraction

$$ x = \frac{1|}{|a_1} + \frac{1|}{|a_2} + \cdots $$

satisfies the condition $a_n = j$.

We have $\bigcap_{1 \le n < \infty} \bigcup_{1 \le j < \infty} [A_{n,i}] = [W] \neq \Lambda$. On the other hand, for every sequence $\Phi(n)$ of finite sets of positive integers, the intersection $\bigcap_{1 \le n < \infty} A_{n, \Phi(n)}$ is a compact subset of W, therefore it is nowhere dense which implies $\bigcap_{1 \le n < \infty} [A_{n, \Phi(n)}] = \Lambda$. Thus $\mathfrak{F}/\Delta$ does not have property (w_1).

C) If $\overline{\overline{X}} = \mathfrak{m}^+$ and $2^{\mathfrak{m}} = \mathfrak{m}^{+\,2}$, $\mathfrak{F}$ is the field of all subsets of X and Δ is a proper $\mathfrak{m}$-ideal containing all one-point subsets of X, then the Boolean $\mathfrak{m}$-algebra $\mathfrak{F}/\Delta$ is not weakly $\mathfrak{m}$-distributive[3].

It suffices to prove the existence of an $\mathfrak{m}$-indexed set $\{B_{t,s}\}_{t \in T, s \in S}$ of subsets of X such that

$$ X = \bigcup_{s \in S} B_{t,s} \quad \text{for every} \quad t \in T , $$

and, for every $\Phi \in S^T$, the set

$$ \bigcap_{t \in T} B_{t, \Phi(t)} $$

has power $\le \mathfrak{m}$. In fact, we have then in $\mathfrak{F}/\Delta$ for $A_{t,s} = [B_{t,s}]$

$$ \bigcap_{t \in T} \bigcup_{s \in S} A_{t,s} = V \neq \Lambda $$

and, for every $\Phi \in S^T$,

$$ \bigcap_{t \in T} A_{t, \Phi(t)} = \Lambda . $$

Since the existence of such sets $B_{t,s}$ in X is invariant under one-to-one transformations of X, it suffices to prove it in the case of a set X of power $\mathfrak{m}^+$ which will be defined below. It is convenient to assume that $T = S =$ the set of all ordinals of cardinality $< \mathfrak{m}$. Let U be the set of all ordinals of cardinality $\le \mathfrak{m}$. By the hypothesis $2^{\mathfrak{m}} = \mathfrak{m}^+$, there exists an indexed set $\{f_u\}_{u \in U}$ containing all the elements of S^T.

For any functions $f, g \in S^T$ we shall write $g \le f$ if $g(t) \le f(t)$ for every $t \in T$. Observe that for any set $X_0 \subset S^T$ with $\overline{\overline{X}}_0 \le \mathfrak{m}$ we can construct, by a diagonal method, a $g \in S^T$ such that, for every $f \in X_0$, the

---

[1] See von Neumann [2].

[2] $\mathfrak{m}^+$ denotes the smallest cardinal greater than $\mathfrak{m}$.

[3] This theorem is due to Banach and Kuratowski [1]. See also Banach [1].

inequality $g \leq f$ does not hold. Consequently, for every ordinal $u \subset U$ there exists a $g = f_{\varphi(u)}$ such that for every $u' \leq u$, the inequality $f_{\varphi(u)} \leq f_{u'}$ does not hold ($\varphi \in U^U$). Moreover we may assume that $f_{\varphi(u)} \neq f_{\varphi(u_1)}$ for $u \neq u_1$.

Let X be the set of all $f_{\varphi(u)}$, $u \in U$. Clearly X has the required power and

(8) for every $f \in S^T$, the set of all $g \in X$ such that $g \leq f$ has power $\leq$ m.

Let $B_{t,s}$ be the set of all $f \in X$ such that $f(t) = s$. Clearly, for every $t \in T$, X is the union of all the sets $B_{t,s}$, $s \in S$. Let $\Phi \in S^T$ and let $f(t)$ = the greatest ordinal in the finite non-void set $\Phi(t)$. By definition, $f \in S^T$, and the intersection $\bigcap_{t \in T} B_{t,\Phi(t)}$ is a subset of the set of all $g \in X$ such that $g \leq f$. By (8), it has power $\leq$ m.

D) Let $\mathfrak{A}$ be a Boolean algebra satisfying the σ-chain condition. $\mathfrak{A}$ is weakly σ-distributive if and only if, in the Stone space of $\mathfrak{A}$, every set of the first category is nowhere dense[1].

In fact, it follows from § 22 K) that if $\mathfrak{A}$ satisfies the σ-chain condition, then the notions of σ-nowhere dense and of nowhere dense coincide in the Stone space of $\mathfrak{A}$. Consequently sets of the σ-category coincide with the sets of the first category. Therefore our remark follows immediately from 30.1 (w$_4'$).

§ 31. Free Boolean m-algebras

Let n be any cardinal. A Boolean m-algebra $\mathfrak{A}$ is said to be a *free Boolean* m-*algebra with* n *free* m-*generators* provided $\mathfrak{A}$ contains a subset $\mathfrak{S}$ such that

(a) the cardinal of $\mathfrak{S}$ is n,

(b) $\mathfrak{S}$ m-generates $\mathfrak{A}$ (i.e. $\mathfrak{A}$ is the least m-subalgebra containing $\mathfrak{S}$),

(c) every mapping f of $\mathfrak{S}$ into any Boolean m-algebra $\mathfrak{A}'$ can be extended to an m-homomorphism of $\mathfrak{A}$ into $\mathfrak{A}'$.

The elements belonging to $\mathfrak{S}$ are said to be the *free* m-*generators of* $\mathfrak{A}$.

It follows from 23.3 that all free Boolean m-algebras with n free m-generators are isomorphic.

31.1. *For any cardinals* m $\geq \aleph_0$, n > 0 *there exists a free Boolean* m-*algebra* $\mathfrak{A}_{m,n}$ *with* n *free* m-*generators*[2].

Let N be a set of power n. By the general definition on p. 2, if $\mathfrak{B}$ is a Boolean algebra, then $\mathfrak{B}^N$ denotes the set of all mappings v from N into $\mathfrak{B}$.

---

[1] This theorem is due to KELLEY [2] and J. OXTOBY.

[2] RIEGER [5]. The above proof differs from that of RIEGER.

We shall now consider mappings α such that

(d) α assigns to every Boolean $\mathfrak{m}$-algebra $\mathfrak{B}$ a mapping $\alpha_\mathfrak{B}$ from $\mathfrak{B}^N$ into $\mathfrak{B}$.

An example of such a mapping α is given by the formula

$$\alpha_\mathfrak{B}(v) = v(n) \quad \text{for} \quad v \in \mathfrak{B}^N$$

where n is a fixed element in N. We shall denote this mapping by n^*. By definition, if $n \in N$, then n^* is the mapping which assigns to every Boolean $\mathfrak{m}$-algebra $\mathfrak{B}$ the mapping $n_\mathfrak{B}^*$ from $\mathfrak{B}^N$ into $\mathfrak{B}$, defined as follows

$$n_\mathfrak{B}^*(v) = v(n) \quad \text{for every} \quad v \in \mathfrak{B}^N .$$

It is easy to see that

(1) $\qquad\qquad\qquad$ if $\quad n_1 \neq n_2 , \quad$ then $\quad n_1^* \neq n_2^* \qquad\qquad (n_1, n_2 \in N) .$

If α is a mapping with property (d), then $-\alpha$ will denote the mapping which assigns to every Boolean $\mathfrak{m}$-algebra $\mathfrak{B}$ the mapping $(-\alpha)_\mathfrak{B}$ from $\mathfrak{B}^N$ into $\mathfrak{B}$ defined by

(2) $\qquad\qquad\qquad (-\alpha)_\mathfrak{B}(v) = -(\alpha_\mathfrak{B}(v)) \quad \text{for} \quad v \in \mathfrak{B}^N .$

Of course $-\alpha$ also has property (d). Similarly, if α and α' have property (d), then the mappings $\alpha \cup \alpha'$ and $\alpha \cap \alpha'$ defined as follows

(3) $\qquad\qquad (\alpha \cup \alpha')_\mathfrak{B}(v) = \alpha_\mathfrak{B}(v) \cup \alpha'_\mathfrak{B}(v) \quad \text{for} \quad v \in \mathfrak{B}^N ,$

(4) $\qquad\qquad (\alpha \cap \alpha')_\mathfrak{B}(v) = \alpha_\mathfrak{B}(v) \cap \alpha'_\mathfrak{B}(v) \quad \text{for} \quad v \in \mathfrak{B}^N$

also have property (d). More generally, if α_t has property (d) for every $t \in T$ and $\overline{\overline{T}} \leq \mathfrak{m}$, then the mappings $\bigcup_{t \in T} \alpha_t$ and $\bigcap_{t \in T} \alpha_t$ defined as follows

(5) $\qquad\qquad (\bigcup_{t \in T} \alpha_t)_\mathfrak{B}(v) = \bigcup_{t \in T} \alpha_{t\mathfrak{B}}(v) \quad \text{for} \quad v \in \mathfrak{B}^N ,$

(6) $\qquad\qquad (\bigcap_{t \in T} \alpha_t)_\mathfrak{B}(v) = \bigcap_{t \in T} \alpha_{t\mathfrak{B}}(v) \quad \text{for} \quad v \in \mathfrak{B}^n$

also have property (d).

Let $\mathfrak{A}_{\mathfrak{m},n}$ be the smallest set of mappings α with property (d), such that

(7) $\qquad\qquad\qquad n^* \in \mathfrak{A}_{\mathfrak{m},n} \quad \text{for every} \quad n \in N;$

(8) $\quad$ if α, α' are in $\mathfrak{A}_{\mathfrak{m},n}$, then $-\alpha$, $\alpha \cup \alpha'$ and $\alpha \cap \alpha'$ are in $\mathfrak{A}_{\mathfrak{m},n}$;

(9) $\qquad\qquad$ if $\overline{\overline{T}} \leq \mathfrak{m}$ and α_t is in $\mathfrak{A}_{\mathfrak{m},n}$ for every $t \in T$, then

$$\bigcup_{t \in T} \alpha_t \quad \text{and} \quad \bigcap_{t \in T} \alpha_t \quad \text{are in} \quad \mathfrak{A}_{\mathfrak{m},n} .$$

It follows directly from (2), (3), (4) and (8) that $\mathfrak{A}_{\mathfrak{m},n}$ is a Boolean algebra with respect to the operations $\cup, \cap, -$ defined by (3), (4), (2). It follows from (5), (6), (9) that $\mathfrak{A}_{\mathfrak{m},n}$ is $\mathfrak{m}$-complete. Namely, if $\alpha_t \in \mathfrak{A}_{\mathfrak{m},n}$ for every $t \in T$ and $\overline{\overline{T}} \leq \mathfrak{m}$, then the elements $\bigcup_{t \in T} \alpha_t$ and $\bigcap_{t \in T} \alpha_t$

defined by (5) and (6) are respectively the join and the meet of all the α_i in the Boolean algebra $\mathfrak{A}_{m,n}$. It follows from the definition of $\mathfrak{A}_{m,n}$ that the set $\mathfrak{S}$ of all n^*, where $n \in N$, m-generates $\mathfrak{A}_{m,n}$. By (1), $\overline{\overline{\mathfrak{S}}} = \mathfrak{n}$. We recall that, by § 23 (4),

(10) $$\overline{\overline{\mathfrak{A}}}_{m,n} \leq \mathfrak{n}^{\mathfrak{m}} .$$

We shall show that $\mathfrak{A}_{m,n}$ is a free Boolean m-algebra and $\mathfrak{S}$ is a set of free m-generators of $\mathfrak{A}_{m,n}$. Let f be any mapping from $\mathfrak{S}$ into a Boolean m-algebra $\mathfrak{A}'$. Then the formula

$$v'(n) = f(n^*) \quad \text{for} \quad n \in N$$

defines an element $v' \in (\mathfrak{A}')^N$. It follows directly from (2), (3), (4), (5), (6) that the formula

$$h(\alpha) = \alpha_{\mathfrak{A}'}(v') \quad \text{for} \quad \alpha \in \mathfrak{A}_{m,n}$$

defines an m-homomorphism from $\mathfrak{A}_{m,n}$ into $\mathfrak{A}'$. By definition,

$$h(n^*) = n^*_{\mathfrak{A}'}(v') = v'(n) = f(n^*) ,$$

i.e. h is an extension of f.

Theorem 31.1 is proved[1].

31.2. *For every Boolean m-algebra $\mathfrak{A}$ with at most $\mathfrak{n}$ m-generators there exists an m-ideal Δ in the free Boolean m-algebra $\mathfrak{A}_{m,n}$ with $\mathfrak{n}$ free m-generators such that $\mathfrak{A}$ is isomorphic to $\mathfrak{A}_{m,n}/\Delta$. Moreover, if $\mathfrak{S}$ is a set of free m-generators in $\mathfrak{A}_{m,n}$, $\mathfrak{S}_0$ is a set of m-generators in $\mathfrak{A}$, $\overline{\overline{\mathfrak{S}}} = \mathfrak{n}$, $\overline{\overline{\mathfrak{S}}}_0 \leq \mathfrak{n}$, we can assume that the isomorphism from $\mathfrak{A}_{m,n}/\Delta$ onto $\mathfrak{A}$ maps the set of all elements $[A] \in \mathfrak{A}_{m,n}/\Delta$ with $A \in \mathfrak{S}$ onto the set $\mathfrak{S}_0$.*

Let f be a mapping from $\mathfrak{S}$ onto $\mathfrak{S}_0$. By (c), f can be extended to an m-homomorphism h_0 of $\mathfrak{A}_{m,n}$ into $\mathfrak{A}$. Since $h_0(\mathfrak{A}_{m,n})$ is an m-subalgebra of $\mathfrak{A}$ containing the set $\mathfrak{S}_0$ of m-generators, we have $\mathfrak{A} = h_0(\mathfrak{A}_{m,n})$, i.e. h_0 is an m-homomorphism from $\mathfrak{A}_{m,n}$ onto $\mathfrak{A}$. The m-ideal Δ of all $A \in \mathfrak{A}_{m,n}$ such that $h_0(A) = \Lambda$ and the isomorphism h defined by the formula

$$h([A]) = h_0(A) \quad \text{for} \quad A \in \mathfrak{A}_{m,n}$$

have the required properties.

---

[1] If, in the proof of 31.1, one wishes to avoid the logical difficulties connected with the claim that α is a mapping defined on the set of *all* Boolean m-algebras $\mathfrak{B}$, it suffices to consider only all those Boolean m-algebras $\mathfrak{B}$ whose elements belong to a fixed set M of power $\mathfrak{n}^{\mathfrak{m}}$. It suffices also to prove property (c) only in the case when all elements of $\mathfrak{A}'$ belong to M.

This follows from the fact that, by § 23 (4), every Boolean m-algebra with at most $\mathfrak{n}$ m-generators is isomorphic to a Boolean algebra whose elements are in M. On the other hand, a Boolean m-algebra $\mathfrak{A}$ is free if it satisfies conditions (a), (b) and condition (c) wehre $\mathfrak{A}'$ has at most $\mathfrak{n}$ m-generators (replace $\mathfrak{A}'$ by its m-subalgebra m-generated by $f(\mathfrak{S})$!). By isomorphism, it suffices to show that $\mathfrak{A}$ satisfies conditions (a), (b) and condition (c) where $\mathfrak{A}'$ is composed of elements of M.

Theorem 31.1 and its proof are valid for a large class of universal algebras with finite operations and infinite operations of a bounded power.

We do not know anything at all about the structure of the free Boolean $\mathfrak{m}$-algebra $\mathfrak{A}_{\mathfrak{m},\mathfrak{n}}$ with $\mathfrak{n}$ free $\mathfrak{m}$-generators, except in some particular cases. If $\mathfrak{n}$ is finite, then the finite (and therefore complete) free Boolean algebra $\mathfrak{A}_{0,\mathfrak{n}}$ defined in § 14 is $\mathfrak{m}$-free for every cardinal $\mathfrak{m}$ (with the same set of free generators). If $\mathfrak{m} = \aleph_0$ and $\mathfrak{n}$ is arbitrary, the structure of $\mathfrak{A}_{\mathfrak{m},\mathfrak{n}}$ is well explained by theorem 31.6 below. About the remaining cases we know only that

31.3. *If* $\mathfrak{m} \geq 2^{\aleph_0}$ *and* $\mathfrak{n} \geq \aleph_0$, *then the free Boolean $\mathfrak{m}$-algebra* $\mathfrak{A}_{\mathfrak{m},\mathfrak{n}}$ *is not $\mathfrak{m}$-representable.*

The complete algebra $\mathfrak{A}$ of all Borel sets (of real numbers) modulo sets of the first category has a countable number of σ-generators (take the elements determined by intervals with rational endpoints!) which, consequently, $\mathfrak{m}$-generate $\mathfrak{A}$ for every infinite $\mathfrak{m}$. By 31.2 there exists an $\mathfrak{m}$-ideal $\varDelta'$ in $\mathfrak{A}_{\mathfrak{m},\mathfrak{n}}$ such that $\mathfrak{A}$ is isomorphic to $\mathfrak{A}_{\mathfrak{m},\mathfrak{n}}/\varDelta'$. Suppose that $\mathfrak{A}_{\mathfrak{m},\mathfrak{n}} = \mathfrak{F}/\varDelta$ where $\mathfrak{F}$ is an $\mathfrak{m}$-field of sets and $\varDelta$ is an $\mathfrak{m}$-ideal of $\mathfrak{F}$. By the remark at the end of § 10, $\mathfrak{A}_{\mathfrak{m},\mathfrak{n}}/\varDelta'$ is isomorphic to $\mathfrak{F}/\varDelta''$ where $\varDelta''$ is the $\mathfrak{m}$-ideal of all sets $A \in \mathfrak{F}$ such that $[A]_\varDelta \in \varDelta'$. Consequently $\mathfrak{A}$ is isomorphic to $\mathfrak{F}/\varDelta''$. This is impossible for $\mathfrak{m} \geq 2^{\aleph_0}$ since $\mathfrak{A}$ is not $\mathfrak{m}$-representable for $\mathfrak{m} \geq 2^{\aleph_0}$ (see the remark on the beginning of § 29).

It can be proved that

$$(11) \qquad\qquad \overline{\overline{\mathfrak{A}_{\mathfrak{m},\mathfrak{n}}}} \geq \mathfrak{m} \quad \text{for} \quad \mathfrak{m}, \mathfrak{n} \geq \aleph_0{}^1 .$$

A proof of (11) will be given in § 35 L).

Sometimes it is useful to relativize the definition of free $\mathfrak{m}$-algebra to a smaller class of Boolean $\mathfrak{m}$-algebras as follows. Let $\mathbf{K}$ be a class of Boolean $\mathfrak{m}$-algebras. An algebra $\mathfrak{A} \in \mathbf{K}$ is said to be a $\mathbf{K}$-*free $\mathfrak{m}$-algebra* (or *free in the class $\mathbf{K}$ of $\mathfrak{m}$-algebras*) *with $\mathfrak{n}$ free $\mathfrak{m}$-generators* provided $\mathfrak{A}$ contains a subset $\mathfrak{S}$ satisfying (a), (b) and condition (c) for every algebra $\mathfrak{A}' \in \mathbf{K}$. The only difference between this definition and the definition on p. 131 is that now we require that $\mathfrak{A}$ and $\mathfrak{A}'$ be in $\mathbf{K}$. Just as before, we infer from 23.3 that all $\mathbf{K}$-free $\mathfrak{m}$-algebras with $\mathfrak{n}$ free $\mathfrak{m}$-generators are isomorphic (provided they exist).

---

[1] This result was recently obtained by GAIFMAN [1, 3] and HALES [1, 2]. This result has the following interesting consequence. Replacing everywhere in the definition of free Boolean $\mathfrak{m}$-algebra the words "$\mathfrak{m}$-algebra" by "complete algebra" and "$\mathfrak{m}$-generates" by "completely generates" we get the analogous definition of free complete Boolean algebra $\mathfrak{A}$ with a set $\mathfrak{S}$ of $\mathfrak{n}$ free complete generators. The free complete Boolean algebra with $\mathfrak{n}$ generators exists if $\mathfrak{n}$ is finite (viz. it coincides then with $\mathfrak{A}_{0,\mathfrak{n}}$) but does not exist if $\mathfrak{n}$ is infinite. For it can be proved (by the same method as a similar statement in § 35 G)) that the smallest $\mathfrak{m}$-subalgebra containing $\mathfrak{S}$ would then be a free Boolean $\mathfrak{m}$-algebra with $\mathfrak{n}$ generators. Hence if $\mathfrak{n}$ is infinite, we get $\overline{\overline{\mathfrak{A}}} \geq \mathfrak{m}$ for every cardinal $\mathfrak{m}$ which is impossible.

In this section we shall discuss only the case where **K** is the class of all m-representable m-algebras. Then a **K**-free m-algebra with n free m-generators will be called a *free m-representable algebra with n free m-generators*.

To explain the structure of free m-representable algebras, we recall the notation from § 14 (p. 43) and § 24 C): $\mathscr{D}_n$ is the Cantor n-space, i.e. the Cartesian product $\mathscr{D}_n = P_{t \in T_0} H_t = H^{T_0}$ where $H_t = H = (-1, 1)$ and $\overline{\overline{T}}_0 = n$. For every $t \in T_0$, D_t is the set of all points (in $\mathscr{D}_n$) whose t^{th} coordinate is equal to 1. $\mathfrak{F}_{m,n}$ is the m-field m-generated by the sets D_t, i.e. m-generated by the field $\mathfrak{F}_{0,n}$ of all open-closed subsets of $\mathscr{D}_n$.

31.4. *The least m-field $\mathfrak{F}_{m,n}$ containing all open-closed subsets of the Cantor space $\mathscr{D}_n$ is a free m-representable algebra with n free m-generators. The sets D_t are the free m-generators of $\mathfrak{F}_{m,n}$[1].*

The proof is a repetition of the method used in the proof of 15.1.

It suffices to prove that every mapping f of the class $\mathfrak{S}$ of all generators D_t ($t \in T_0$) into any Boolean algebra $\mathfrak{F}/\varDelta$, where $\mathfrak{F}$ is an m-field of subsets of a set X and $\varDelta$ is an m-ideal of $\mathfrak{F}$, can be extended to a homomorphism h of $\mathfrak{F}_{m,n}$ into $\mathfrak{F}/\varDelta$.

For every $t \in T_0$, let $X_t \in \mathfrak{F}$ be a set such that

$$f(D_t) = [X_t] .$$

Let

$$\varphi_t(x) = \begin{cases} 1 & \text{if } x \in X_t \\ -1 & \text{if } x \in -X_t = X - X_t . \end{cases}$$

The mapping $\varphi(x) = \{\varphi_t(x)\}$ of X into $\mathscr{D}_n$ induces an m-homomorphism h, given by

$$h(A) = [\varphi^{-1}(A)]_\varDelta \in \mathfrak{F}/\varDelta \quad \text{for} \quad A \in \mathfrak{F}_{m,n} ,$$

which is the required extension of f.

In a manner similar to that used in proving 31.2, we prove the following theorem.

31.5. *For every m-representable m-algebra $\mathfrak{A}$ with a set $\mathfrak{S}_0$ of m-generators, $\overline{\overline{\mathfrak{S}}}_0 \leq n$, there exists an m-ideal $\varDelta$ in $\mathfrak{F}_{m,n}$ such that $\mathfrak{A}$ is isomorphic to $\mathfrak{F}_{m,n}/\varDelta$. Moreover, we can assume that this isomorphism maps the set of all elements $[D_t]$ onto $\mathfrak{S}_0$.*

In the case where $m = \aleph_0$, the class of all σ-algebras coincides with the class of all σ-representable σ-algebras by 29.1. Consequently, by 31.4 and 31.2 (or 31.5),

---

[1] In the case $m = n = \aleph_0$, theorems 31.4−31.6 were proved by SIKORSKI [14]. The case of $n > \aleph_0$ was examined first by RIEGER [5]. The proof quoted here was given by SIKORSKI [17].

31.6. *The least σ-field $\mathfrak{F}_{\sigma,n}$ (of subsets of the Cantor space $\mathscr{D}_n$) containing all open-closed subsets of $\mathscr{D}_n$ is a free Boolean σ-algebra. The sets D_t are free σ-generators of $\mathfrak{F}_{\sigma,n}$.*

Every Boolean σ-algebra with at most $\mathfrak{n}$ generators is isomorphic to $\mathfrak{F}_{\sigma,n}/\varDelta$ for some σ-ideal $\varDelta$ of $\mathfrak{F}_{\sigma,n}$.

Modifying the definition from § 14, p. 43, we shall say that an indexed set $\{A_t\}_{t \in T}$ of elements of a Boolean algebra $\mathfrak{A}$ is $\mathfrak{m}$-*independent* provided that, for every set $T' \subset T$, $\overline{\overline{T}}' \leq \mathfrak{m}$ and for every function $\varepsilon(t) = \pm 1$, the meet $\bigcap_{t \in T'} \varepsilon(t) \cdot A_t$ exists and is not equal to $\wedge$. Of course, if $\{A_t\}_{t \in T}$ is $\mathfrak{m}$-independent, it is also independent in the sense defined on p. 43.

31.7. *The free $\mathfrak{m}$-generators of $\mathfrak{A}_{m,n}$ are $\mathfrak{m}$-independent. The free $\mathfrak{m}$-generators of $\mathfrak{F}_{m,n}$ are also $\mathfrak{m}$-independent.*

The second part of 31.7 follows immediately from the definition of the sets D_t. The proof is similar to the proof of an analogous statement on p. 43.

To obtain the first part of 31.7, let f denote any one-to-one mapping from the set $\mathfrak{S}$ of free $\mathfrak{m}$-generators of $\mathfrak{A}_{m,n}$ onto the class of all the sets D_t, and let A_t denote the element in $\mathfrak{S}$ whose image is D_t. By (c), f can be extended to an $\mathfrak{m}$-homomorphism h from $\mathfrak{A}_{m,n}$ into $\mathfrak{F}_{m,n}$. Since for $T' \subset T_0$, $\overline{\overline{T}}' \leq \mathfrak{m}$,

$$h\left(\bigcap_{t \in T'} \varepsilon(t) \cdot A_t \right) = \bigcap_{t \in T'} \varepsilon(t) \cdot D_t \neq \wedge,$$

we infer that $\bigcap_{t \in T'} \varepsilon(t) \cdot A_t \neq \wedge$.

The next remark follows directly from 31.7 and 14.2.

31.8. *The subalgebra $\mathfrak{A}_{0,n}$ (of $\mathfrak{A}_{m,n}$) generated by the set $\mathfrak{S}$ of free $\mathfrak{m}$-generators of $\mathfrak{A}_{m,n}$ is a free Boolean algebra with $\mathfrak{n}$ free generators.*

The analogous remark concerning $\mathfrak{F}_{m,n}$ is also true because by definition, the algebra generated by free $\mathfrak{m}$-generators D_t of $\mathfrak{F}_{m,n}$ is the free Boolean algebra $\mathfrak{F}_{0,n}$ defined on p. 43.

Since $\mathfrak{F}_{m,n}$ is an $\mathfrak{m}$-field of sets, and since every $\mathfrak{m}$-field of sets is an $\mathfrak{m}$-representable $\mathfrak{m}$-algebra, it follows directly from 31.4 that $\mathfrak{F}_{m,n}$ is a free algebra in the class of all $\mathfrak{m}$-fields of sets[1].

§ 32. Homomorphisms induced by point mappings

We recall (see § 5 A) and § 11) that a homomorphism h of a field $\mathfrak{F}$ (of subsets of a space X) into a field $\mathfrak{F}'$ (of subsets of a space X') is induced by a point mapping φ of X' into X if

(1) $h(A) = \varphi^{-1}(A)$ for every $A \in \mathfrak{F}$.

---

[1] This also follows from a general remark on free algebras, due to PIERCE [8].

Of course, if $\mathfrak{F}$ and $\mathfrak{F}'$ are m-fields of sets, and h is induced by a point mapping φ, then h is an m-homomorphism. However, not every m-homomorphism of an m-field $\mathfrak{F}$ into an m-field $\mathfrak{F}'$ is induced by a point mapping.

Example. A) The following example is an easy modification of that in § 8 D) and § 11 B).

Let X be a set of power $> \mathfrak{m}$ and let $\mathfrak{F}$ be the m-field composed of all subsets of power $\leq \mathfrak{m}$ and of their complements in X. Suppose $x_0 \notin X$. Let $X' = X \cup (x_0)$ and let

$$h(A) = \begin{cases} A & \text{if } \overline{\overline{A}} \leq \mathfrak{m} \\ A \cup (x_0) & \text{if } \overline{\overline{X-A}} \leq \mathfrak{m}. \end{cases}$$

h is an m-isomorphism of $\mathfrak{F}$ into the field $\mathfrak{F}'$ of all subsets of X'. No mapping φ of X' into X induces h.

The following condition is necessary and sufficient for an m-homomorphism h of an m-field $\mathfrak{F}$ (of subsets of a space X) into an m-field $\mathfrak{F}'$ (of subsets of a space X') to be induced by a point mapping: if a maximal m-filter V' of $\mathfrak{F}'$ is determined by a point x' in X', then the maximal m-filter $V = h^{-1}(V')$ is also determined by a point x in X. The proof of this statement is exactly the same as that of the analogous statement in § 11.

The proof of the following theorem is analogous to that of 11.1.

32.1. *If the m-field $\mathfrak{F}$ is m-perfect, then every m-homomorphism h of $\mathfrak{F}$ into any m-field $\mathfrak{F}'$ is induced by a point mapping.*

Conversely, if the m-isomorphism h (mentioned in 24.3) of $\mathfrak{F}$ onto an m-perfect reduced m-field of sets is induced by a point mapping, then $\mathfrak{F}$ is m-perfect[1].

The following theorem is an immediate consequence of 32.1 (see also 27.1).

32.2. *If the cardinal of X is m-perfect, then every m-homomorphism of the field $\mathfrak{F}$ of all subsets of X into any m-field $\mathfrak{F}'$ is induced by a point mapping.*

If the cardinal of a metric space X is σ-perfect, then every σ-homomorphism h of the field $\mathfrak{F}$ of all Borel subsets of X into any σ-field $\mathfrak{F}'$ is induced by a point mapping.

The following theorem explains the structure of m-homomorphisms from a Boolean m-algebra $\mathfrak{A}$ into any m-field of sets. To formulate it, let us denote by $h_0(A)$ the set of all maximal m-filters V in $\mathfrak{A}$ such that $A \in V$ ($A \in \mathfrak{A}$). Let X be the set of all maximal m-filters in $\mathfrak{A}$.

---

[1] Theorems 32.1−32.2 and 32.4−32.6 are slight modifications of analogous theorems proved for $\mathfrak{m} = \aleph_0$ by SIKORSKI [6, 8, 18].

32.3. *For every* $\mathfrak{m}$*-homomorphism* h *from a Boolean* $\mathfrak{m}$*-algebra* $\mathfrak{A}$ *into an* $\mathfrak{m}$*-field* $\mathfrak{F}'$ *of subsets of a space* X' *there exists a mapping* φ *from* X' *into* X *such that*

$$h(A) = \varphi^{-1}(h_0(A)) \quad \text{for every} \quad A \in \mathfrak{A}.$$

First, observe that for any $A_1, A_2 \in \mathfrak{A}$,

$$\text{if} \quad h(A_1) \neq h(A_2), \quad \text{then} \quad h_0(A_1) \neq h_0(A_2),$$

for if there is a point $x_0 \in h(A_1) - h(A_2)$, then the set V of all $A \in \mathfrak{A}$ such that $x_0 \in h(A)$ is a maximal $\mathfrak{m}$-filter such that $V \in h_0(A_1) - h_0(A_2)$.

Consequently the equation

$$h_1(h_0(A)) = h(A) \quad \text{for} \quad A \in \mathfrak{A}$$

defines a mapping h_1 from the $\mathfrak{m}$-field $\mathfrak{F} = h_0(\mathfrak{A})$ (see 22.1) into $\mathfrak{F}'$. It is easy to verify that h_1 is an $\mathfrak{m}$-homomorphism. By § 24 B), $\mathfrak{F}$ is $\mathfrak{m}$-perfect. By 32.1, h_1 is induced by a point mapping φ. The mapping φ has the required property.

We recall that if Δ and Δ' are ideals of fields $\mathfrak{F}$ and $\mathfrak{F}'$ (of subsets of X and X') respectively, then a homomorphism h of $\mathfrak{F}/\Delta$ into $\mathfrak{F}'/\Delta'$ is said to be induced by a point mapping φ of X' into X provided

(2) $$h([A]_\Delta) = [\varphi^{-1}(A)]_{\Delta'}$$

for every $A \in \mathfrak{F}$ (see § 15). In particular (the case where Δ is the zero ideal), a homomorphism h of $\mathfrak{F}$ into $\mathfrak{F}'/\Delta'$ is said to be induced by a point mapping φ of X' into X provided

(3) $$h(A) = [\varphi^{-1}(A)]_{\Delta'}$$

for every $A \in \mathfrak{F}$.

Of course, if $\mathfrak{F}$ and $\mathfrak{F}'$ are $\mathfrak{m}$-fields, Δ and Δ' are $\mathfrak{m}$-ideals and the homomorphism h of $\mathfrak{F}/\Delta$ into $\mathfrak{F}'/\Delta'$ is induced by a point mapping φ, then h is an $\mathfrak{m}$-homomorphism of $\mathfrak{F}/\Delta$ into $\mathfrak{F}'/\Delta'$.

Let $\mathfrak{F}_{\mathfrak{m},\mathfrak{n}}$ have the same meaning as in § 31, i.e. it is the least $\mathfrak{m}$-field containing all open-closed subsets of the Cantor $\mathfrak{n}$-space $\mathscr{D}_\mathfrak{n}$. The proof of the following theorem is the same as that of 15.1.

32.4. *If* Δ *is an* $\mathfrak{m}$*-ideal of* $\mathfrak{F}_{\mathfrak{m},\mathfrak{n}}$ *and* Δ' *is any* $\mathfrak{m}$*-ideal of an* $\mathfrak{m}$*-field* $\mathfrak{F}'$, *then every* $\mathfrak{m}$*-homomorphism* h *of* $\mathfrak{F}_{\mathfrak{m},\mathfrak{n}}/\Delta$ *into* $\mathfrak{F}'/\Delta'$ *is induced by a point mapping*.

As it is known, every separable metric space X is homeomorphic to a subset of the Hilbert cube $\mathscr{H}$ (i.e. to the Cartesian product of an enumerable sequence of unit intervals). The space X is said to be an *absolute Borel space* provided it is homeomorphic to a Borel subset of $\mathscr{H}$. For instance, every complete separable metric space is an absolute Borel space[1].

---

[1] See e.g. KURATOWSKI [3], p. 337.

32.5. *If X is an absolute Borel space, $\mathfrak{F}$ is the σ-field of all Borel sub-sets of X, and Δ is a σ-ideal of $\mathfrak{F}$, then every σ-homomorphism h of $\mathfrak{F}/\Delta$ into $\mathfrak{F}'/\Delta'$, where $\mathfrak{F}'$ is a σ-field and Δ' is a σ-ideal of $\mathfrak{F}'$, is induced by a point mapping.*

Consider first the case where the space X is at most enumerable, $X = (x_1, x_2, \ldots)$. Let $B_n \in \mathfrak{F}'$ be a set such that $h([(x_n)]_\Delta) = [B_n]_{\Delta'}$ and let $C_1 = B_1 \cup (X - (B_1 \cup B_2 \cup \cdots))$, $C_n = B_n - (B_1 \cup \cdots \cup B_{n-1})$ for $n > 1$. The sets C_n are disjoint, their union is equal to X and $h([(x_n)]_\Delta) = [C_n]_{\Delta'}$. The mapping φ defined by the formula

$$\varphi(x') = x_n \quad \text{for} \quad x' \in C_n$$

induces h.

Suppose now that X is not enumerable. In this case, the proof is based on the following topological theorem: If X is a non-enumerable absolute Borel space, then there exists[1] a one-to-one mapping ψ of $\mathscr{D}_{\aleph_0}$ onto X such that, for every set $A \subset \mathscr{D}_{\aleph_0}$, $\psi(A)$ is a Borel subset of X if and only if A is a Borel subset of $\mathscr{D}_{\aleph_0}$ (i.e. if $A \in \mathfrak{F}_{\sigma,\sigma}$).

Let Δ'' be the class of all sets $A \in \mathfrak{F}_{\sigma,\sigma}$ such that $\psi(A) \in \Delta$. The mapping h' defined by

$$h'([A]_{\Delta''}) = h([\psi(A)]_\Delta) \qquad\qquad (A \in \mathfrak{F}_{\sigma,\sigma})$$

is a σ-homomorphism of $\mathfrak{F}_{\sigma,\sigma}/\Delta''$ into $\mathfrak{F}'/\Delta'$. By 32.4, the σ-homomorphism h' is induced by a point mapping φ'. It is easy to verify that the mapping $\varphi(x') = \psi(\varphi'(x'))$ induces h.

Example. B) If a separable metric space X is not an absolute Borel space, then there exists a σ-field $\mathfrak{F}'$, a σ-ideal Δ' of $\mathfrak{F}'$ and a σ-homomorphism h of the field $\mathfrak{F}$ of all Borel subsets of X into $\mathfrak{F}'/\Delta'$ such that h is not induced by any point mapping[2].

In fact, we may suppose that X is a non-Borel subset of the Hilbert cube $\mathscr{H}$. Let $X' = \mathscr{H}$, let $\mathfrak{F}'$ be the σ-field of all Borel subsets of $\mathscr{H}$ and let Δ' be the σ-ideal of all sets $A \in \mathfrak{F}'$ which do not intersect X. By definition, $\mathfrak{F} = \mathfrak{F}'|X$ (see § 10 B)). The σ-isomorphism h of $\mathfrak{F}$ onto $\mathfrak{F}'/\Delta'$ defined by the formula

(4) $h(A \cap X) = [A]_{\Delta'} \quad \text{for} \quad A \in \mathfrak{F}'$

is not induced by any point mapping φ of X' into X. For suppose such a mapping φ exists. We have $\varphi^{-1}(B) \in \mathfrak{F}'$ for $B \in \mathfrak{F}$, i.e. φ is a Baire mapping of X' into X. Replacing in (4) the set A by a one-point sub-set (x) of X, we obtain easily that $\varphi(x) = x$ for $x \in X$. Consequently, X is the set of all points $x' \in X'$ such that $\varphi(x') = x'$. This implies that X is a Borel subset of X'. Contradiction.

---

[1] See e.g. KURATOWSKI [3], p. 358.
[2] SIKORSKI [6].

Now let $\mathfrak{A}$ and $\mathfrak{A}'$ be $\mathfrak{m}$-representable algebras, let X and X' be their Stone spaces, and let $\mathfrak{F}/\varDelta$ and $\mathfrak{F}'/\varDelta'$ be their canonical $\mathfrak{m}$-representations respectively (see § 29 p. 119). Since $\mathfrak{A}$ is isomorphic to $\mathfrak{F}/\varDelta$ and $\mathfrak{A}'$ is isomorphic to $\mathfrak{F}'/\varDelta'$, the investigation of $\mathfrak{m}$-homomorphisms of $\mathfrak{A}$ into $\mathfrak{A}'$ can be always reduced to the investigation of $\mathfrak{m}$-homomorphisms of $\mathfrak{F}/\varDelta$ into $\mathfrak{F}'/\varDelta'$. The following theorem asserts that this investigation can always be reduced to the investigation of $\mathfrak{m}$-continuous mappings of X' into X.

32.6. *Let $\mathfrak{A}$ and $\mathfrak{A}'$ be $\mathfrak{m}$-representable Boolean algebras. Every $\mathfrak{m}$-homomorphism h of the canonical $\mathfrak{m}$-representation $\mathfrak{F}/\varDelta$ of $\mathfrak{A}$ into the canonical $\mathfrak{m}$-representation $\mathfrak{F}'/\varDelta'$ of $\mathfrak{A}'$ is induced by an $\mathfrak{m}$-continuous mapping φ of the Stone space X' of $\mathfrak{A}'$ into the Stone space X of $\mathfrak{A}$.*

Let $\mathfrak{F}_0$ and $\mathfrak{F}'_0$ be the fields of open-closed subsets of the spaces X and X' respectively. Since $\mathfrak{F}_0$ is isomorphic to $\mathfrak{F}/\varDelta$, and $\mathfrak{F}'_0$ is isomorphic to $\mathfrak{F}'/\varDelta'$, the homomorphism h determines the corresponding homomorphism h' of $\mathfrak{F}_0$ into $\mathfrak{F}'_0$ [see § 22 (9)]. By definition

$$h([A_0]_\varDelta) = [h'(A_0)]_{\varDelta'} \quad \text{for every} \quad A_0 \in \mathfrak{F}_0 .$$

By theorem 11.1, the homomorphism h' is induced by a continuous mapping φ of X' into X, i.e.

$$h'(A_0) = \varphi^{-1}(A_0) \quad \text{for every} \quad A_0 \in \mathfrak{F}_0 .$$

We have

(5) $$h([A_0]_\varDelta) = [\varphi^{-1}(A_0)]_{\varDelta'} \quad \text{for every} \quad A_0 \in \mathfrak{F}_0 .$$

Since $\varphi^{-1}(A_0) \in \mathfrak{F}'_0$ for all $A_0 \in \mathfrak{F}_0$, for every set A in the least $\mathfrak{m}$-field containing $\mathfrak{F}_0$ the set $\varphi^{-1}(A)$ belongs to the least $\mathfrak{m}$-field containing $\mathfrak{F}'_0$.

By 22.5 φ is $\mathfrak{m}$-continuous. Thus $\varphi^{-1}(B) \in \varDelta'$ for every nowhere dense $\mathfrak{m}$-closed subset of X and, consequently, for every set $B \in \varDelta$. This implies that $\varphi^{-1}(A) \in \mathfrak{F}'$ for every $A \in \mathfrak{F}$ and that the formula

(6) $$h''([A]_\varDelta) = [\varphi^{-1}(A)]_{\varDelta'} \quad \text{for} \quad A \in \mathfrak{F}$$

defines a homomorphism h'' of $\mathfrak{F}/\varDelta$ into $\mathfrak{F}'/\varDelta'$. Since every element $A \in \mathfrak{F}$ is of the form

$$A = (A_0 \cup B_1) - B_2 \quad \text{where} \quad A_0 \in \mathfrak{F}_0 \quad \text{and} \quad B_1, B_2 \in \varDelta ,$$

it follows from (5) and (6) that h'' is identical with h, i.e. φ induces h.

Example. C) In all of the theorems on induced homomorphisms proved in §§ 11, 15, 32, except theorem 32.6, only a restriction on the domain $\mathfrak{F}/\varDelta$ (or $\mathfrak{F}$) of the homomorphisms in question was necessary, the structure of the counterdomain $\mathfrak{F}'/\varDelta'$ (or $\mathfrak{F}'$) being arbitrary. In theorem 32.6 a special structure of $\mathfrak{F}'/\varDelta'$ was also supposed. The question arises as to whether theorem 32.6 remains true (of course without the continuity of φ) if we assume only that $\mathfrak{F}'$ is an $\mathfrak{m}$-field of sets and that $\varDelta'$ is an $\mathfrak{m}$-ideal of $\mathfrak{F}'$. The answer to this question is negative.

For instance, let $\mathfrak{A} = \mathfrak{F}'/\varDelta'$ be the quotient algebra defined in § 29 B), i.e. $\mathfrak{F}'$ is an $\mathfrak{m}$-field, $\varDelta'$ is an $\mathfrak{m}$-ideal of $\mathfrak{F}'$, and there exists no isomorphism g of $\mathfrak{F}'/\varDelta'$ into $\mathfrak{F}'$ such that $[g(A)]_{\varDelta'} = A$ for every $A \in \mathfrak{F}'/\varDelta'$. Let $\mathfrak{F}/\varDelta$ be the canonical representation of $\mathfrak{A}$. Then the natural isomorphism h of $\mathfrak{F}/\varDelta$ onto $\mathfrak{F}'/\varDelta'$ is not induced by any point mapping[1]. For suppose such a mapping φ exists. Let $A \in \mathfrak{F}'/\varDelta'$. The element $h^{-1}(A) \in \mathfrak{F}/\varDelta$ is of the form

$$h^{-1}(A) = [A_0]_\varDelta$$

where A_0 is both open and closed. The set A_0 is uniquely determined by A and the conditions mentioned above. The mapping $g(A) = \varphi^{-1}(A_0)$ is an isomorphism of $\mathfrak{F}'/\varDelta'$ into $\mathfrak{F}'$ such that $[g(A)]_{\varDelta'} = A$ for every $A \in \mathfrak{F}'/\varDelta'$. Contradiction.

By analogy with the final part of § 15 the following problem, which is closely related to the existence of inducing mappings, should be investigated. Under what conditions has a given Boolean $\mathfrak{m}$-algebra $\mathfrak{A}$ the following property:

(a) for every $\mathfrak{m}$-homomorphism h of $\mathfrak{A}$ into any Boolean algebra $\mathfrak{A}'/\varDelta'$ (where $\mathfrak{A}'$ is a Boolean $\mathfrak{m}$-algebra and $\varDelta'$ is an $\mathfrak{m}$-ideal of $\mathfrak{A}'$) there exists an $\mathfrak{m}$-homomorphism h' of $\mathfrak{A}$ into $\mathfrak{A}'$ such that

$$h(A) = [h'(A)]_{\varDelta'} \quad \text{for every} \quad A \in \mathfrak{A} \ ?$$

The answer to this problem is given by a theorem analogous to 15.3. The exact formulation is left to the reader.

§ 33. Theorems on extension of homomorphisms

The following theorem will be useful in § 35.

33.1. *Let $\mathfrak{A}_0$ be a subalgebra of a Boolean algebra $\mathfrak{A}$. Every homomorphism of $\mathfrak{A}_0$ into a complete Boolean algebra $\mathfrak{A}'$ can be extended to a homomorphism of $\mathfrak{A}$ into $\mathfrak{A}'$*[2].

It suffices to prove that if h_0 is a homomorphism of a subalgebra $\mathfrak{A}_0$ of $\mathfrak{A}$ into a complete Boolean algebra $\mathfrak{A}'$ and if $A_0 \in \mathfrak{A}$, then h_0 can be extended to a homomorphism h of the subalgebra $\mathfrak{A}_1$ generated by $\mathfrak{A}_0$ and A_0. In fact, theorem 33.1 follows immediately from this lemma. It suffices to order all elements of $\mathfrak{A}$ into a transfinite sequence $\{A_\alpha\}_{\alpha < \beta}$ and to extend the homomorphism, step by step, onto the least subalgebras containing $\mathfrak{A}_0$ and elements $A_0, A_1, \ldots, A_\alpha, \ldots$ respectively.

---

[1] TRACZYK [1].

[2] SIKORSKI [5].

To prove the lemma, let us recall that $\mathfrak{A}_1$ is the set of all elements $A \in \mathfrak{A}$ which can be represented in the form

(1) $$A = (A_1 \cap A_0) \cup (A_2 - A_0)$$

where $A_1, A_2 \in \mathfrak{A}_0$ [see § 4 (3)]. Let B_1 be the join (in the complet) algebra $\mathfrak{A}'$) of all elements $h_0(A)$ where $A \in \mathfrak{A}_0$ and $A \subset A_0$. Similarlye let B_2 be the meet (in the complete algebra $\mathfrak{A}'$) of all elements $h_0(A,$ where $A \in \mathfrak{A}_0$ and $A_0 \subset A$. By definition, $B_1 \subset B_2$. Choose an element $B \in \mathfrak{A}'$ such that $B_1 \subset B \subset B_2$. By definition,

(2)   if $C, D \in \mathfrak{A}_0$ and $C \subset A_0 \subset D$,   then $h_0(C) \subset B \subset h_0(D)$.

If $A \in \mathfrak{A}_1$ is an element of the form (1), we define

(3) $$h(A) = (h_0(A_1) \cap B) \cup (h_0(A_2) - B) .$$

To verify the unambiguity of this definition, we have to show that the element on the right side of (3) does not depend on the representation of $A \in \mathfrak{A}_1$ in the form (1)[1]. Suppose that (1) holds and that simultaneously

(1') $$A = (A_1' \cap A_0) \cup (A_2' - A_0)$$

where $A_1', A_2' \in \mathfrak{A}_0$. It follows from (1) and (1') that

$$A_2 - A_2' \subset A_0 , \quad A_2' - A_2 \subset A_0 ,$$
$$A_0 \subset -A_1 \cup A_1' , \quad A_0 \subset A_1 \cup -A_1' .$$

Consequently, by (2),

$$h_0(A_2) - h_0(A_2') \subset B , \quad h_0(A_2') - h_0(A_2) \subset B ,$$
$$B \subset -h_0(A_1) \cup h_0(A_1') , \quad B \subset h_0(A_1) \cup -h_0(A_1') ,$$

which implies

$$(h_0(A_1) \cap B) \cup (h_0(A_2) - B) = (h_0(A_1') \cap B) \cup (h_0(A_2') - B) .$$

Thus (3) defines uniquely a mapping h of $\mathfrak{A}_1$ into $\mathfrak{A}'$. It is easy to verify that h is a homomorphism. If $A \in \mathfrak{A}_0$, then $A = (A \cap A_0) \cup (A - A_0)$ and by (3)

$$h(A) = (h_0(A) \cap B) \cup (h_0(A) - B) = h_0(A) ,$$

i.e. h is an extension of h_0.

Observe that theorem 33.1 is a generalization of theorem 6.1 (iii). To obtain theorem 6.1 (iii) it suffices to assume in 33.1 that $\mathfrak{A}'$ is the two-element Boolean algebra, that $\mathfrak{A}_0$ is the subalgebra composed of all elements $A \in \varDelta_0$ and their complements, and that h_0 is the homomorphism defined by the formula

$$h_0(A) = \begin{cases} \wedge & \text{if } A \in \varDelta_0 , \\ \vee & \text{if } -A \in \varDelta_0 . \end{cases}$$

---

[1] This part of the proof can be replaced by verification that the mapping f which coincides with h_0 on $\mathfrak{A}_0$ and assumes the value B at A_0 satisfies condition § 12 (4).

Theorem 33.1 can also be formulated as follows. Let h be a homomorphism of a Boolean algebra $\mathfrak{A}_0$ into a complete Boolean algebra $\mathfrak{A}'$ and let g be an isomorphism of $\mathfrak{A}_0$ into a Boolean algebra $\mathfrak{A}$. Then there exists a homomorphism h' of $\mathfrak{A}$ into $\mathfrak{A}'$ such that $h = h'g$.

Passing to the corresponding Stone spaces X_0, X, X' of $\mathfrak{A}_0$, $\mathfrak{A}$, $\mathfrak{A}'$ respectively, and to the continuous mappings φ, φ', ψ inducing respectively h, h', g, we obtain the following topological formulation of 33.1. If φ is a continuous mapping of an extremally disconnected compact space X' into a totally disconnected compact space X_0, and if ψ is a continuous mapping of a totally disconnected compact space X onto X_0, then there exists a continuous mapping φ' of X' into X such that $\varphi = \psi\varphi'$.[1]

Observe that the hypothesis that $\mathfrak{A}'$ is complete can be replaced in 33.1 by the hypothesis that $\mathfrak{A}'$ is an $\mathfrak{m}$-complete Boolean algebra for $\mathfrak{m} = \overline{\overline{\mathfrak{A}}}$. If $\overline{\overline{\mathfrak{A}}}_0 < \overline{\overline{\mathfrak{A}}}$, it can also be replaced by the hypothesis that $\mathfrak{A}'$ is $\mathfrak{m}$-complete for every $\mathfrak{m} < \overline{\overline{\mathfrak{A}}}$. The proof remains unchanged.

33.2. *Let $\mathfrak{A}_0$ be a dense subalgebra of a Boolean algebra $\mathfrak{A}$ and let h_0 be an isomorphism of $\mathfrak{A}_0$ into a complete Boolean algebra $\mathfrak{A}'$. The isomorphism h_0 can be extended to an isomorphism h of $\mathfrak{A}$ into $\mathfrak{A}'$*[2].

By 33.1, the isomorphism h_0 can be extended to a homomorphism h of $\mathfrak{A}$ into $\mathfrak{A}'$. If $A \in \mathfrak{A}$, $A \neq \wedge$, then there exists an $A_0 \in \mathfrak{A}_0$ such that $\wedge \neq A_0 \subset A$. Consequently $\wedge \neq h_0(A_0) = h(A_0) \subset h(A)$. Thus $h(A) \neq \wedge$ which proves that h is an isomorphism.

33.3. *Let $\mathfrak{A}$ and $\mathfrak{A}'$ be two complete Boolean algebras and let h_0 be an isomorphism of a subalgebra $\mathfrak{A}_0$ of $\mathfrak{A}$ onto a dense subalgebra $\mathfrak{A}_0'$ of $\mathfrak{A}'$. Every homomorphism h of $\mathfrak{A}$ into $\mathfrak{A}'$ that is an extension of h_0 maps $\mathfrak{A}$ onto $\mathfrak{A}'$.*

Let B be an element in $\mathfrak{A}'$. Let A be the Boolean join (in the complete Boolean algebra $\mathfrak{A}$) of all elements $h_0^{-1}(B_1)$ where $B_1 \in \mathfrak{A}_0'$ and $B_1 \subset B$. If B_1, $B_2 \in \mathfrak{A}_0'$ and $B_1 \subset B \subset B_2$, then

$$h_0^{-1}(B_1) \subset A \subset h_0^{-1}(B_2).$$

Consequently

$$B_1 = h(h_0^{-1}(B_1)) \subset h(A) \subset h(h_0^{-1}(B_2)) = B_2.$$

Since $\mathfrak{A}_0'$ is dense in $\mathfrak{A}'$, the element B is the meet of all $B_2 \in \mathfrak{A}_0'$ such that $B \subset B_2$, and simultaneously B is the join of all $B_1 \in \mathfrak{A}_0'$ such that $B_1 \subset B$ (see 23.1). Thus $B = h(A)$ which proves the theorem.

---

[1] This theorem is a particular case of a more general topological theorem proved by GLEASON [1]. For a discussion of the connection between theorem 33.1 and the Gleason theorem and for application of the Gleason theorem and other related questions, see HALMOS [8], ISBELL and SEMADENI [1], RAINWATER [1], SEMADENI [4, 5].

[2] Lemmas 33.2—33.4 were proved by SIKORSKI [13].

33.4. *Let $\mathfrak{A}_0$ and $\mathfrak{A}'_0$ be dense subalgebras of complete Boolean algebras $\mathfrak{A}$ and $\mathfrak{A}'$ respectively. Every isomorphism of $\mathfrak{A}_0$ onto $\mathfrak{A}'_0$ can be extended to an isomorphism of $\mathfrak{A}$ onto $\mathfrak{A}'$.*

This follows immediately from 33.2 and 33.3.

§ 34. Theorems on extending to homomorphisms

Let $\mathfrak{A}$ and $\mathfrak{A}'$ be Boolean $\mathfrak{m}$-algebras and let $\mathfrak{S}$ be a subset (of $\mathfrak{A}$) which $\mathfrak{m}$-generates $\mathfrak{A}$ (see § 23). We shall examine the problem of determining under what conditions a mapping f of $\mathfrak{S}$ into $\mathfrak{A}'$ can be extended to an $\mathfrak{m}$-homomorphism h of $\mathfrak{A}$ into $\mathfrak{A}'$. We recall that if this extension exists, it is unique [see § 23 (5)].

Obviously the following condition is necessary for a mapping f to have an extension to an $\mathfrak{m}$-homomorphism:

(a)   if $\bigcap_{t \in T}^{\mathfrak{A}} \varepsilon(t) \cdot A_t = \bigwedge_{\mathfrak{A}}$,   then $\bigcap_{t \in T}^{\mathfrak{A}'} \varepsilon(t) \cdot f(A_t) = \bigwedge_{\mathfrak{A}'}$

for every $\mathfrak{m}$-indexed set $A_t \in \mathfrak{S}$, and for every function $\varepsilon(t) = \pm 1$.

This follows from the commutativity of the extension h with $\bigcap$ and $-$.

In general, condition (a) is not sufficient for the existence of the extension h (see examples A), B), D) below).

We shall say that a Boolean $\mathfrak{m}$-algebra $\mathfrak{A}'$ has the *strong $\mathfrak{m}$-extension property* if, for every Boolean $\mathfrak{m}$-algebra $\mathfrak{A}$, every mapping f (from a set $\mathfrak{S}$ $\mathfrak{m}$-generating $\mathfrak{A}$, into $\mathfrak{A}'$) satisfying condition (a) can be extended to an $\mathfrak{m}$-homomorphism h of $\mathfrak{A}$ into $\mathfrak{A}'$.

34.1. *Every $\mathfrak{m}$-field of sets has the strong $\mathfrak{m}$-extension property.*

Let $\mathfrak{A}'$ be an $\mathfrak{m}$-field of subsets of a space X, and let $\mathfrak{A}$, $\mathfrak{S}$, f have the same meaning as above.

Observe first that if a proper $\mathfrak{m}$-filter V of $\mathfrak{A}$ has the property that

(1)   for every $A \in \mathfrak{S}$, either $A \in V$ or $-A \in V$,

then V is maximal. For the natural homomorphism of $\mathfrak{A}$ onto $\mathfrak{A}/V$ maps $\mathfrak{S}$ into the subalgebra composed of the zero and unit only. Consequently it maps the whole algebra $\mathfrak{A}$ onto this two-element subalgebra, i.e. $\mathfrak{A}/V$ coincides with this subalgebra.

Now let $\{A_t\}_{t \in T_0}$ be an indexed set composed of all elements in $\mathfrak{S}$. For every point $x \in X$, let

(2) $\varepsilon_x(t) = \begin{cases} 1 & \text{if} \quad x \in f(A_t) , \\ -1 & \text{if} \quad x \in -f(A_t) . \end{cases}$

Let V_x be the $\mathfrak{m}$-filter $\mathfrak{m}$-generated by all elements $\varepsilon_x(t) \cdot A_t$, $t \in T_0$.

The $\mathfrak{m}$-filter V_x is proper. For if not, then $\bigcap_{t \in T} \varepsilon_x(t) \cdot A_t = \bigwedge$ for a set $T \subset T_0$, $\overline{\overline{T}} \leqq \mathfrak{m}$. This would imply, by (a), that the set $\bigcap_{t \in T} \varepsilon_x(t) \cdot f(A_t)$ is empty, which is impossible since this set contains the point x.

The $\mathfrak{m}$-filter V_x is maximal since it has property (1).

Let X' be the set of all maximal $\mathfrak{m}$-filters V in $\mathfrak{A}$. The formula

$$\varphi(x) = V_x \quad \text{for} \quad x \in X$$

defines a mapping of X into X'. The mapping

$$h_0(A) = \text{the set of all } V \in X' \text{ such that } A \in V$$

is an $\mathfrak{m}$-homomorphism of $\mathfrak{A}$ into the field of all subsets of X' (see 22.1). Thus the mapping

$$h(A) = \varphi^{-1}(h_0(A)) \quad \text{for} \quad A \in \mathfrak{A}$$

is an $\mathfrak{m}$-homomorphism of $\mathfrak{A}$ into the field of all subsets of X.

h is an extension of f. In fact, it follows immediately from (2) and the definition of V_x that $x \in f(A_t)$ if and only if $A_t \in V_x$, i.e. if $\varphi(x) \in h_0(A_t)$. This proves that $f(A_t) = \varphi^{-1}(h_0(A_t)) = h(A_t)$.

Since f maps $\mathfrak{S}$ into $\mathfrak{A}'$, the extension h maps $\mathfrak{A}$ into $\mathfrak{A}'$. Thus h has all the required properties.

Observe that incidentally we proved here theorem 32.3 for the second time.

Theorem 34.1 can be generalized as follows.

34.2. *Every $\mathfrak{m}$-distributive Boolean $\mathfrak{m}$-algebra has the strong $\mathfrak{m}$-extension property*[1].

Suppose $\mathfrak{A}'$ is an $\mathfrak{m}$-distributive $\mathfrak{m}$-algebra, and let $\mathfrak{A}$, $\mathfrak{S}$ and f mean the same as above.

If $\mathfrak{S}_1$ is any subset of $\mathfrak{S}$, $\overline{\overline{\mathfrak{S}}}_1 \leq \mathfrak{m}$, let $\mathfrak{A}(\mathfrak{S}_1)$ denote the $\mathfrak{m}$-subalgebra of $\mathfrak{A}$ $\mathfrak{m}$-generated by $\mathfrak{S}_1$, and let $\mathfrak{A}'(\mathfrak{S}_1)$ denote the $\mathfrak{m}$-subalgebra of $\mathfrak{A}'$ $\mathfrak{m}$-generated by $f(\mathfrak{S}_1)$. Since $\overline{\overline{f(\mathfrak{S}_1)}} \leq \mathfrak{m}$, $\mathfrak{A}'(\mathfrak{S}_1)$ is isomorphic to an $\mathfrak{m}$-field of sets by 24.5. Since the strong $\mathfrak{m}$-extension property is invariant under isomorphisms, the Boolean $\mathfrak{m}$-algebra $\mathfrak{A}'(\mathfrak{S})$ has the strong $\mathfrak{m}$-extension property. Thus there exists an $\mathfrak{m}$-homomorphism $h_{\mathfrak{S}_1}$ from $\mathfrak{A}(\mathfrak{S}_1)$ into $\mathfrak{A}'(\mathfrak{S}_1)$ which is an extension of the mapping $f|\mathfrak{S}_1$, i.e. the mapping f restricted to $\mathfrak{S}_1$.

Now let $\mathfrak{S}_1$ and $\mathfrak{S}_2$ be two subsets of $\mathfrak{S}$ such that $\overline{\overline{\mathfrak{S}}}_1 \leq \mathfrak{m}$ and $\overline{\overline{\mathfrak{S}}}_2 \leq \mathfrak{m}$. We shall prove that

(3)   if $A \in \mathfrak{A}(\mathfrak{S}_1) \cap \mathfrak{A}(\mathfrak{S}_2)$, then $h_{\mathfrak{S}_1}(A) = h_{\mathfrak{S}_2}(A)$.

Indeed, let $\mathfrak{S}_3$ be the union of $\mathfrak{S}_1$ and $\mathfrak{S}_2$. It follows from the uniqueness of extensions that $h_{\mathfrak{S}_3}|\mathfrak{A}(\mathfrak{S}_1) = h_{\mathfrak{S}_1}$ and $h_{\mathfrak{S}_3}|\mathfrak{A}(\mathfrak{S}_2) = h_{\mathfrak{S}_2}$, i.e. that the homomorphism $h_{\mathfrak{S}_3}$ restricted to $\mathfrak{A}(\mathfrak{S}_1)$ or $\mathfrak{A}(\mathfrak{S}_2)$ coincides with $h_{\mathfrak{S}_1}$ and $h_{\mathfrak{S}_2}$, respectively. Hence $h_{\mathfrak{S}_1}(A) = h_{\mathfrak{S}_3}(A) = h_{\mathfrak{S}_2}(A)$ for all A belonging simultaneously to $\mathfrak{A}(\mathfrak{S}_1)$ and $\mathfrak{A}(\mathfrak{S}_2)$.

Since $\mathfrak{S}$ $\mathfrak{m}$-generates $\mathfrak{A}$, $\mathfrak{A}$ is the union of all the subalgebras $\mathfrak{A}(\mathfrak{S}_1)$ where $\mathfrak{S}_1 \subset \mathfrak{S}$, $\overline{\overline{\mathfrak{S}}}_1 \leq \mathfrak{m}$. It follows from (3) that all the $\mathfrak{m}$-homomorphisms $h_{\mathfrak{S}_1}$ on $\mathfrak{A}(\mathfrak{S}_1)$ determine together a mapping h from $\mathfrak{A}$ into $\mathfrak{A}'$. It is easy to verify that h is an $\mathfrak{m}$-homomorphism.

---

[1] SIKORSKI [32], SIKORSKI and TRACZYK [2].

The problem of whether there exist non-$\mathfrak{m}$-distributive Boolean $\mathfrak{m}$-algebras with the strong $\mathfrak{m}$-extension property is not solved. We shall prove below (theorem 34.5) that every $\mathfrak{m}$-algebra with the strong $\mathfrak{m}$-extension property is $\mathfrak{m}$-representable. On the other hand, there exist $\mathfrak{m}$-representable $\mathfrak{m}$-algebras which do not have the strong $\mathfrak{m}$-extension property. See example A) and § 37 A).

We finish the examination of the strong $\mathfrak{m}$-extension property with the following theorem.

34.3. *For every $t \in T$, let $\mathfrak{A}_t$ be an $\mathfrak{m}$-subalgebra of a Boolean $\mathfrak{m}$-algebra $\mathfrak{A}$ and let h_t be an $\mathfrak{m}$-homomorphism of $\mathfrak{A}_t$ into an $\mathfrak{m}$-algebra $\mathfrak{A}'$ having the strong $\mathfrak{m}$-extension property (in particular, into an $\mathfrak{m}$-distributive $\mathfrak{m}$-algebra $\mathfrak{A}'$, or into an $\mathfrak{m}$-field of sets $\mathfrak{A}'$). Suppose that the set-theoretical union of all $\mathfrak{A}_t$ $\mathfrak{m}$-generates $\mathfrak{A}$. Then, in order that there exist an $\mathfrak{m}$-hormomophism h of $\mathfrak{A}$ into $\mathfrak{A}'$ which is a common extension of all the h_t, i.e.*

$$h(A) = h_t(A) \quad for \quad A \in \mathfrak{A}_t, \quad t \in T$$

it is necessary and sufficient that, for every subset $T' \subset T$ with $\overline{\overline{T'}} \le \mathfrak{m}$, and for every indexed set $\{A_t\}_{t \in T'}$ such that $A_t \in \mathfrak{A}_t$ for every $t \in T'$, the condition

$$\bigcap_{t \in T'}^{\mathfrak{A}} A_t = \wedge \quad imply \quad \bigcap_{t \in T'}^{\mathfrak{A}'} h_t(A_t) = \wedge .$$

The proof of 34.3 is the same as the proof of the analogous theorem 12.4.

Now let $\mathfrak{A}_0$ be a subalgebra of a Boolean $\mathfrak{m}$-algebra $\mathfrak{A}$ such that $\mathfrak{A}_0$ $\mathfrak{m}$-generates $\mathfrak{A}$, and let f be a homomorphism from $\mathfrak{A}_0$ into a Boolean $\mathfrak{m}$-algebra $\mathfrak{A}'$. Then the following condition is necessary for the homomorphism f to have an extension to an $\mathfrak{m}$-homomorphism h from $\mathfrak{A}$ into $\mathfrak{A}'$:

(a')   if $\bigcap_{t \in T}^{\mathfrak{A}} A_t = \wedge_{\mathfrak{A}}$, then $\bigcap_{t \in T}^{\mathfrak{A}'} f(A_t) = \wedge_{\mathfrak{A}'}$

for every $\mathfrak{m}$-indexed set $\{A_t\}_{t \in T}$ of elements of $\mathfrak{A}_0$.

Condition (a') is an equivalent formulation of (a) in the case where $\mathfrak{S}$ is the subalgebra $\mathfrak{A}_0$. Condition (a') is also equivalent to the following condition:

(a'') $f(\bigcap_{t \in T}^{\mathfrak{A}} A_t) = \bigcap_{t \in T}^{\mathfrak{A}'} f(A_t)$ and $f(\bigcup_{t \in T}^{\mathfrak{A}} B_t) = \bigcup_{t \in T}^{\mathfrak{A}'} f(B_t)$

for any $\mathfrak{m}$-indexed sets $\{A_t\}_{t \in T}$, $\{B_t\}_{t \in T}$ of elements of $\mathfrak{A}_0$ such that $\bigcap_{t \in T}^{\mathfrak{A}} A_t \in \mathfrak{A}_0$, $\bigcup_{t \in T}^{\mathfrak{A}} B_t \in \mathfrak{A}_0$.

In general, condition (a') is not sufficient for the existence of the extension h (see examples A) and B) below).

A Boolean $\mathfrak{m}$-algebra $\mathfrak{A}'$ is said to have the *weak $\mathfrak{m}$-extension property* if, for every Boolean $\mathfrak{m}$-algebra $\mathfrak{A}$ and for every subalgebra $\mathfrak{A}_0$ $\mathfrak{m}$-generating $\mathfrak{A}$, every homomorphism f of $\mathfrak{A}_0$ into $\mathfrak{A}'$ satisfying condition (a') can be extended to an $\mathfrak{m}$-homomorphism h of $\mathfrak{A}$ into $\mathfrak{A}'$.

The strong $\mathfrak{m}$-extension property always implies the weak $\mathfrak{m}$-extension property.

However, the strong $\mathfrak{m}$-extension property and the weak $\mathfrak{m}$-extension property are not equivalent; there exist Boolean $\mathfrak{m}$-algebras which have the weak $\mathfrak{m}$-extension property but do not have the strong $\mathfrak{m}$-extension property (and consequently are not $\mathfrak{m}$-distributive, by 34.2). See example E) below.

34.4. *Every weakly $\mathfrak{m}$-distributive Boolean algebra has the weak $\mathfrak{m}$-extension property*[1].

Suppose that $\mathfrak{A}_0$ is a subalgebra of an $\mathfrak{m}$-algebra $\mathfrak{A}$, $\mathfrak{A}_0$ $\mathfrak{m}$-generates $\mathfrak{A}$ and f is a homomorphism of $\mathfrak{A}_0$ into a weakly $\mathfrak{m}$-distributive Boolean $\mathfrak{m}$-algebra $\mathfrak{A}'$ such that (a') holds. We have to prove that f can be extended to an $\mathfrak{m}$-homomorphism h from $\mathfrak{A}$ into $\mathfrak{A}'$. In the proof below the letters T (with indices) and S will always denote sets of power $\mathfrak{m}$. According to the convention on p. 127, $\boldsymbol{S}$ will denote the class of all finite non-void subsets of S and, for any $\Phi \in S^T$ and any indexed set $\{A_{t,s}\}_{t \in T, s \in S}$ of elements of $\mathfrak{A}'$, $A_{t, \Phi(t)} = \bigcup_{s \in \Phi(t)} A_{t,s}$.

Let $\mathfrak{K}$ be the set of all elements A in $\mathfrak{A}$ having the following property:

(b) There exist an element $A' \in \mathfrak{A}'$ and an $\mathfrak{m}$-indexed set $\{A_{t,s}\}_{t \in T, s \in S}$ of elements of $\mathfrak{A}'$ such that

(4) $$\bigcup_{s \in S} A_{t,s} = V \quad \text{for every} \quad t \in T$$

and, moreover, for every $\Phi \in S^T$ there exists a non-void set $\mathfrak{K}_\Phi \subset \mathfrak{A}_0$ such that

(5) $$\bar{\bar{\mathfrak{K}}}_\Phi \leq \mathfrak{m},$$

(6) $$\bigcap_{B \in \mathfrak{K}_\Phi} B \subset A \subset \bigcup_{B \in \mathfrak{K}_\Phi} B,$$

(7) $$A' \cap A_\Phi = f(B) \cap A_\Phi \text{ for every } B \in \mathfrak{K}_\Phi$$

where, for brevity,

(8) $$A_\Phi = \bigcap_{t \in T} A_{t, \Phi(t)}.$$

Suppose A^* is another element in $\mathfrak{K}$, i.e. there exist

$$A'^*, \{A^*_{t,s}\}_{t \in T, s \in S} \text{ and } \mathfrak{K}^*_{\Phi^*} \text{ (for } \Phi^* \in S^T)$$

satisfying conditions (4), (5), (6), (7). We shall prove that

(9) $$\text{if} \quad A \subset A^*, \quad \text{then} \quad A' \subset A'^*.$$

It follows from (6) and the analogous statement for $\mathfrak{K}^*_{\Phi^*}$ that for any $\Phi, \Phi^* \in S^T$,

$$\bigcap_{B \in \mathfrak{K}_\Phi} B \subset \bigcup_{B^* \in \mathfrak{K}^*_{\Phi^*}} B^*,$$

i.e.

$$\bigcap_{B \in \mathfrak{K}_\Phi} \bigcap_{B^* \in \mathfrak{K}^*_{\Phi^*}} (B \cap -B^*) = \wedge_{\mathfrak{A}}.$$

---

[1] Matthes [1]. The proof given below is a slight modification of a proof (not published) communicated by K. Matthes to the author.

Hence by (a') (see (5) and the analogous statement for $\Re^*_{\Phi*}$)

$$\bigcap_{B \in \Re_\Phi} \bigcap_{B^* \in \Re^*_{\Phi*}} (f(B) \cap -f(B^*)) = \wedge_{\mathfrak{A}'} ,$$

i.e.

(10) $\bigcap_{B \in \Re_\Phi} f(B) \subset \bigcup_{B^* \in \Re^*_{\Phi*}} f(B^*) .$

By (7), $A' \cap A_\Phi \subset f(B)$ for $B \in \Re_\Phi$. Thus

(11) $A' \cap A_\Phi \subset \bigcap_{B \in \Re_\Phi} f(B) .$

By condition (7) applied to A'^*, we have $f(B^*) \subset A'^* \cup -A^*_{\Phi*}$ for $B^* \in \Re^*_{\Phi*}$, where analogously to (8)

$$A^*_{\Phi*} = \bigcap_{t \in T} A^*_{t, \Phi^*(t)} .$$

Hence it follows that

(12) $\bigcup_{B^* \in \Re^*_{\Phi*}} f(B^*) \subset A'^* \cup -A^*_{\Phi*} .$

It follows from (10), (11), (12) that

$$A' \cap A_\Phi \subset A'^* \cup -A^*_{\Phi*} \quad \text{for any} \quad \Phi, \Phi^* \in S^T .$$

Since $\mathfrak{A}'$ is weakly m-distributive we have

$$\bigcup_{\Phi \in S^T} A_\Phi = V \quad \text{and} \quad \bigcup_{\Phi^* \in S^T} A^*_{\Phi*} = V$$

by (4) and the analogous statement for $\{A^*_{t,s}\}_{t \in T, s \in S}$. Therefore

$$A' \subset A'^* \cup -A^*_{\Phi*} \quad \text{for every} \quad \Phi^* \in S^T$$

and finally $A' \subset A'^*$, i.e. (9) is proved.

It follows from (9) that if $A = A^* \in \Re$, then $A' = A'^*$, i.e. for every $A \in \Re$ there exists exactly one element A' such that (b) holds. Thus the formula

$$h(A) = A'$$

defines a mapping of $\Re$ into $\mathfrak{A}'$.

Observe now that

(13) $\mathfrak{A}_0 \subset \Re \quad \text{and} \quad h(A) = f(A) \quad \text{for} \quad A \in \mathfrak{A}_0 .$

In fact, to show this it suffices to assume $A' = f(A)$, $A_{t,s} = V$ for all t and s, and $\Re_\Phi = (A)$ for every $\Phi \in S^T$. All the conditions (4), (5), (6), (7) are satisfied.

It is evident that if $A, A', \{A_{t,s}\}_{t \in T, s \in S}$ and $\Re_\Phi$ satisfy (4), (5), (6), (7), then $-A, -A', \{A_{t,s}\}_{t \in T, s \in S}$ and

$$\Re_{-\Phi} = \text{the set of all elements} -B \text{ where } B \in \Re_\Phi$$

also satisfy those conditions. This proves that

(14)   if $A \in \Re$, then $-A \in \Re$ and $h(-A) = -h(A) .$

To complete the proof of 34.4 it suffices to show that

(15) $\bigcap_{r \in S} A_r \in \Re \quad \text{and} \quad h(\bigcap_{r \in S} A_r) = \bigcap_{r \in S} h(A_r)$

for any monotonic m-indexed set $\{A_r\}_{r \in S}$ of elements of $\Re$.

In fact, it follows from (14) and (15) that also

(15') $\bigcup_{r \in S} A_r \in \Re$ and $h(\bigcup_{r \in S} A_r) = \bigcup_{r \in S} h(A_r)$

for every monotonic $\mathfrak{m}$-indexed set $\{A_r\}_{r \in S}$ of elements of $\Re$. By (13), (15) and (15'), $\Re$ satisfies all the hypotheses of theorem 23.4. Since $\mathfrak{A}_0$ $\mathfrak{m}$-generates $\mathfrak{A}$, by 23.4 $\Re = \mathfrak{A}$ and h is an $\mathfrak{m}$-homomorphism from $\mathfrak{A}$ into $\mathfrak{A}'$. By (13), h is an extension of f.

To prove (15), let $\{A_r\}_{r \in S}$ be a monotonic $\mathfrak{m}$-indexed set of elements in $\Re$. Thus for every $r \in S$ there exists an element $A_r' \in \mathfrak{A}'$ and an $\mathfrak{m}$-indexed set $\{A_{t,s}\}_{t \in T_r, s \in S}$ such that

(16) $\bigcup_{s \in S} A_{t,s} = V$ for $t \in T_r$,

and moreover, for every $\Phi \in S^{T_r}$ there exists a non-void set $\Re_{r,\Phi} \subset \mathfrak{A}_0$ such that

(17) $\overline{\overline{\Re}}_{r,\Phi} \leq \mathfrak{m}$,

(18) $\bigcap_{B \in \Re_{r,\Phi}} B \subset A_r \subset \bigcup_{B \in \Re_{r,\Phi}} B$,

(19) $A_r' \cap \bigcap_{t \in T_r} A_{t, \Phi(t)} = f(B) \cap \bigcap_{t \in T_r} A_{t, \Phi(t)}$ for $B \in \Re_{r,\Phi}$.

We may suppose that all the sets T_r are disjoint.

Let $T = \bigcup_{r \in S} T_r \cup (t_0)$ where t_0 is an element which does not belong to the union $\bigcup_{r \in S} T_r$. Let

(20) $A = \bigcap_{r \in S} A_r$, $A' = \bigcap_{r \in S} A_r'$,

(21) $A_{t_0, s} = A' \cup -A_s'$ for all $s \in S$,

and for every $\Phi \in S^T$ let

(22) $\Re_\Phi = \bigcup_{r \in S_\Phi} \Re_{r, \Phi | T_r}$

where S_Φ is the set of all $r \in S$ such that

(23) $A_r \subset \bigcap_{s \in \Phi(t_0)} A_s$.

Since $\{A_r\}_{r \in S}$ is monotonic, (23) is equivalent to

(23') $A_r \subset A_{s_0}$

where A_{s_0} is the smallest of all the elements A_s where $s \in \Phi(t_0)$ (the set $\Phi(t_0)$ is finite and non-void!).

The elements A, A', $\{A_{t,s}\}_{t \in T, s \in S}$ and $\Re_\Phi$ $(\Phi \in S^T)$ just defined satisfy conditions (4), (5), (6), (7). In fact, (16), (20) and (21) imply (4). (17) and (22) imply (5). (18), (22) and (23)-(23') imply (6). To prove (7), suppose that $B \in \Re_\Phi$, $\Phi \in S^T$. Thus there exists an $r \in S_\Phi$ such that $B \in \Re_{r, \Phi | T_r}$ $(\Phi | T_r \in S^{T_r})$. By (19),

$$A_r' \cap \bigcap_{t \in T_r} A_{t, \Phi(t)} = f(B) \cap \bigcap_{t \in T_r} A_{t, \Phi(t)}.$$

Hence

$$A_r' \cap A_{t_0, \Phi(t_0)} \cap \bigcap_{t \in T} A_{t, \Phi(t)} = f(B) \cap \bigcap_{t \in T} A_{t, \Phi(t)}.$$

Since $r \in S_\Phi$, (23) holds and consequently $A_r' \subset \bigcap_{s \in \Phi(t_0)} A_s'$ by (9). Thus

$$A_r' \cap A_{t_0, \Phi(t_0)} = A_r' \cap (A' \cup - \bigcap_{s \in \Phi(t_0)} A_s') = A'$$

by (20) and (21). This proves (7) and completes the proof of (15).

It is not known whether there exist Boolean algebras having the weak $\mathfrak{m}$-extension property which are not weakly $\mathfrak{m}$-distributive.

34.5. *Every Boolean $\mathfrak{m}$-algebra $\mathfrak{A}$ with the weak $\mathfrak{m}$-extension property (and, consequently, every Boolean $\mathfrak{m}$-algebra $\mathfrak{A}$ with the strong $\mathfrak{m}$-extension property) is $\mathfrak{m}$-representable.*

Let $\mathscr{D}_\mathfrak{n}$, $\mathfrak{F}_{\mathfrak{m},\mathfrak{n}}$ and D_t have the same meaning as in § 31. Assume that $\mathfrak{n}$ is the cardinal of $\mathfrak{A}$. Thus there exists a one-to-one mapping f_0 from the set of all the sets D_t (the free $\mathfrak{m}$-generators of $\mathfrak{F}_{\mathfrak{m},\mathfrak{n}}$) onto $\mathfrak{A}$. By 14.3 f_0 can be extended to a homomorphism f of the field $\mathfrak{F}_{0,\mathfrak{n}}$ of all open-closed subsets of $\mathscr{D}_\mathfrak{n}$ onto $\mathfrak{A}$. The subalgebra $\mathfrak{F}_{0,\mathfrak{n}}$ $\mathfrak{m}$-generates $\mathfrak{F}_{\mathfrak{m},\mathfrak{n}}$. The homomorphism f_0 satisfies condition (a'). Indeed, if $\bigcap_{t\in T}^{\mathfrak{F}_{\mathfrak{m},\mathfrak{n}}} A_t = \wedge$ ($\overline{\overline{T}} \leq \mathfrak{m}$, $A_t \in \mathfrak{F}_{0,\mathfrak{n}}$), then by the compactness of $\mathscr{D}_\mathfrak{n}$ we have $\bigcap_{t\in T'}^{\mathfrak{F}_{\mathfrak{m},\mathfrak{n}}} A_t = \wedge$ for a finite set $T' \subset T$. Since f is a homomorphism, we get $\bigcap_{t\in T'}^{\mathfrak{A}} f(A_t) = \wedge_A$ and consequently $\bigcap_{t\in T}^{\mathfrak{A}} f(A_t) = \wedge_A$.

Since $\mathfrak{A}$ has the weak $\mathfrak{m}$-extension property, f_0 can be extended to an $\mathfrak{m}$-homomorphism h from the $\mathfrak{m}$-field $\mathfrak{F}_{\mathfrak{m}.\mathfrak{n}}$ onto $\mathfrak{A}$. Thus $\mathfrak{A}$ is $\mathfrak{m}$-representable.

Examples. A) Let E be a Borel subset of the Cantor space $\mathscr{D}_\sigma$ (see § 14, p. 43) such that E is not a G_δ-set. Let $\mathfrak{A}_0 = \mathfrak{F}_{0,\sigma}|E$, $\mathfrak{A} = \mathfrak{F}_{\sigma,\sigma}|E$ and let $\varDelta$ be the σ-ideal σ-generated by all closed sets disjoint from E, i.e. a set $A \in \mathfrak{F}_{\sigma,\sigma}$ belongs to $\varDelta$ if and only if it is a subset of an F_σ-set disjoint from E. The mapping f defined by the formula

$$(24) \qquad\qquad f(A \cap E) = [A]_\varDelta \qquad\qquad (A \in \mathfrak{F}_{0,\sigma})$$

is an isomorphism of $\mathfrak{A}_0$ into $\mathfrak{A}' = \mathfrak{F}_{\sigma,\sigma}/\varDelta$ and satisfies condition (a'). In fact, if $A_n \in \mathfrak{F}_{0,\sigma}$ and $\bigcap_{1 \leq n < \infty}(E \cap A_n) = E \cap (\bigcap_{1 \leq n < \infty} A_n) = \wedge$, then $\bigcap_{1 \leq n < \infty} A_n \in \varDelta$ since it is closed and disjoint from E; consequently $\bigcap_{1 \leq n < \infty} f(E \cap A_n) = [\bigcap_{1 \leq n < \infty} A_n]_\varDelta = \wedge$.

However, f cannot be extended to a σ-homomorphism h of $\mathfrak{A}$ into $\mathfrak{A}'$. For suppose such an extension h exists. Then $h_1(A) = h(A \cap E)$ and $h_2(A) = [A]_\varDelta$ ($A \in \mathfrak{F}_{\sigma,\sigma}$) are two σ-homomorphisms of $\mathfrak{F}_{\sigma,\sigma}$ into $\mathfrak{A}'$ and, by (24), $h_1(A) = h_2(A)$ for all $A \in \mathfrak{F}_{0,\sigma}$. Since $\mathfrak{F}_{0,\sigma}$ σ-generates $\mathfrak{F}_{\sigma,\sigma}$, we infer that $h_1 = h_2$. This is impossible since, on the one hand, $h_1(-E) = \wedge$ and, on the other hand, $-E \notin \varDelta$ and consequently $h_2(-E) \neq \wedge$. [1]

This proves that condition (a') (and consequently also condition (a)) is not sufficient, even in the case where $\mathfrak{m} = \aleph_0$ and f is supposed to be an isomorphism. The Boolean σ-algebra $\mathfrak{A}'$ has not the weak σ-extension property and, consequently, $\mathfrak{A}'$ has not the strong σ-extension property.

Let $\{\mathfrak{A}_t\}_{t\in T}$ be the indexed set of all finite subalgebras of $\mathfrak{A}_0 = \mathfrak{F}_{0,\sigma}|E$, and let h_t be the isomorphism f (in the above example) restricted to $\mathfrak{A}_t$. There exists no σ-homomorphism h which is a common extension of all h_t

---

[1] Example A) was given by SIKORSKI [14].

(since h would be an extension of f). This proves that the hypothesis that $\mathfrak{A}'$ has the strong $\mathfrak{m}$-extension property is also essential in 34.3 even if the h_t are supposed to be isomorphisms and $\mathfrak{A}$ is an $\mathfrak{m}$-field. In this example we have $\overline{\overline{T}} = \aleph_0$. Another example with $\overline{\overline{T}} = 2$ will be given in § 37 A).

B) Suppose that $\mathfrak{m} \geq 2^{\aleph_0}$ and $\mathfrak{n} \geq \aleph_0$. Let $\mathfrak{A}_{\mathfrak{m},\mathfrak{n}}$, $\mathfrak{A}_{0,\mathfrak{n}}$, $\mathscr{D}_{\mathfrak{n}}$, $\mathfrak{F}_{\mathfrak{m},\mathfrak{n}}$ and D_t have the same meaning as in § 31, p. 135.

Let f be a one-to-one mapping of the class of free $\mathfrak{m}$-generators D_t of the field $\mathfrak{F}_{\mathfrak{m},\mathfrak{n}}$ onto a set $\mathfrak{S}$ of $\mathfrak{m}$-generators of the free $\mathfrak{m}$-algebra $\mathfrak{A}_{\mathfrak{m},\mathfrak{n}}$. The mapping f satisfies condition (a) since the sets D_t are $\mathfrak{m}$-independent. However f cannot be extended to an $\mathfrak{m}$-homomorphism h from $\mathfrak{F}_{\mathfrak{m},\mathfrak{n}}$ into $\mathfrak{A}_{\mathfrak{m},\mathfrak{n}}$. For suppose that such an extension h existed. Then h would map $\mathfrak{F}_{\mathfrak{m},\mathfrak{n}}$ onto $\mathfrak{A}_{\mathfrak{m},\mathfrak{n}}$, and $\mathfrak{A}_{\mathfrak{m},\mathfrak{n}}$ could be $\mathfrak{m}$-representable, which contradicts theorem 31.3.

By 14.3 f can be extended to a homomorphism of the field $\mathfrak{F}_{0,\mathfrak{n}}$ of all open-closed subsets of $\mathscr{D}_{\mathfrak{n}}$ into $\mathfrak{A}_{\mathfrak{m},\mathfrak{n}}$ (it is easy to verify by means of 12.1 that f_0 is an isomorphism from $\mathfrak{F}_{0,\mathfrak{n}}$ onto $\mathfrak{A}_{0,\mathfrak{n}}$). The subalgebra $\mathfrak{F}_{0,\mathfrak{n}}$ $\mathfrak{m}$-generates $\mathfrak{F}_{\mathfrak{m},\mathfrak{n}}$. The homomorphism f_0 satisfies condition (a') (by the same argument as in the proof of 34.5). However the homomorphism f_0 cannot be extended to an $\mathfrak{m}$-homomorphism h from $\mathfrak{F}_{\mathfrak{m},\mathfrak{n}}$ into $\mathfrak{A}_{\mathfrak{m},\mathfrak{n}}$ since such an extension h would be an extension of f and f cannot be extended to an $\mathfrak{m}$-homomorphism.

Thus $\mathfrak{A}_{\mathfrak{m},\mathfrak{n}}$ has neither the strong $\mathfrak{m}$-extension property nor the weak $\mathfrak{m}$-extension property for $\mathfrak{m} \geq 2^{\aleph_0}$ and $\mathfrak{n} \geq \aleph_0$.

C) Every Boolean σ-algebra $\mathfrak{A}'$ having a strictly positive σ-finite σ-measure has the weak σ-extension property[1].

This follows directly from 34.4 and § 30 A).

Observe that the algebra $\mathfrak{A}'$ having a strictly positive σ-finite σ-measure need not be σ-distributive (take e.g. the algebra of Borel sets of real numbers modulo the sets of Lebesgue measure zero — see § 19 A)).

D) Let f be a mapping from a set $\mathfrak{S}$ of $\mathfrak{m}$-generators of an $\mathfrak{m}$-algebra $\mathfrak{A}$ into an $\mathfrak{m}$-algebra $\mathfrak{A}'$. Let $\mathfrak{A}_0$ be the subalgebra of $\mathfrak{A}$ generated by $\mathfrak{S}$. Suppose f satisfies condition (a). Then, by 12.2, f can be uniquely extended to a homomorphism f_0 from $\mathfrak{A}_0$ into $\mathfrak{A}'$. However f_0 need not satisfy condition (a), i.e. (a'). If f_0 does not satisfy this condition, then neither f nor f_0 can be extended to an $\mathfrak{m}$-homomorphism from $\mathfrak{A}$ into $\mathfrak{A}'$.

An example (for $\mathfrak{m} = \aleph_0$) can be constructed as follows. Let $\mathfrak{F}$ be the σ-field of all Borel sets of real numbers, let $\varDelta$ be the σ-ideal of all Borel sets of the first category, let $\varDelta'$ be the ideal of all Borel sets of Lebesgue measure zero, and let $\mathfrak{A} = \mathfrak{F}/\varDelta$, $\mathfrak{A}' = \mathfrak{F}/\varDelta'$. Let $\mathfrak{S}$ be the class of all

---

[1] DUBINS [1].

elements $[A]_\Delta$ where A is a half line, and let $f([A]_\Delta) = [A]_{\Delta'}$. It is easy to verify that f satisfies condition (a). Its extension f_0 over the subalgebra $\mathfrak{A}_0$ generated by $\mathfrak{S}$ does not have property (a'). For suppose f_0 satisfies (a'). Then, by C), f_0 can be extended to a σ-homomorphism h from $\mathfrak{A}$ into $\mathfrak{A}'$. The Lebesgue σ-measure on $\mathfrak{F}$ determines a non-vanishing strictly positive σ-finite σ-measure m on $\mathfrak{A}'$. Thus the formula $m'(B) = m(h(B))$ $(B \in \mathfrak{A})$ defines a non-vanishing σ-finite σ-measure m' on A. On the other hand, it is known that every σ-finite measure on $\mathfrak{A}$ vanishes identically (see § 21 F)).

E) The algebra of Borel sets (of real numbers) modulo sets of Lebesgue measure zero has the weak σ-extension property but it does not have the strong σ-extension property. This follows directly from C) and the properties of the mapping f constructed in D).

§ 35. Completions and $\mathfrak{m}$-completions

The representation problem examined in § 24 can be formulated as follows: does there exist an isomorphism that maps a given Boolean algebra $\mathfrak{A}$ into a complete atomic Boolean algebra (i.e. onto a field of sets — see 25.1) and preserves some given infinite joins and meets? Now we shall discuss another problem: does there exist an isomorphism that maps a given Boolean algebra $\mathfrak{A}$ into a complete (but, in general, not atomic) Boolean algebra and preserves all infinite joins and meets?

The following theorem shows that the answer to the last problem is always affirmative.

35.1. *Let h_0 be the isomorphism of a Boolean algebra $\mathfrak{A}$ onto the field $\mathfrak{F}_0$ of all open-closed subsets of the Stone space X of $\mathfrak{A}$. Let $\mathfrak{F}$ be the σ-field of all Borel subsets of X (or the σ-field of all subsets having the Baire property) and let Δ be the ideal of all sets $A \in \mathfrak{F}$ of the first category in X. The mapping i_0 given by*

(1) $$i_0(A) = [h_0(A)]_\Delta \qquad (A \in \mathfrak{A})$$

is a complete isomorphism of $\mathfrak{A}$ into the complete Boolean algebra $\mathfrak{F}/\Delta$. Moreover $i_0(\mathfrak{A})$ is a dense regular subalgebra of $\mathfrak{F}/\Delta$.

The homomorphism i_0 is an isomorphism since no non-empty open subset of a compact Hausdorff space is of the first category. The Boolean algebra $\mathfrak{F}/\Delta$ is complete on account of § 21 C). Every set $A \in \mathfrak{F}$ is of the form

$$A = (G - A_1) \cup A_2$$

where G is open and $A_1, A_2 \in \Delta$. If $[A]_\Delta \neq \wedge$, then G is not empty, i.e. there exists an element $A_0 \in \mathfrak{A}$ such that $A_0 \neq \wedge$ and $h_0(A_0) \subset G$. Consequently $\wedge \neq i_0(A_0) = [h_0(A_0)] \subset [G] = [A]$ which proves that $\mathfrak{B}$ is dense in $\mathfrak{F}/\Delta$. By 23.1, $i_0(\mathfrak{A})$ is a regular subalgebra of $\mathfrak{F}/\Delta$. If

$A = \bigcup_{t \in T}^{\mathfrak{A}} A_t$, then $i_0(A) = \bigcup_{t \in T}^{i_0(\mathfrak{A})} i_0(A_t)$ since i_0 is an isomorphism of $\mathfrak{A}$ onto $i_0(\mathfrak{A})$. Since $i_0(\mathfrak{A})$ is a regular subalgebra of $\mathfrak{F}/\Delta$, we can write the last equality in the form $i_0(A) = \bigcup_{t \in T}^{\mathfrak{F}/\Delta} i_0(A_t)$ which proves that i_0 is a complete isomorphism of $\mathfrak{A}$ into $\mathfrak{F}/\Delta$.

Let $\mathfrak{A}$ be a fixed Boolean algebra.

In this section and in § 36 we shall examine ordered pairs $\{i, \mathfrak{B}\}$ where i is an isomorphism from $\mathfrak{A}$ into $\mathfrak{B}$. A pair $\{i', \mathfrak{B}'\}$ is said to be *isomorphic* to $\{i, \mathfrak{B}\}$ if there exists an isomorphism h from $\mathfrak{B}$ onto $\mathfrak{B}'$ such that $i' = hi$ (in other words, if the isomorphism $i'i^{-1}$ from $i(\mathfrak{A})$ onto $i'(\mathfrak{A})$ can be extended to an isomorphism h from $\mathfrak{B}$ onto $\mathfrak{B}'$). Note that h^{-1} is then an isomorphism from $\mathfrak{B}'$ onto $\mathfrak{B}$ such that $i = h^{-1}i'$ and consequently $\{i, \mathfrak{B}\}$ is isomorphic to $\{i', \mathfrak{B}'\}$. Thus we can then say simply that $\{i, \mathfrak{B}\}$ and $\{i', \mathfrak{B}'\}$ are isomorphic.

A pair $\{i, \mathfrak{B}\}$ is said to be a *completion*[1] *of* $\mathfrak{A}$ provided

(a) $\mathfrak{B}$ is a complete Boolean algebra,

(b) i is a complete isomorphism from $\mathfrak{A}$ into $\mathfrak{B}$ (i.e. $i(\mathfrak{A})$ is a regular subalgebra of $\mathfrak{B}$),

(c) $i(\mathfrak{A})$ completely generates $\mathfrak{B}$.

For instance, it follows from 35.1 that the pair

$$(2) \qquad\qquad \{i_0, \mathfrak{F}/\Delta\}$$

defined in 35.1 is a completion of $\mathfrak{A}$. It is easy to see that if $\{i, \mathfrak{B}\}$ is a completion of $\mathfrak{A}$ and $\{i', \mathfrak{B}'\}$ is isomorphic to $\{i, \mathfrak{B}\}$ then $\{i', \mathfrak{B}'\}$ is also a completion of $\mathfrak{A}$. Thus all isomorphs of (2) are completions of $\mathfrak{A}$. The converse statement is also true. It is a part of the following theorem which implies that all completions of $\mathfrak{A}$ are isomorphic.

35.2. *Let i be an isomorphism from $\mathfrak{A}$ into a complete Boolean algebra $\mathfrak{B}$. The following conditions are equivalent:*

(i) $\{i, \mathfrak{B}\}$ *is a completion of* $\mathfrak{A}$;

(ii) $\{i, \mathfrak{B}\}$ *is isomorphic to* (2);

(iii) $i(\mathfrak{A})$ *is a dense subalgebra of* $\mathfrak{B}$;

(iv) *for every isomorphism i' of $\mathfrak{A}$ into any complete Boolean algebra $\mathfrak{B}'$ there exists an isomorphism h of $\mathfrak{B}$ into $\mathfrak{B}'$ such that $i' = hi$ (in other words, the isomorphism $i'i^{-1}$ from $i(\mathfrak{A})$ into $\mathfrak{B}'$ can be extended to an isomorphism h from $\mathfrak{B}$ into $\mathfrak{B}'$)*[2].

(ii) implies (iii). This follows from 35.1.

(iii) implies (iv). This follows from 33.2.

(iv) implies (ii). Indeed, $i_0 i^{-1}$ can be extended to an isomorphism h from $\mathfrak{B}$ into $\mathfrak{F}/\Delta$. By 35.1 and 33.3, h maps $\mathfrak{B}$ onto $\mathfrak{F}/\Delta$. Thus $\{i, \mathfrak{B}\}$ and $\{i_0, \mathfrak{F}/\Delta\}$ are isomorphic.

---

[1] For an examination of completions of Boolean algebras, see GLIVENKO [1], MacNEILLE [1], SIKORSKI [13] and STONE [5]. See also DILWORTH [3], GLEASON [1], RAINWATER [1], SEMADENI [2].

[2] SIKORSKI [13].

(i) implies (iv). For brevity, let $\mathfrak{B}_0 = \mathfrak{F}/\varDelta$ (see (2)). It follows from the equivalence of (ii), (iii) and (iv) that $\{i_0, \mathfrak{B}_0\}$ has property (iv). Thus the isomorphism $i i_0^{-1}$ from $i_0(\mathfrak{A})$ into $\mathfrak{B}$ can be extended to an isomorphism h from $\mathfrak{B}_0$ into $\mathfrak{B}$. The subalgebra $i(\mathfrak{A}) = h(i_0(\mathfrak{A}))$ is dense in $h(\mathfrak{B}_0)$ by 35.1, since h is an isomorphism. By (c), $i(\mathfrak{A})$ is a regular subalgebra of $\mathfrak{B}$. Consequently, by 23.2, $h(\mathfrak{B}_0)$ is a regular subalgebra of $\mathfrak{B}$. On the other hand, $h(\mathfrak{B}_0)$ is a complete Boolean algebra since it is isomorphic to $\mathfrak{B}_0$. This implies that $h(\mathfrak{B}_0)$ is a complete subalgebra of $\mathfrak{B}$. Hence $h(\mathfrak{B}_0) = \mathfrak{B}$ by (b), i.e. $\{i_0, \mathfrak{B}_0\}$ is isomorphic to $\{i, \mathfrak{B}\}$. Since property (iv) is invariant under isomorphisms and $\{i_0, \mathfrak{B}_0\}$ has this property, $\{i, \mathfrak{B}\}$ also has property (iv).

(iii) implies (i). This follows from 23.1.

If $\mathfrak{A}$ is a subalgebra of $\mathfrak{B}$, i is the identity mapping and $\{i, \mathfrak{B}\}$ is a completion of $\mathfrak{A}$, then $\mathfrak{B}$ itself will also be called a *completion of* $\mathfrak{A}$. The completion $\mathfrak{B}$ of $\mathfrak{A}$ always exists (identify $\mathfrak{A}$ with $i_0(\mathfrak{A})$ by means of the isomorphism i_0!), and is determined by $\mathfrak{A}$ uniquely up to an isomorphism. By 35.2, each of the following conditions is necessary and sufficient for a complete Boolean algebra $\mathfrak{B}$ to be a completion of its subalgebra $\mathfrak{A}$:

(c_1) $\mathfrak{A}$ is a regular subalgebra and completely generates $\mathfrak{B}$;

(c_2) $\mathfrak{A}$ is a dense subalgebra of $\mathfrak{B}$;

(c_3) every isomorphism from $\mathfrak{A}$ into a complete Boolean algebra $\mathfrak{B}'$ can be extended to an isomorphism from $\mathfrak{B}$ into $\mathfrak{B}'$.

Examples. A) If a Boolean algebra $\mathfrak{A}$ is atomic, $\mathfrak{F}$ is the field of all subsets of the set of all atoms of $\mathfrak{A}$ and $i(A) = $ the set of all atoms $a \subset A$ for $A \in \mathfrak{A}$, then $\{i, \mathfrak{F}\}$ is a completion of $\mathfrak{A}$. This follows immediately from 35.2 (i), (iii) and 24.4.

B) If $\mathfrak{A}$ is atomless and $\{i, \mathfrak{B}\}$ is a completion of $\mathfrak{A}$, then $\mathfrak{B}$ is also atomless. This follows immediately from 35.2 (i), (iii).

C) The Boolean algebra of all regular open subsets (see § 1 B)) of a zero-dimensional space is a completion of the Boolean algebra of all open-closed subsets of this space[1]. This remark follows directly from 35.2 (i), (iii) and § 20 C).

An analogous remark is true also for the Boolean algebra of all regular closed subsets (see § 5 C)).

D) The hypothesis in 33.1 that the Boolean algebra $\mathfrak{A}'$ is complete is essential. More exactly, if $\mathfrak{A}'$ is a non-complete Boolean algebra, then there exist a Boolean algebra $\mathfrak{A}$, a subalgebra $\mathfrak{A}_0$ of $\mathfrak{A}$ and a homomorphism h of $\mathfrak{A}_0$ into $\mathfrak{A}'$ such that h cannot be extended to any homomorphism of $\mathfrak{A}$ into $\mathfrak{A}'$.

Viz. let $\mathfrak{A}$ be a completion of $\mathfrak{A}'$, let $\mathfrak{A}_0 = \mathfrak{A}'$ and let h be the identity mapping of $\mathfrak{A}_0$ onto $\mathfrak{A}'$. Suppose that h can be extended to a homo-

---

[1] See MacNeille [1].

morphism h' of $\mathfrak{A}$ into $\mathfrak{A}'$. Then h' maps $\mathfrak{A}$ onto $\mathfrak{A}'$. By the same argument as in the proof of 33.2 we verify that h' is an isomorphism. Thus $\mathfrak{A}'$ is complete since it is isomorphic to the complete algebra $\mathfrak{A}$. Contradiction.

E) All separable atomless complete Boolean algebras are isomorphic to the algebra $\mathfrak{F}/\Delta$ where $\mathfrak{F}$ is the field of all Borel sets of real numbers and Δ is the ideal of all sets of the first category[1].

This is a consequence of the following statements: (a_0) Completions of isomorphic Boolean algebras are isomorphic. (b_0) All enumerable atomless Boolean algebras are isomorphic (see § 9 C)). (c_0) Every atomless complete separable Boolean algebra is a completion of an enumerable atomless Boolean algebra.

To prove (c_0) it suffices to remark that if $\mathfrak{S}$ is an enumerable dense subset of an atomless Boolean algebra, then the subalgebra generated by $\mathfrak{S}$ is enumerable, atomless and dense in the whole algebra.

It follows from § 21 F) that the algebra of Borel sets modulo sets of Lebesgue measure zero is not separable.

F) A pair $\{\mathfrak{S}_1, \mathfrak{S}_2\}$ of subsets of a Boolean algebra $\mathfrak{A}$ is said to be a *cut* in $\mathfrak{A}$ provided

(d') if $A_1 \subset A_2$ for every $A_2 \in \mathfrak{S}_2$, then $A_1 \in \mathfrak{S}_1$;

(d'') if $A_1 \subset A_2$ for every $A_1 \in \mathfrak{S}_1$, then $A_2 \in \mathfrak{S}_2$.

Let $\mathfrak{B}$ be a completion of its subalgebra $\mathfrak{A}$. If B is any element in $\mathfrak{B}$, $\mathfrak{S}_1$ is the set of all $A_1 \in \mathfrak{A}$ such that $A_1 \subset B$ and $\mathfrak{S}_2$ is the set of all $A_2 \in \mathfrak{A}$ such that $B \subset A_2$, then $\{\mathfrak{S}_1, \mathfrak{S}_2\}$ is a cut in $\mathfrak{A}$. Conversely, every cut in $\mathfrak{A}$ is determined in this way by an element $B \in \mathfrak{B}$, viz. by the element $B = \bigcup_{A_1 \in \mathfrak{S}_1}^{\mathfrak{B}} A_1 = \bigcap_{A_2 \in \mathfrak{S}_2}^{\mathfrak{B}} A_2$. Thus elements in any completion $\mathfrak{B}$ of $\mathfrak{A}$ can be identified with cuts in $\mathfrak{A}$, i.e. $\mathfrak{B}$ can be defined as the set of all cuts in $\mathfrak{A}$[2].

G) As an application of the existence of completions of Boolean algebras we shall prove the following supplementary remark to 31.8. Let $\mathfrak{S}$ be a set of $\mathfrak{n}$ free m-generators of the free Boolean m-algebra $\mathfrak{A}_{\mathfrak{m},\mathfrak{n}}$ ($\overline{\overline{\mathfrak{S}}} = \mathfrak{n}$) and let $\aleph_0 \leq \mathfrak{m}' \leq \mathfrak{m}$. The m'-subalgebra $\mathfrak{C}$ (of $\mathfrak{A}_{\mathfrak{m},\mathfrak{n}}$) m'-generated by $\mathfrak{S}$ is isomorphic to the free Boolean m'-algebra $\mathfrak{A}_{\mathfrak{m}',\mathfrak{n}}$ with $\mathfrak{n}$ free m'-generators.

To prove this, consider $\mathfrak{A}_{\mathfrak{m}',\mathfrak{n}}$ as a regular subalgebra of its completion $\mathfrak{B}$. Let f be a one-to-one mapping of the set $\mathfrak{S}$ onto a set $\mathfrak{S}'$ of free m-generators of $\mathfrak{A}_{\mathfrak{m}',\mathfrak{n}}$. The mapping f can be extended to an m-homomorphism h from $\mathfrak{A}_{\mathfrak{m},\mathfrak{n}}$ into $\mathfrak{B}$. The extension h of f, if considered only on $\mathfrak{C}$, is an m'-homomorphism from $\mathfrak{C}$ into $\mathfrak{A}_{\mathfrak{m}',\mathfrak{n}}$. On the other hand, the mapping f^{-1} can be extended to an m'-homomorphism g from $\mathfrak{A}_{\mathfrak{m}',\mathfrak{n}}$ into $\mathfrak{C}$. By 23.3, $\mathfrak{C}$ is isomorphic to $\mathfrak{A}_{\mathfrak{m}',\mathfrak{n}}$.

---

[1] This result is due to S. JAŚKOWSKI (not published). See TARSKI [3], p. 199.

[2] This method of construction of completions of Boolean algebras was applied by MACNEILLE [1].

As before, let i be an isomorphism of a Boolean algebra $\mathfrak{A}$ into a Boolean algebra $\mathfrak{B}$. The pair $\{i, \mathfrak{B}\}$ is said to be an $\mathfrak{m}$-*completion of* $\mathfrak{A}$ provided

(a') $\mathfrak{B}$ is an $\mathfrak{m}$-algebra,

(b') i is a complete isomorphism from $\mathfrak{A}$ into $\mathfrak{B}$ (i.e. $i(\mathfrak{A})$ is a regular subalgebra of $\mathfrak{B}$),

(c') $i(\mathfrak{A})$ $\mathfrak{m}$-generates $\mathfrak{B}$.

For instance, if $\{i_0, \mathfrak{C}\}$ is a completion of $\mathfrak{A}$, and $\mathfrak{B}_0$ is the subalgebra $\mathfrak{m}$-generated by $i(\mathfrak{A})$, then

(3) $$\{i_0, \mathfrak{B}_0\}$$

is an $\mathfrak{m}$-completion of $\mathfrak{A}$. Since $\{i_0, \mathfrak{C}\}$ is determined by $\mathfrak{A}$ uniquely up to an isomorphism, (3) is also uniquely determined by $\mathfrak{A}$ up to an isomorphism. Note that if $\{i, \mathfrak{B}\}$ is an $\mathfrak{m}$-completion of $\mathfrak{A}$ and $\{i', \mathfrak{B}'\}$ is isomorphic to $\{i, \mathfrak{B}\}$, then $\{i', \mathfrak{B}'\}$ is also an $\mathfrak{m}$-completion of $\mathfrak{A}$. Thus all isomorphs of (3) are $\mathfrak{m}$-completions of $\mathfrak{A}$. The converse statement is also true; it is a part of the following theorem, which implies that all $\mathfrak{m}$-completions of $\mathfrak{A}$ are isomorphic.

35.3. *Let i be an isomorphism from $\mathfrak{A}$ into a Boolean $\mathfrak{m}$-algebra $\mathfrak{B}$. The following conditions are equivalent:*

(i) $\{i, \mathfrak{B}\}$ *is an $\mathfrak{m}$-completion of* $\mathfrak{A}$;

(ii) $\{i, \mathfrak{B}\}$ *is isomorphic to* (3);

(iii) $i(\mathfrak{A})$ *is a dense subalgebra of $\mathfrak{B}$ and $\mathfrak{m}$-generates $\mathfrak{B}$*[1].

(i) implies (ii). In fact, let $\{h, \mathfrak{C}\}$ be a completion of $\mathfrak{B}$, $i_0 = hi$ and $\mathfrak{B}_0 = h(\mathfrak{B})$. Then $\{i_0, \mathfrak{C}\}$ is a completion of $\mathfrak{A}$, $\mathfrak{B}_0$ is the subalgebra $\mathfrak{m}$-generated by $i_0(\mathfrak{A})$ and h is an isomorphism from $\{i, \mathfrak{B}\}$ onto $\{i_0, \mathfrak{B}_0\}$.

(ii) implies (iii). This follows from the corresponding implication (ii) $\rightarrow$ (iii) in 35.2.

(iii) implies (i). This follows from 23.1.

If $\mathfrak{A}$ is a subalgebra of $\mathfrak{B}$, i is the identity mapping and $\{i, \mathfrak{B}\}$ is an $\mathfrak{m}$-completion of $\mathfrak{A}$, then $\mathfrak{B}$ itself will also be called an $\mathfrak{m}$-*completion of* $\mathfrak{A}$. The $\mathfrak{m}$-completion $\mathfrak{B}$ of $\mathfrak{A}$ always exists (identify $\mathfrak{A}$ with $i_0(\mathfrak{A})$ in (3) by means of the isomorphism i_0!) and is determined by $\mathfrak{A}$ uniquely up to an isomorphism. By 35.3, each of the following conditions is necessary and sufficient for an $\mathfrak{m}$-complete Boolean algebra $\mathfrak{B}$ to be an $\mathfrak{m}$-completion of its subalgebra $\mathfrak{A}$:

(c'_1) $\mathfrak{A}$ is a regular subalgebra and $\mathfrak{m}$-generates $\mathfrak{B}$;

(c'_2) $\mathfrak{A}$ is a dense subalgebra and $\mathfrak{m}$-generates $\mathfrak{B}$.

E x a m p l e s. H) If a Boolean algebra $\mathfrak{A}$ satisfies the $\mathfrak{n}$-chain condition, then its completion and $\mathfrak{m}$-completion also satisfy the $\mathfrak{n}$-chain condition. Thus for $\mathfrak{m} \geq \mathfrak{n}$, the $\mathfrak{m}$-completion of $\mathfrak{A}$ is a complete Boolean algebra by 20.5 and, consequently, it coincides with the completion of $\mathfrak{A}$.

---

[1] For the case $\mathfrak{m} = \aleph_0$, see SIKORSKI [13].

I) If $\{i, \mathfrak{B}\}$ is an m-completion of an atomic Boolean algebra $\mathfrak{A}$, then $\mathfrak{B}$ is also atomic. This follows easily e.g. from A).

J) Let $\mathfrak{A}$, h_0, X and $\mathfrak{F}_0$ have the same meaning as in 35.1, let $\mathfrak{F}_\sigma$ be the σ-field σ-generated by $\mathfrak{F}_0$, let $\varDelta_0$ be the σ-ideal of all sets $B \in \mathfrak{F}_\sigma$ of the first category, and let

$$i_0(A) = [h_0(A)]_{\varDelta_0} \quad \text{for} \quad A \in \mathfrak{A}.$$

Then $\{i_0, \mathfrak{F}_\sigma/\varDelta_0\}$ is a σ-completion of $\mathfrak{A}$.

K) As an application of the existence of m-completions we shall prove the following statement: for every infinite cardinal $\mathfrak{m}$ there exists a Boolean m-algebra $\mathfrak{B}$ such that $\overline{\overline{\mathfrak{B}}} \geq \mathfrak{m}$ and $\mathfrak{B}$ is m-generated by a countable set $\mathfrak{S}_0$ of elements[1].

First we shall construct a field $\mathfrak{F}$ of subsets of the Cantor space $\mathscr{D}_\mathfrak{m} = H^{T_0}$ where, as usual $H = (1, -1)$, and T_0 is a set of power $\mathfrak{m}$. According to the convention on p. 43, D_t will denote the set of all points $x = \{x_t\}_{t \in T_0} \in \mathscr{D}_\mathfrak{m}$ whose t^{th} coordinate x_t is equal to 1. It is convenient to assume that

$$T_0 = (S \cup U) \times S$$

where U is the set of all ordinals with cardinality $< \mathfrak{m}$ (0 included) and S is a countable set disjoint from U. The letters j, k, l, n, m, p, q and s (with indices if necessary) will denote elements of S, and the letters α, β and γ (with indices) will denote ordinals in U. Thus elements t of T_0 are pairs

$$t = \{j, s\} \quad \text{and} \quad t = \{\alpha, j\}.$$

For simplicity we shall write $D_{j,s}$ and $D_{\alpha,j}$ instead of $D_{\{j,s\}}$ and $D_{\{\alpha,j\}}$ respectively. For any j in S, let

$$S_j = S - (j).$$

The symbols $\cup$, $\cap$ without any superscript will denote the set-theoretical union and intersection, respectively.

By transfinite induction we now define sets $A_{\alpha,j}$ as follows:

(4) $$A_{0,j} = D_{0,j},$$

(4') $$A_{\alpha+1,j} = \bigcup_{s \in S_j} (A_{\alpha,j} \cap A_{\alpha,s} \cap D_{j,s}) \cup \bigcap_{\beta \leq \alpha+1} D_{\beta,j},$$

(4'') $$A_{\alpha,j} = \bigcap_{\beta < \alpha} A_{\beta,j} \quad \text{if } \alpha \text{ is a limit ordinal}.$$

By definition

(5) $$\bigcap_{\beta \leq \alpha} D_{\beta,j} \subset A_{\alpha,j}$$

and

(5') $$A_{\alpha+1,j} \subset A_{\alpha,j}.$$

---

[1] GAIFMAN [1, 3] and HALES [1, 2]. The construction of $\mathfrak{B}$ quoted here is that of HALES [1, 2].

The element $x = \{x_t\}_{t \in T_0} \in \mathscr{D}_{\mathrm{m}}$ defined as follows:

$$x_t = \begin{cases} 1 & \text{if } t = \{\beta, j\} \text{ where } \beta \leq \alpha, \\ -1 & \text{otherwise} \end{cases}$$

belongs to $A_{\alpha, j}$ by (5) but it does not belong to $A_{\alpha+1, j}$ because $x \notin D_{j, s}$ and $x \notin D_{\alpha+1, j}$. Thus $A_{\alpha+1, j} \neq A_{\alpha, j}$ and consequently, by (4') and (5'),

(6) $$A_{\beta, j} \supset A_{\alpha, j} \neq A_{\beta, j} \quad \text{for} \quad \beta < \alpha.$$

Let $\mathfrak{F}$ be the field (of subsets of $\mathscr{D}_{\mathrm{m}}$) generated by all the sets $A_{\alpha, j}$ and $D_{i, j}$. It follows from (6) (see also § 4 (1) or (2)) that

(7) $$\overline{\overline{\mathfrak{F}}} = \mathrm{m}.$$

We shall prove now that

(8) $$A_{\alpha+1, j} = \bigcup{}^{\mathfrak{F}}_{s \in S_j} (A_{\alpha, j} \cap A_{\alpha, s} \cap D_{j, s}),$$

i.e. that

(8')   if $A_{\alpha, j} \cap A_{\alpha, s} \cap D_{j, s} \subset A \in \mathfrak{F}$ for all $s \in S_j$, then $A_{\alpha+1, j} \subset A$.

By § 4 (2), it suffices to prove (8') only in the case where

(9) $A = \bigcup_{r \in R_1} A_{\alpha_r, j_r} \cup \bigcup_{r \in R_2} - A_{\gamma_r, k_r} \cup \bigcup_{r \in R_3} D_{m_r, n_r} \cup \bigcup_{r \in R_4} - D_{p_r, q_r}$,

where R_1, R_2, R_3, R_4 are finite sets of indices. Moreover it suffices to prove (8') only in the case when the set $-A_{\alpha+1, j}$ appears on the right side of (9) (since by adding the set $-A_{\alpha+1, j}$ to set (9) we obtain an equivalent statement). In order to prove (8) it suffices therefore to prove the following lemma:

(L) If A is a set (9) and

(10) $$\gamma_{r_0} = \alpha + 1, \quad k_{r_0} = j \quad \text{for an} \quad r_0 \in R_2$$

and if

(11) $$A_{\alpha, j} \cap A_{\alpha, s} \cap D_{j, s} \subset A \quad \text{for every} \quad s \in S_j,$$

then

(12) $$A = \mathscr{D}_{\mathrm{m}}.$$

Let $l \in S$ be an element such that

(13) $l \neq j, \quad l \neq j_r$ for $r \in R_1, \quad l \neq k_r$ for $r \in R_2,$

         $l \neq n_r$ for $r \in R_3, \quad p_r \neq l \neq q_r$ for $r \in R_4.$

Consider the point $x = \{x_t\}_{t \in T_0} \in \mathscr{D}_{\mathrm{m}}$ defined as follows:

(14a) $x_t = 1$ if $t = \{\beta, k_r\}$ where $r \in R_2$ and $\beta \leq \gamma_r$;

(14b) $x_t = 1$ if $t = \{p_r, q_r\}$ where $r \in R_4$;

(14c) $x_t = 1$ if $t = \{\beta, l\}$ and $\beta \leq \alpha$;

(14d) $x_t = 1$ if $t = \{j, l\}$;

(14e) $x_t = -1$ otherwise.

By (14d), $x \in D_{j,l}$. By (14c) and (5), $x \in A_{\alpha,l}$. By (14a) (the case where $r = r_0$ — see (10)) and (5), $x \in A_{\alpha,j}$. Thus

$$(15) \qquad\qquad x \in A_{\alpha,j} \cap A_{\alpha,l} \cap D_{j,l} .$$

Since $l \neq j$ by (13), we infer from (11) that

$$x \in A .$$

On the other hand, $x \notin \bigcup_{r \in R_2} - A_{\gamma_r, k_r}$ by (14a) and (5), and $x \notin \bigcup_{r \in R_4} -$ $- D_{p_r q_r}$ by (14b). Thus there are two cases to be considered:

1) $x \in \bigcup_{r \in R_3} D_{m_r, n_r}$, i.e. $x \in D_{m_{r'}, n_{r'}}$ for an $r' \in R_3$. Then, by (13), (14b), (14d), (14e), there exists an $r \in R_4$ such that $\{m_{r'}, n_{r'}\} = \{p_r, q_r\}$. This proves (12) since on the right side of (9) there appear the set $D_{m_{r'}, n_{r'}}$ and its complement $- D_{p_r, q_r}$.

2) $x \in \bigcup_{r \in R_1} A_{\alpha_r, j_r}$. Then $x \in A_{\alpha_r, j_r}$ for some $r \in R_1$. If $- A_{\alpha_r, j_r} \subset A$, then (12) holds. Thus to prove lemma (L) it suffices to show that the hypothesis

$$(16) \qquad\qquad x \in A_{\alpha_r, j_r} \quad \text{and} \quad - A_{\alpha_r, j_r} \not\subset A$$

leads to a contradiction.

Suppose (16) is true. Note that $j_r \neq l$ by (13). Let α' be the smallest ordinal such that there exists a $j' \in S$ with the properties that

$$(17) \qquad\qquad j' \neq l, \quad x \in A_{\alpha',j'} \quad \text{and} \quad - A_{\alpha',j'} \not\subset A .$$

If $\alpha' = 0$, then by (4) and (14a), (14e) we have $k_{r'} = j'$ for some $r' \in R_2$. Thus (see (6))

$$- A_{\alpha',j'} = - A_{0,j'} \subset - A_{\gamma_{r'},j'} \subset A .$$

Contradiction with (17).

If α' is a limit number, then by (4") there exists a $\beta < \alpha'$ such that $x \in A_{\beta,j'}$ and $- A_{\beta,j'} \not\subset A$. This contradicts the minimality of α'.

Suppose now that $\alpha' = \alpha'' + 1$. Then, by (4'), the following two cases have to be considered.

2') $x \in \bigcap_{\beta \leq \alpha'} D_{\beta,j'}$. Then, by (13), (14a), (14e), there exists an $r \in R_2$ such that $\gamma_r \geq \alpha'$ and $k_r = j'$. Then $- A_{\alpha',j'} \subset - A_{\gamma_r, k_r} \subset A$. Contradiction with (17).

2") $x \in \bigcup_{s \in S_{j'}} (A_{\alpha'',j'} \cap A_{\alpha'',s} \cap D_{j',s})$, i.e. $x \in A_{\alpha'',j'} \cap A_{\alpha'',s} \cap D_{j',s}$ for an $s \neq j'$.

Consider first the case where $s \neq l$. Then, by the minimality of α', we have $- A_{\alpha'',j'} \subset A$ and $- A_{\alpha'',s} \subset A$. Since $x \in D_{j',s}$, we infer from (13), (14b), (14e) that $\{j', s\} = \{p_r, q_r\}$ for some $r \in R_4$, i.e. $D_{j',s} = D_{p_r, q_r}$. Thus $- D_{j',s} \subset A$. Consequently

$$- A_{\alpha',j'} \subset - A_{\alpha'',j'} \cup - A_{\alpha'',s} \cup - D_{j',s} \subset A .$$

Contradiction with (17).

Consider now the case where $s = l$, i.e.

$$x \in A_{\alpha'',j'} \cap A_{\alpha'',l} \cap D_{j',l} .$$

Since $x \in D_{j',l}$ it follows from (14b), (14d), (14e) and (13) that $j' = j$. Thus $-A_{\alpha',j} \not\subset A$. Since $-A_{\alpha+1,j} \subset A$ by (10), we have $\alpha'' \geq \alpha + 1$ by (6). Then, since x is in $A_{\alpha'',l}$, x is in $A_{\alpha+1,l}$ by (6). From (4') there are two remaining possibilities:

$$x \in \bigcap_{\beta \leq \alpha+1} D_{\beta,1} \quad \text{or} \quad x \in \bigcup_{s \in S_l} (A_{\alpha,1} \cap A_{\alpha,s} \cap D_{l,s}) .$$

In the first case $x_t = 1$ for $t = \{\alpha + 1, l\}$, which is impossible in view of (13), (14a), (14c) and (14e). In the second case there exists an $s \in S_l$ (i.e. $s \neq l$) such that $x_t = 1$ for $t = \{l, s\}$; this is impossible in view of (13), (14b), (14d) and (14e). The proof of lemma (L) is now complete. Thus (8) holds.

Now let $\mathfrak{B}$ be an m-completion of the Boolean algebra $\mathfrak{F}$, let $\mathfrak{S}_0$ be the countable set of all the elements $D_{j,s}$ ($j, s \in S$) and let $\mathfrak{B}_0$ be the m-subalgebra (of $\mathfrak{B}$) m-generated by $\mathfrak{S}$. Since $\mathfrak{F}$ is a regular subalgebra of $\mathfrak{B}$, we have by (8) and (4'')

$$A_{\alpha+1,j} = \bigcup_{s \in S_j}^{\mathfrak{B}} (A_{\alpha,j} \cap A_{\alpha,s} \cap D_{j,s})$$

for any ordinal α in U, and (see the remark at the end of § 18 A))

$$A_{\alpha,j} = \bigcap_{\beta < \alpha}^{\mathfrak{F}} A_{\beta,j} = \bigcap_{\beta < \alpha}^{\mathfrak{B}} A_{\beta,j}$$

for any limit ordinal α in U. Hence it follows by transfinite induction (see also (4)) that all the sets $A_{\alpha,j}$ are elements of $\mathfrak{B}_0$. Since $\mathfrak{F}$ is the subalgebra generated by $\mathfrak{S}_0$ and all $A_{\alpha,j}$, we infer that $\mathfrak{F} \subset \mathfrak{B}_0$. This proves that $\mathfrak{B} = \mathfrak{B}_0$; i.e. $\mathfrak{B}$ is m-generated by the countable set $\mathfrak{S}_0$. It follows from (7) that $\overline{\overline{\mathfrak{B}}} \geq m$.

L) Now we are able to prove inequality (11) from § 31 (p. 134). Let $\mathfrak{A}_{m,n}$ be a free Boolean m-algebra with a set $\mathfrak{S}$ of n free m-generators, $m, n \geq \aleph_0$. Let f be any mapping of $\mathfrak{S}$ onto the countable set $\mathfrak{S}_0$ of m-generators of the Boolean m-algebra $\mathfrak{B}$ constructed in K). The mapping f can be extended to an m-homomorphism h of $\mathfrak{A}_{m,n}$ into $\mathfrak{B}$. Since $\mathfrak{S}_0$ m-generates $\mathfrak{B}$, h is a mapping from $\mathfrak{A}_{m,n}$ onto $\mathfrak{B}$. Since $\overline{\overline{\mathfrak{B}}} \geq m$, we infer that $\overline{\overline{\mathfrak{A}}}_{m,n} \geq m$.

Now we shall give some applications of completions and m-completions to an examination of the (m, n)-distributivity of Boolean algebras.

The definition of (m, n)-distributivity on p. 62 is a little complicated because the existence of various infinite joins and meets has to be postulated. Theorem 35.4 suggests another definition which avoids this difficulty.

35.4. *Let $\mathfrak{B}$ be a completion (or an m'-completion where $m' \geq m$ and $m' \geq n$) of its subalgebra $\mathfrak{A}$. The following conditions are equivalent:*
(d) $\mathfrak{A}$ is (m, n)-distributive;

(d$_5$) *for every* (m, n)-*indexed set* $\{A_{t,s}\}_{t \in T, s \in S}$ *of elements in* $\mathfrak{A}$,

$$(18) \qquad \bigcap_{t \in T}^{\mathfrak{B}} \bigcup_{s \in S}^{\mathfrak{B}} A_{t,s} = \bigcup_{\varphi \in S^T}^{\mathfrak{B}} \bigcap_{t \in T}^{\mathfrak{B}} A_{t, \varphi(t)} .$$

Consider first the case where $\mathfrak{B}$ is a completion of $\mathfrak{A}$. Then all the meets and joins in question exist because $\mathfrak{B}$ is complete. Since $A_{t, \varphi(t)} \subset \subset \bigcup_{s \in S}^{\mathfrak{B}} A_{t,s}$, we have

$$\bigcap_{t \in T}^{\mathfrak{B}} A_{t, \varphi(t)} \subset \bigcap_{t \in T}^{\mathfrak{B}} \bigcup_{s \in S}^{\mathfrak{B}} A_{t,s} .$$

Thus (18) holds if and only if the element

$$B = \bigcap_{t \in T}^{\mathfrak{B}} \bigcup_{s \in S}^{\mathfrak{B}} A_{t,s} - \bigcup_{\varphi \in S^T}^{\mathfrak{B}} \bigcap_{t \in T}^{\mathfrak{B}} A_{t, \varphi(t)} \in \mathfrak{B}$$

is equal to $\wedge$.

Suppose $B \neq \wedge$. Since $\mathfrak{A}$ is dense in $\mathfrak{B}$, there is an element $A \in \mathfrak{A}$ such that $\wedge \neq A \subset B$. Then (see § 18 (5))

$$\bigcup_{s \in S}^{\mathfrak{A}} A \cap A_{t,s} = A$$

and consequently

$$\bigcap_{t \in T}^{\mathfrak{A}} \bigcup_{s \in S}^{\mathfrak{A}} A \cap A_{t,s} = A \neq \wedge$$

but

$$\bigcap_{t \in T}^{\mathfrak{A}} A \cap A_{t, \varphi(t)} = \wedge$$

for every $\varphi \in S^T$. Thus the (m, n)-indexed set $\{A \cap A_{t,s}\}_{t \in T, s \in S}$ of elements of $\mathfrak{A}$ does not satisfy condition (d$_1$) in 19.2, i.e. $\mathfrak{A}$ is not (m,n)-distributive. This proves that (d) implies (d$_5$).

On the other hand, it is evident that (d$_5$) implies (d) since $\mathfrak{A}$ is a regular subalgebra of $\mathfrak{B}$.

Let now $\mathfrak{B}'$ be the subalgebra (of $\mathfrak{B}$) m'-generated by $\mathfrak{A}$. Since $\mathfrak{A}$ is dense in $\mathfrak{B}$, so is $\mathfrak{B}'$. Consequently $\mathfrak{B}'$ is a regular subalgebra of $\mathfrak{B}$. Since m' $\geq$ m, m' $\geq$ n and $\mathfrak{B}'$ is an m'-subalgebra of $\mathfrak{B}$, we have

$$\bigcap_{t \in T}^{\mathfrak{B}} \bigcup_{s \in S}^{\mathfrak{B}} A_{t,s} = \bigcap_{t \in T}^{\mathfrak{B}'} \bigcup_{s \in S}^{\mathfrak{B}'} A_{t,s} \in \mathfrak{B}',$$

$$\bigcap_{t \in T}^{\mathfrak{B}} A_{t, \varphi(t)} = \bigcap_{t \in T}^{\mathfrak{B}'} A_{t, \varphi(t)} \in \mathfrak{B}' .$$

If (18) holds, then also

$$(18') \qquad \bigcap_{t \in T}^{\mathfrak{B}'} \bigcup_{s \in S}^{\mathfrak{B}'} A_{t,s} = \bigcup_{\varphi \in S^T}^{\mathfrak{B}'} \bigcap_{t \in T}^{\mathfrak{B}'} A_{t, \varphi(t)}$$

by § 18 (5). On the other hand, (18') implies (18) because $\mathfrak{B}'$ is a regular subalgebra of $\mathfrak{B}$. Thus (18) is equivalent to (18') which proves 35.4 in the version for m'-completions.

35.5. *A Boolean algebra* $\mathfrak{A}$ *satisfying the* n-*chain condition is* (m, n)-*distributive if and only if its completion is* (m, n)-*distributive.*

Suppose $\mathfrak{B}$ is a completion of its subalgebra $\mathfrak{A}$ which is (m, n)-distributive and satisfies the n-chain condition. By 19.2 (d), (d$_1$), to prove that $\mathfrak{B}$ is (m,n)-distributive, it suffices to show that if $\{B_{t,s}\}_{t \in T, s \in S}$ is an (m, n)-indexed set of elements of $\mathfrak{B}$ such that

$$B = \bigcap_{t \in T}^{\mathfrak{B}} \bigcup_{s \in S}^{\mathfrak{B}} B_{t,s} \neq \wedge ,$$

then there exists a $\varphi \in S^T$ such that $\bigcap_{t \in T}^{\mathfrak{B}} B_{t, \varphi(t)} \neq \wedge$.

Since $\mathfrak{A}$ is dense in $\mathfrak{B}$, there exists an element $A \in \mathfrak{A}$ such that $\wedge \neq A \subset B$. By § 23 F),

$$B_{t,s} = \bigcup_{s' \in S'}^{\mathfrak{B}} B_{t,s,s'} \quad \text{where} \quad B_{t,s,s'} \in \mathfrak{A}, \quad \overline{S}' \leq \mathfrak{n}.$$

Let $A_{t,s,s'} = A \cap B_{t,s,s'}$. Since

$$\bigcap_{t \in T}^{\mathfrak{A}} \bigcup_{(s,s') \in S \times S'}^{\mathfrak{A}} A_{t,s,s'} = A \neq \wedge,$$

by 19.2 (d_1) there exists a mapping $\psi(t) = (\varphi(t), \varphi'(t))$ defined on T with $\varphi(t) \in S$, $\varphi'(t) \in S'$ such that $\bigcap_{t \in T}^{\mathfrak{A}} A_{t,\varphi(t),\varphi'(t)} \neq \wedge$. Since $A_{t,\varphi(t),\varphi'(t)} \subset B_{t,\varphi(t)}$, we have $\bigcap_{t \in T}^{\mathfrak{B}} B_{t,\varphi(t)} \neq \wedge$.

Thus the $(\mathfrak{m}, \mathfrak{n})$-distributivity of $\mathfrak{A}$ implies the $(\mathfrak{m}, \mathfrak{n})$-distributivity of its completion $\mathfrak{B}$. The converse implication results from the following statement which comes directly from the definition of $(\mathfrak{m}, \mathfrak{n})$-distributivity: every regular subalgebra of an $(\mathfrak{m}, \mathfrak{n})$-distributive algebra is $(\mathfrak{m}, \mathfrak{n})$-distributive.

By the last remark, the word "completion" can be replaced in 35.5 by "$\mathfrak{m}'$-completion" where $\mathfrak{m}'$ is any infinite cardinal.

35.6. *A Boolean algebra $\mathfrak{A}$ is $\mathfrak{m}$-distributive if and only if its $\mathfrak{m}$-completion is $\mathfrak{m}$-distributive*[1].

Suppose $\mathfrak{B}$ is an $\mathfrak{m}$-completion of its subalgebra $\mathfrak{A}$. If $\mathfrak{A}$ is $\mathfrak{m}$-distributive, every subalgebra $\mathfrak{B}_0$ of $\mathfrak{B}$ $\mathfrak{m}$-generated by an $\mathfrak{m}$-indexed set $\{A_t\}_{t \in T}$ of elements in $\mathfrak{A}$ is atomic and, consequently, isomorphic to an $\mathfrak{m}$-field of sets. In fact, the element

$$(19) \qquad\qquad \bigcap_{t \in T}^{\mathfrak{B}} \varepsilon(t) \cdot A_t \qquad\qquad (\varepsilon(t) = \pm 1)$$

is an atom of $\mathfrak{B}_0$ or is equal to $\wedge$. By 35.4, where $S = (1, -1)$ and $A_{t,s} = s \cdot A_t$, the unit element is the union of all the elements (19) (see the similar argument in the proof of 24.5).

Every $\mathfrak{m}$-subalgebra of $\mathfrak{B}$, $\mathfrak{m}$-generated by at most $\mathfrak{m}$ generators, is contained in a subalgebra $\mathfrak{B}_0$ $\mathfrak{m}$-generated by an $\mathfrak{m}$-indexed set $\{A_t\}_{t \in T}$ of elements of $\mathfrak{A}$, and consequently is isomorphic to an $\mathfrak{m}$-field of sets. By 24.6, $\mathfrak{B}$ is $\mathfrak{m}$-distributive.

Thus the $\mathfrak{m}$-distributivity of $\mathfrak{A}$ implies the $\mathfrak{m}$-distributivity of its $\mathfrak{m}$-completion $\mathfrak{B}$. The converse implication results from the remark at the end of the proof of 35.5.

35.7. *Let $\mathfrak{B}$ be a completion (or $\mathfrak{m}'$-completion where $\mathfrak{m}' \geq \mathfrak{m}$ and $\mathfrak{m}' \geq \mathfrak{n}$) of its subalgebra $\mathfrak{A}$. The following conditions are equivalent:*

(w) *$\mathfrak{A}$ is weakly $(\mathfrak{m}, \mathfrak{n})$-distributive;*

(w_6) *for every $(\mathfrak{m}, \mathfrak{n})$-indexed set $\{A_{t,s}\}_{t \in T, s \in S}$ of elements in $\mathfrak{A}$,*

$$(20) \qquad\qquad \bigcap_{t \in T}^{\mathfrak{B}} \bigcup_{s \in S}^{\mathfrak{B}} A_{t,s} = \bigcup_{\Phi \in S^T}^{\mathfrak{B}} \bigcap_{t \in T}^{\mathfrak{B}} A_{t,\Phi(t)} .$$

---

[1] PIERCE [3]. The above proof of 35.5 differs from that of PIERCE.

S and $A_{t,\Phi(t)}$ have here the same meaning as in the definition of weak $(\mathfrak{m}, \mathfrak{n})$-distributivity on p. 127.

The proof of 35.6 is analogous to that of 35.4 (instead of 19.2 (d_1) we have now to apply 30.1 (w_1)). Condition (w_6) sounds like the original definition of weak $(\mathfrak{m}, \mathfrak{n})$-distributivity but is simpler because the existence of infinite joins and meets need not be additionally postulated here.

35.8. *A Boolean algebra* $\mathfrak{A}$ *satisfying the* $\mathfrak{n}$-*chain condition is weakly* $(\mathfrak{m}, \mathfrak{n})$-*distributive if and only if its completion is weakly* $(\mathfrak{m}, \mathfrak{n})$-*distributive*[1].

The proof is analogous to that of 35.5. Since every regular subalgebra of a weakly $(\mathfrak{m}, \mathfrak{n})$-distributive algebra is also weakly $(\mathfrak{m}, \mathfrak{n})$-distributive, the word "completion" can be replaced in 35.8 by "$\mathfrak{m}'$-completion" where $\mathfrak{m}'$ is any infinite cardinal.

Examples. M) There exists an $\mathfrak{m}$-distributive Boolean $\mathfrak{m}$-algebra (and even an $\mathfrak{m}$-field of sets) whose completion and all $\mathfrak{m}'$-completions for $\mathfrak{m}' \geq 2^{\mathfrak{m}}$ are not $\mathfrak{m}$-distributive[2].

Indeed, in § 20 J) we have shown that there exists an $\mathfrak{m}$-field $\mathfrak{F}$ of subsets of a space X such that $\mathfrak{F}$ is not $(\mathfrak{m}, 2^{\mathfrak{m}})$-distributive. Let $\mathfrak{B}$ be any $2^{\mathfrak{m}}$-complete Boolean algebra such that $\mathfrak{F}$ is a regular subalgebra of $\mathfrak{B}$. The algebra $\mathfrak{B}$ is not $\mathfrak{m}$-distributive. For suppose $\mathfrak{B}$ is $\mathfrak{m}$-distributive. By 20.4 $\mathfrak{B}$ would also be $(\mathfrak{m}, 2^{\mathfrak{m}})$-distributive. Consequently $\mathfrak{F}$ would be $(\mathfrak{m}, 2^{\mathfrak{m}})$-distributive as a regular subalgebra of $\mathfrak{B}$.

N) If $2^{\mathfrak{m}} = \mathfrak{m}^{+}$[3], then there exists a weakly $\mathfrak{m}$-distributive Boolean $\mathfrak{m}$-algebra (and even an $\mathfrak{m}$-field of sets) whose completion and all $\mathfrak{m}'$-completions for $\mathfrak{m}' \geq \mathfrak{m}^{+}$ are not weakly $\mathfrak{m}$-distributive[4].

Let T and S be two sets of power $\mathfrak{m}$ and $\mathfrak{m}^{+}$ respectively, and let $\boldsymbol{T}$ be the set of all finite non-void subsets of T. By § 30 C) there exist sets $S_{t',t''} \subset S$ ($t', t'' \in T$) such that

(21) $\qquad S = \mathsf{U}_{t'' \in T} S_{t',t''}$ for every $t' \in T$,

(22) $\qquad$ the set $S_{\Phi} = \mathsf{\cap}_{t' \in T} S_{t',\Phi(t')}$ has power $\leq \mathfrak{m}$ for every $\Phi \in \boldsymbol{T}^{T}$.

Here, according to the convention on p. 127,

(23) $\qquad S_{t',\Phi(t')} = \mathsf{U}_{t'' \in \Phi(t')} S_{t',t''}$.

First we shall prove that if $\{B_{t,s}\}_{t \in T, s \in S}$ is any $(\mathfrak{m}, \mathfrak{m}^{+})$-indexed set of elements of a weakly $\mathfrak{m}$-distributive Boolean $\mathfrak{m}^{+}$-algebra such that

(24) $\qquad \mathsf{U}_{s \in S} B_{t,s} = V$ for every $t \in T$,

(25) $\qquad B_{t,s} \cap B_{t,s'} = \wedge$ for $s \neq s'$,

then

(26) $\qquad \mathsf{U}_{\Psi \in \boldsymbol{T}^{T \times T}} \mathsf{\cap}_{t \in T} \mathsf{U}_{s \in S_{\Psi_t}} B_{t,s} = V$.

---

[1] Traczyk [5].
[2] Pierce [3].
[3] $\mathfrak{m}^{+}$ denotes the smallest cardinal greater than $\mathfrak{m}$.
[4] Traczyk [5].

Here, for every $\Psi \in T^{T \times T}$ and $t \in T$, Ψ_t is the element of T^T defined as follows

$$\Psi_t(t') = \Psi(t, t') \quad \text{for} \quad t' \in T,$$

and S_{Ψ_t} is defined as in (23), i.e.

$$S_{\Psi_t} = \bigcap\nolimits_{t' \in T} S_{t', \Psi_t(t')} = \bigcap\nolimits_{t' \in T} \bigcup\nolimits_{t'' \in \Psi_t(t')} S_{t', t''}.$$

Let

(27) $$B_{t, t', t''} = \bigcup\nolimits_{s \in S_{t', t''}} B_{t, s} \quad \text{for} \quad t, t', t'' \in T.$$

By (21) and (24),

$$\bigcup\nolimits_{t'' \in T} B_{t, t', t''} = V \quad \text{for every pair} \quad (t, t') \in T \times T.$$

Since $\mathfrak{B}$ is weakly m-distributive, we infer that

(28) $$\bigcup\nolimits_{\Psi \in T^{T \times T}} \bigcap\nolimits_{(t, t') \in T \times T} B_{t, t', \Psi(t, t')} = V$$

where, according to the convention on p. 127 (see also (23)),

$$B_{t, t', \Psi(t, t')} = \bigcup\nolimits_{t'' \in \Psi(t, t')} B_{t, t', t''}.$$

(28) can be also written in the following form

(29) $$\bigcup\nolimits_{\Psi \in T^{T \times T}} \bigcap\nolimits_{t \in T} \bigcap\nolimits_{t' \in T} B_{t, t', \Psi_t(t')} = V.$$

By (24) and (25) for every fixed $t \in T$, the mapping

$$h(S') = \bigcup\nolimits_{s \in S'} B_{t, s} \qquad\qquad (S' \subset S)$$

is an m^+-isomorphism from the field of all subsets of S into $\mathfrak{B}$. Hence it follows that

$$\bigcap\nolimits_{t' \in T} B_{t, t', \Psi_t(t')} = \bigcup\nolimits_{s \in S_{\Psi_t}} B_{t, s}$$

which, together with (29), yields (26). The exact proof of the last identity is as follows:

$$\bigcap\nolimits_{t' \in T} B_{t, t', \Psi_t(t')} = \bigcap\nolimits_{t' \in T} \bigcup\nolimits_{t'' \in \Psi_t(t')} h(S_{t', t})$$
$$= h\left(\bigcap\nolimits_{t' \in T} \bigcup\nolimits_{t'' \in \Psi_t(t')} S_{t', t''}\right) = h(S_{\Psi_t}) = \bigcup\nolimits_{s \in S_{\Psi_t}} B_{t, s}.$$

Now let $\mathfrak{F}$ be the m-field defined in § 20 J). We recall the fundamental properties of $\mathfrak{F}$. It contains an (m, m^+)-indexed set $\{A_{t, s}\}_{t \in T, s \in S}$ of sets such that

(30) $$\bigcup\nolimits_{s \in S}^{\mathfrak{F}} A_{t, s} = V,$$

(31) $$\bigcap\nolimits_{t \in T} \bigcup\nolimits_{s \in S} A_{t, s} = \Lambda$$

where $\bigcup$ and $\bigcap$ denote the set-theoretical union and intersection. By (30) and 20.2 there exist sets $B_{t, s} \in \mathfrak{F}$ such that

(32) $B_{t, s} \subset A_{t, s}$, $B_{t, s} \cap B_{t, s'} = \Lambda$ for $s \neq s'$ and $\bigcup\nolimits_{s \in S}^{\mathfrak{F}} B_{t, s} = V$ for $t \in T$.

Suppose now that $\mathfrak{B}$ is any m^+-algebra such that $\mathfrak{F}$ is a regular subalgebra of $\mathfrak{B}$. Then $\mathfrak{B}$ is not weakly m-distributive. For suppose $\mathfrak{B}$ is weakly $\mathfrak{m}$-distributive. By (32), hypotheses (24) and (25) are satisfied. Thus (26) holds. On the other hand, (31) and (32) imply that $\bigcap_{t \in T} \bigcup_{s \in S_{\Psi_t}} B_{t,s} = \Lambda$ for every $\Psi \in T^{T \times T}$. Contradiction.

Taking as $\mathfrak{B}$ a completion or an $\mathfrak{m}'$-completion of $\mathfrak{F}$ ($\mathfrak{m}' \geq \mathfrak{m}^+$) we get statement J).

§ 36. Extensions of Boolean algebras

Let $\mathfrak{A}$ be a Boolean algebra, $\mathfrak{m}$ a fixed infinite cardinal and let $\mathbf{J}$ and $\mathbf{M}$ be two fixed sets of non-empty subsets of $\mathfrak{A}$ such that

(1) $\overline{\overline{\mathfrak{S}}} \leq \mathfrak{m}$ for every $\mathfrak{S} \in \mathbf{J}$ and for every $\mathfrak{S} \in \mathbf{M}$,

and all the joins

(2) $\bigcup_{A \in \mathfrak{S}} A$ $(\mathfrak{S} \in \mathbf{J})$

and meets

(2') $\bigcap_{A \in \mathfrak{S}} A$ $(\mathfrak{S} \in \mathbf{M})$

exist.

A pair $\{i, \mathfrak{B}\}$ is said to be a $(\mathbf{J}, \mathbf{M}, \mathfrak{m})$-*extension*[1] *of* $\mathfrak{A}$ provided

(a) $\mathfrak{B}$ is a Boolean $\mathfrak{m}$-algebra;

(b) i is a $(\mathbf{J}, \mathbf{M})$-isomorphism of $\mathfrak{A}$ into $\mathfrak{B}$ (see p. 88);

(c) $i(\mathfrak{A})$ $\mathfrak{m}$-generates $\mathfrak{B}$.

In the case where $\mathbf{J}$ and $\mathbf{M}$ are the sets of all non-void subsets $\mathfrak{S}$ of $\mathfrak{A}$ of power $\leq \mathfrak{m}$ such that the joins (2) or meets (3) exist respectively, then $(\mathbf{J}, \mathbf{M}, \mathfrak{m})$-extensions of $\mathfrak{A}$ will be called simply $\mathfrak{m}$-*extensions of* $\mathfrak{A}$. By § 22 (p. 88), $(\mathbf{J}, \mathbf{M})$-isomorphisms coincide then with $\mathfrak{m}$-isomorphisms and consequently condition (b) can be formulated in this case as follows:

(b$_0$) i is an $\mathfrak{m}$-isomorphism of $\mathfrak{A}$ into $\mathfrak{B}$.

Conditions (a), (c) coincide with conditions (a'), (c') in the definition of $\mathfrak{m}$-completions of $\mathfrak{A}$ in § 35, p. 156, but (b) and (b$_0$) are weaker than (b'). This causes that, in contrast to $\mathfrak{m}$-completions, $\mathfrak{A}$ can have non-isomorphic $(\mathbf{J}, \mathbf{M}, \mathfrak{m})$-extensions or non-isomorphic $\mathfrak{m}$-extensions (see Examples B) and C)).

Let $\{i, \mathfrak{B}\}$ and $\{i', \mathfrak{B}'\}$ be $(\mathbf{J}, \mathbf{M}, \mathfrak{m})$-extensions of $\mathfrak{A}$. Then $\{i', \mathfrak{B}'\}$ is said to be an $\mathfrak{m}$-*homomorphic image of* $\{i, \mathfrak{B}\}$ if there exists an $\mathfrak{m}$-homomorphism h from $\mathfrak{B}$ into $\mathfrak{B}'$ such that

(3) $i' = hi$.

---

[1] σ-extensions were first examined by SIKORSKI [13]. Later KERSTAN [1] proved the existence of free $(\mathbf{J}, \mathbf{M}, \mathfrak{m})$-extensions for any $\mathfrak{m}$. Another proof of the existence of free $(\mathbf{J}, \mathbf{M}, \mathfrak{m})$-extensions was given by SIKORSKI [32]. Independently the same proof of the existence of free $\mathfrak{m}$-extensions was found by YAQUB [1] who also examined the case where $\mathbf{J}$ and $\mathbf{M}$ are empty. See also DAY [1] and DAY and YAQUB [1]. The exposition in this section is a slight modification of that in SIKORSKI [32].

We then write

(4) $\{i', \mathfrak{B}'\} \leq \{i, \mathfrak{B}\}$.

Observe that condition (3) is equivalent to the following condition

(5) h is an extension of the isomorphism $i' i^{-1}$ of $i(\mathfrak{A})$ onto $i'(\mathfrak{A})$.

Hence it follows that if the $\mathfrak{m}$-homomorphism h with the required properties exists, it is unique. Moreover h maps $\mathfrak{B}$ onto $\mathfrak{B}'$. The two last statements are direct consequences of (c).

If the homomorphism h is an isomorphism, then $\{i, \mathfrak{B}\}$ and $\{i', \mathfrak{B}'\}$ are isomorphic in the sense defined in § 35, p. 153.

Observe that if $\{i, \mathfrak{B}\}$ is an $(\mathbf{J}, \mathbf{M}, \mathfrak{m})$-extension of $\mathfrak{A}$ and h is any isomorphism of $\mathfrak{B}$ into another Boolean algebra $\mathfrak{B}'$, then $\{hi, h(\mathfrak{B})\}$ is a $(\mathbf{J}, \mathbf{M}, \mathfrak{m})$-extension of $\mathfrak{A}$ isomorphic to $\{i, \mathfrak{B}\}$. In other words, any isomorph of a $(\mathbf{J}, \mathbf{M}, \mathfrak{m})$-extension is also a $(\mathbf{J}, \mathbf{M}, \mathfrak{m})$-extension.

It is easy to verify that relation (4) is a quasi-ordering in the class K of all $(\mathbf{J}, \mathbf{M}, \mathfrak{m})$-extensions of $\mathfrak{A}$, i.e. it is reflexive and transitive. Moreover two elements $\{i, \mathfrak{B}\}$, $\{i', \mathfrak{B}'\}$ in K are isomorphic if and only if simultaneously

$$\{i', \mathfrak{B}'\} \leq \{i, \mathfrak{B}\} \quad \text{and} \quad \{i, \mathfrak{B}\} \leq \{i', \mathfrak{B}'\}$$

(see 23.3). Sometimes it is convenient to identify isomorphic elements of K; after this identification, K is partially ordered by $\leq$.

The class K is not empty, viz. every $\mathfrak{m}$-completion $\{i, \mathfrak{B}\}$ of $\mathfrak{A}$ belongs to K, i.e. every $\mathfrak{m}$-completion of $\mathfrak{A}$ is a $(\mathbf{J}, \mathbf{M}, \mathfrak{m})$-extension of $\mathfrak{A}$.

Let $\mathfrak{n}$ be the cardinal of a set of generators of $\mathfrak{A}$, and let $\mathfrak{A}_{\mathfrak{m}, \mathfrak{n}}$ be a free Boolean $\mathfrak{m}$-algebra with a set of $\mathfrak{n}$ free $\mathfrak{m}$-generators. By 31.8 the smallest subalgebra $\mathfrak{A}_{0, \mathfrak{n}}$ of $\mathfrak{A}_{\mathfrak{m}, \mathfrak{n}}$ containing all the free $\mathfrak{m}$-generators is a free Boolean algebra with $\mathfrak{n}$ free generators. Thus every Boolean algebra with $\mathfrak{n}$ generators is a homomorphic image of $\mathfrak{A}_{0, \mathfrak{n}}$. In particular there exists a homomorphism g_0 of $\mathfrak{A}_{0, \mathfrak{n}}$ onto $\mathfrak{A}$. Since the free $\mathfrak{m}$-generators of $\mathfrak{A}_{\mathfrak{m}, \mathfrak{n}}$ are free generators of $\mathfrak{A}_{0, \mathfrak{n}}$, the algebras $\mathfrak{A}_{0, \mathfrak{n}}$ and $\mathfrak{A}_{\mathfrak{m}, \mathfrak{n}}$ have the property that

(d) every homomorphism from $\mathfrak{A}_{0, \mathfrak{n}}$ into any $\mathfrak{m}$-algebra $\mathfrak{A}'$ can be extended uniquely to an $\mathfrak{m}$-homomorphism from $\mathfrak{A}_{\mathfrak{m}, \mathfrak{n}}$ into $\mathfrak{A}'$.

Let $\varDelta_0$ be the ideal of all elements $A \in \mathfrak{A}_{0, \mathfrak{n}}$ such that $g_0(A) = \varLambda$, and let I be the set of all $\mathfrak{m}$-ideals $\varDelta$ in $\mathfrak{A}_{\mathfrak{m}, \mathfrak{n}}$ such that

(e') $\varDelta \cap \mathfrak{A}_{0, \mathfrak{n}} = \varDelta_0$,

(e'') $\varDelta$ contains all the elements

$$A_0 - \bigcup_{A \in \mathfrak{S}_1} A , \quad \bigcup_{A \in \mathfrak{S}_1} A - A_0$$

$$A_0 - \bigcap_{A \in \mathfrak{S}_2} A , \quad \bigcap_{S \in \mathfrak{S}_2} A - A_0$$

where $A_0 \in \mathfrak{A}_{0, \mathfrak{n}}$ and $\mathfrak{S}_1$ and $\mathfrak{S}_2$ are any subsets of $\mathfrak{A}_{0, \mathfrak{n}}$ of powers $\leq \mathfrak{m}$

such that

$$g_0(\mathfrak{S}_1) \in \mathbf{J}, \quad g_0(A_0) = \bigcup{}_{A \in \mathfrak{S}_1} g_0(A),$$
$$g_0(\mathfrak{S}_2) \in \mathbf{M}, \quad g_0(A_0) = \bigcap{}_{A \in \mathfrak{S}_2} g_0(A).$$

For every $\Delta \in \mathbf{I}$, let $\mathfrak{A}_\Delta = \mathfrak{A}_{m,n}/\Delta$ and let $\mathfrak{A}_{\Delta,0}$ be the subalgebra (of $\mathfrak{A}_\Delta$) composed of all the elements $[A]_\Delta$ with $A \in \mathfrak{A}_{0,n}$.

Condition (e') means that the formula

$$g_\Delta([A]_\Delta) = g_0(A) \quad \text{for} \quad A \in \mathfrak{A}_{0,n}$$

defines an isomorphism g_Δ of $\mathfrak{A}_{\Delta,0}$ onto $\mathfrak{A}$. Let

$$i_\Delta = g_\Delta^{-1}.$$

Condition (e'') means that i_Δ is a $(\mathbf{J}, \mathbf{M})$-isomorphism of $\mathfrak{A}$ into $\mathfrak{A}_\Delta$. Since $\mathfrak{A}_{\Delta,0} = i_\Delta(\mathfrak{A})$ m-generates $\mathfrak{A}_\Delta$, we obtain the following theorem.

36.1. *For every Δ in $\mathbf{I}$, $\{i_\Delta, \mathfrak{A}_\Delta\}$ is a $(\mathbf{J}, \mathbf{M}, \mathfrak{m})$-extension of $\mathfrak{A}$.*

Now we shall prove that, conversely,

36.2. *For every $(\mathbf{J}, \mathbf{M}, \mathfrak{m})$-extension $\{i, \mathfrak{B}\}$ of $\mathfrak{A}$ there exists an ideal $\Delta \in \mathbf{I}$ such that $\{i, \mathfrak{B}\}$ is isomorphic to $\{i_\Delta, \mathfrak{A}_\Delta\}$.*

By (d), the homomorphism $i g_0$ from $\mathfrak{A}_{0,n}$ into $\mathfrak{B}$ can be extended to an m-homomorphism g of $\mathfrak{A}_{m,n}$ onto $\mathfrak{B}$. The m-ideal Δ of all $A \in \mathfrak{A}_{m,n}$ such that $g(A) = \wedge$ has properties (e'), (e''), i.e. $\Delta \in \mathbf{I}$, and the formula

$$h([A]_\Delta) = g(A) \quad \text{for} \quad A \in \mathfrak{A}_{m,n}$$

defines the required isomorphism h of $\{i_\Delta, \mathfrak{A}_\Delta\}$ onto $\{i, \mathfrak{B}\}$.

36.3. *For any Δ', Δ'' in $\mathbf{I}$,*

$$\{i_{\Delta'}, \mathfrak{A}_{\Delta'}\} \leq \{i_{\Delta''}, \mathfrak{A}_{\Delta''}\} \quad \text{if and only if} \quad \Delta'' \subset \Delta'.$$

Consequently $\{i_{\Delta'}, \mathfrak{A}_{\Delta'}\}$ and $\{i_{\Delta''}, \mathfrak{A}_{\Delta''}\}$ are isomorphic if and only if $\Delta' = \Delta''$.

For suppose h is an m-homomorphism of $\mathfrak{A}_{\Delta''}$ into $\mathfrak{A}_{\Delta'}$ that is an extension of $i_{\Delta'} i_{\Delta''}^{-1}$, i.e.

(6) $$h([A]_{\Delta''}) = [A]_{\Delta'}$$

for $A \in \mathfrak{A}_{0,n}$. It is easy to verify that the class of all $A \in \mathfrak{A}_{m,n}$ for which (6) holds is an m-subalgebra of $\mathfrak{A}_{m,n}$. Since the m-subalgebra contains $\mathfrak{A}_{0,n}$, it coincides with $\mathfrak{A}_{m,n}$, i.e. (6) holds for every $A \in \mathfrak{A}_{m,n}$. If $A \in \Delta''$, we have $A \in \Delta'$. Thus $\Delta'' \subset \Delta'$. Conversely, if $\Delta'' \subset \Delta'$, then (6) defines an m-homomorphism h of $\mathfrak{A}_{\Delta''}$ into $\mathfrak{A}_{\Delta'}$ which is an extension of $i_{\Delta'} i_{\Delta''}^{-1}$. This proves the first part of 36.3. The second part follows directly from the first.

Theorems 36.1, 36.2, 36.3 describe exactly the class K of all $(\mathbf{J}, \mathbf{M}, \mathfrak{m})$-extensions of $\mathfrak{A}$. $\{i, \mathfrak{B}\}$ is in K if and only if it is an isomorph of $\{i_\Delta, \mathfrak{A}_\Delta\}$ for some $\Delta \in \mathbf{I}$. After identification of isomorphic elements the set K is isomorphic (in the sense of the theory of partially ordered sets) to the set $\mathbf{I}$ partially ordered by the converse of the set-theoretical inclusion.

The intersection Δ^0 of all ideals Δ in I also belongs to I. By 36.1,

(7) $\{i_{\Delta^0}, \mathfrak{A}_{\Delta^0}\}$

is a $(\mathbf{J}, \mathbf{M}, \mathfrak{m})$-extension of $\mathfrak{A}$. The $(\mathbf{J}, \mathbf{M}, \mathfrak{m})$-extension (7) and all its isomorphs are called *maximal* $(\mathbf{J}, \mathbf{M}, \mathfrak{m})$-*extensions of* $\mathfrak{A}$.

36.4. *In order that a* $(\mathbf{J}, \mathbf{M}, \mathfrak{m})$-*extension* $(i, \mathfrak{B})$ *of* $\mathfrak{A}$ *to be maximal it is necessary and sufficient that for every* $(\mathbf{J}, \mathbf{M})$-*homomorphism* g *from* $\mathfrak{A}$ *into any Boolean* $\mathfrak{m}$-*algebra* $\mathfrak{C}$ *there exist an* $\mathfrak{m}$-*homomorphism* h *from* $\mathfrak{B}$ *into* $\mathfrak{C}$ *such that*

$$g = hi .$$

To explain the last condition, it is convenient to identify $\mathfrak{A}$ with $i(\mathfrak{A})$ by means of the isomorphism i. Then the last condition means that every $(\mathbf{J}, \mathbf{M})$-homomorphism of the subalgebra $\mathfrak{A} = i(\mathfrak{A}) \subset \mathfrak{B}$ into any Boolean $\mathfrak{m}$-algebra $\mathfrak{C}$ can be extended to an $\mathfrak{m}$-homomorphism from $\mathfrak{B}$ into $\mathfrak{C}$.

To prove 36.4 let us observe that the homomorphism gg_0 from $\mathfrak{A}_{0,n}$ into $\mathfrak{C}$ can be extended to an $\mathfrak{m}$-homomorphism h' from $\mathfrak{A}_{m,n}$ into $\mathfrak{C}$ by (d). The $\mathfrak{m}$-ideal Δ' of all $A \in \mathfrak{A}_{m,n}$ such that $h'(A) = \wedge$ has property (e'') since g is a $(\mathbf{J}, \mathbf{M})$-homomorphism. Since $\Delta_0 \subset \Delta'$, we have

$$(\Delta' \cap \Delta^0) \cap \mathfrak{A}_{0,n} = \Delta' \cap (\Delta^0 \cap \mathfrak{A}_{0,n}) = \Delta' \cap \Delta_0 = \Delta_0 ,$$

i.e. the $\mathfrak{m}$-ideal $\Delta' \cap \Delta^0$ satisfies (e'). Since it also satisfies (e'') (as the intersection of two ideals satysfying (e'')), it belongs to I. Hence $\Delta^0 \subset \Delta' \cap \Delta^0$, i.e. $\Delta^0 \subset \Delta'$. Consequently the formula

$$h([A]_{\Delta^0}) = h'(A) \quad \text{for} \quad A \in \mathfrak{A}_{m,n}$$

defines an $\mathfrak{m}$-homomorphism from $\mathfrak{A}_{\Delta^0}$ into $\mathfrak{C}$, and $g = hi$.

Thus we have proved that (7) has the extension property mentioned in 36.4. Consequently every isomorph of (7), i.e. every maximal $(\mathbf{J}, \mathbf{M}, \mathfrak{m})$-extension of $\mathfrak{A}$, has this property.

On the other hand, it follows from 23.3 that all $(\mathbf{J}, \mathbf{M}, \mathfrak{m})$-extensions of $\mathfrak{A}$ with the extension property are isomorphic. Since (7) has the extension property, all $(\mathbf{J}, \mathbf{M}, \mathfrak{m})$-extensions with the extension property are isomorphic to (7), i.e. are maximal.

According to a general definition for partially ordered sets, an element $\{i, \mathfrak{B}\}$ in K is said to be a greatest element in K if

$$\{i', \mathfrak{B}'\} \leq \{i, \mathfrak{B}\} \quad \text{for every} \quad \{i', \mathfrak{B}'\} \text{ in } K .$$

$\{i, \mathfrak{B}\}$ is called a smallest element in K if

$$\{i, \mathfrak{B}\} \leq \{i', \mathfrak{B}'\} \quad \text{for every} \quad \{i', \mathfrak{B}'\} \text{ in } K .$$

An element $\{i, \mathfrak{B}\}$ in K is said to be minimal if, for every $\{i', \mathfrak{B}'\}$ in K

$$\{i', \mathfrak{B}'\} \leq \{i, \mathfrak{B}\} \quad \text{implies} \quad \{i, \mathfrak{B}\} \leq \{i', \mathfrak{B}'\}$$

i.e. it implies that $\{i', \mathfrak{B}'\}$ is isomorphic to $\{i, \mathfrak{B}\}$ and, consequently, can be identified with $\{i, \mathfrak{B}\}$.

36.5. *Any maximal* $(\mathbf{J}, \mathbf{M}, \mathfrak{m})$*-extension of* $\mathfrak{A}$ *is a greatest element in* K*. Any* $\mathfrak{m}$*-completion of* $\mathfrak{A}$ *is a minimal element in* K*.*

The first part of 36.5 follows easily from 36.4. To prove the second part it suffices to show that if $\{i, \mathfrak{B}\}$ is an $\mathfrak{m}$-completion of $\mathfrak{A}$, $\{i', \mathfrak{B}'\}$ is any element in K and h is an $\mathfrak{m}$-homomorphism from $\mathfrak{B}$ onto $\mathfrak{B}'$ such that $i' = hi$, then h is an isomorphism. Indeed, h is an extension of the isomorphism $i'i^{-1}$ from $i(\mathfrak{A})$ onto $i'(\mathfrak{A})$. Thus $h(A) \neq \wedge_{\mathfrak{B}'}$ for $A \neq \wedge_{\mathfrak{B}}$, $A \in i(\mathfrak{A})$. If B is any non-zero element in $\mathfrak{B}$, then there is a non-zero element A in $i(\mathfrak{A})$ with $A \subset B$ since $i(\mathfrak{A})$ is dense in $\mathfrak{B}$, by 35.3. Since $\wedge_{\mathfrak{B}'} \neq h(A) \subset h(B)$, we have $h(B) \neq \wedge_{\mathfrak{B}'}$ which proves that h is an isomorphism.

Theorem 35.5 explains the name "maximal" as used for isomorphs of (7). It also suggests calling $\mathfrak{m}$-completions of $\mathfrak{A}$ *minimal* $\mathfrak{m}$*-extensions of* $\mathfrak{A}$ (consequently completions of $\mathfrak{A}$ are also called *minimal extensions of* $\mathfrak{A}$). The extension property mentioned in 35.4 is similar to a fundamental property of free $\mathfrak{m}$-algebras. Therefore, maximal $(\mathbf{J}, \mathbf{M}, \mathfrak{m})$-extensions of $\mathfrak{A}$ are also called *free* $(\mathbf{J}, \mathbf{M}, \mathfrak{m})$*-extensions of* $\mathfrak{A}$.

It is not known whether $\mathfrak{m}$-completions of $\mathfrak{A}$ are smallest elements in K. A partial answer to this problem is given by the following two theorems.

36.6. *If* $\{i, \mathfrak{B}\} \in K$ *and the* $\mathfrak{m}$*-algebra* $\mathfrak{B}$ *has the weak* $\mathfrak{m}$*-extension property (in particular, if* $\mathfrak{B}$ *is weakly* $\mathfrak{m}$*-distributive or is an* $\mathfrak{m}$*-field of sets), then* $\{i, \mathfrak{B}\}$ *is an* $\mathfrak{m}$*-completion of* $\mathfrak{A}$ *and a smallest element in* K*.*

In fact, let $\{i', \mathfrak{B}'\}$ be another element in K. Since $\mathfrak{B}$ has the weak $\mathfrak{m}$-extension property (see § 34), the isomorphism ii'^{-1} from $i'(\mathfrak{A})$ into $\mathfrak{B}$ can be extended to an $\mathfrak{m}$-homomorphism from $\mathfrak{B}'$ into $\mathfrak{B}$. This proves that $\{i, \mathfrak{B}\} \leq \{i', \mathfrak{B}'\}$. Thus $\{i, \mathfrak{B}\}$ is the smallest element in K. Since in every partially ordered set all minimal elements coincide with the smallest element if it exists, $\{i, \mathfrak{B}\}$ has to be a completion of $\mathfrak{A}$ by the second part of 36.5.

36.7. *If* $\mathfrak{A}$ *is* $\mathfrak{m}$*-distributive, or if* $\mathfrak{A}$ *is weakly* $\mathfrak{m}$*-distributive and satisfies the* $\mathfrak{m}$*-chain condition, then* $\mathfrak{m}$ *completions of* $\mathfrak{A}$ *are smallest elements in* K*.*

This follows directly from 36.6 and 35.6 or 35.8.

Let K_r denote the class of all $(\mathbf{J}, \mathbf{M}, \mathfrak{m})$-extensions $\{i, \mathfrak{B}\}$ of $\mathfrak{A}$ such that $\mathfrak{B}$ is $\mathfrak{m}$-representable. Clearly K_r is not empty if and only if $\mathfrak{A}$ is $(\mathbf{J}, \mathbf{M}, \mathfrak{m})$-representable in the sense defined in § 29, p. 125. Thus we now suppose that $\mathfrak{A}$ is $(\mathbf{J}, \mathbf{M}, \mathfrak{m})$-representable.

To investigate the class K_r we have to recall the notation from § 29, p. 125: h_0 will be an isomorphism of $\mathfrak{A}$ onto the field $\mathfrak{F}$ of all open-closed subsets of a Stone space X of $\mathfrak{A}$, $\mathfrak{F}_{\mathfrak{m}}$ will denote the smallest $\mathfrak{m}$-field containing $\mathfrak{F}$. The $\mathfrak{m}$-ideal of all sets $A \in \mathfrak{F}_{\mathfrak{m}}$ of the $(\mathbf{J}, \mathbf{M}, \mathfrak{m})$-category (see p. 125) will now be denoted by Δ^*.

Let I_r be the class of all $\mathfrak{m}$-ideals Δ in $\mathfrak{F}_{\mathfrak{m}}$, such that

(e_1) no non-void open subset of X belongs to Δ;

(e_2) Δ^* is a subset of Δ.

For every Δ in I_r, the formula

$$i^\Delta(A) = [h_0(A)]_\Delta$$

defines an $\mathfrak{m}$-isomorphism i^Δ from $\mathfrak{A}$ into $\mathfrak{F}_{\mathfrak{m}}/\Delta$, and $i^\Delta(\mathfrak{A})$ $\mathfrak{m}$-generates $\mathfrak{F}_{\mathfrak{m}}/\Delta$. In other words,

36.8. *For every* $\Delta \in I_r$, $\{i^\Delta, \mathfrak{F}_{\mathfrak{m}}/\Delta\}$ *is in* K_r.

Namely i^Δ is an isomorphism by (e_1). It is a $(\mathbf{J}, \mathbf{M})$-homomorphism by (e_2). The subalgebra $i^\Delta(\mathfrak{A})$ $\mathfrak{m}$-generates $\mathfrak{F}_{\mathfrak{m}}/\Delta$ since $\mathfrak{F}$ $\mathfrak{m}$-generates $\mathfrak{F}_{\mathfrak{m}}$. By 29.6,

(8) $\{i^{\Delta^*}, \mathfrak{F}_{\mathfrak{m}}/\Delta^*\}$

is in K_r. The $(\mathbf{J}, \mathbf{M}, \mathfrak{m})$-extension (8) and all its isomorphs will be called *maximal representable* $(\mathbf{J}, \mathbf{M}, \mathfrak{m})$-*extensions of* $\mathfrak{A}$ or *free representable* $(\mathbf{J}, \mathbf{M}, \mathfrak{m})$-*extensions of* $\mathfrak{A}$ for reasons explained in theorems 36.10 and 36.9 below.

36.9. *In order that a* $(\mathbf{J}, \mathbf{M}, \mathfrak{m})$-*extension* $\{i, B\} \in K_r$ *be maximal representable it is necessary and sufficient that for every* $(\mathbf{J}, \mathbf{M})$-*homomorphism g from* $\mathfrak{A}$ *into any* $\mathfrak{m}$-*representable* $\mathfrak{m}$-*algebra* $\mathfrak{C}$, *there exist an* $\mathfrak{m}$-*homomorphism h from* $\mathfrak{B}$ *into* $\mathfrak{C}$ *such that*

$$g = hi .$$

The explanation of the last extension property is the same as in the case of the analogous theorem 36.4.

First we shall prove that (8) has the extension property. It suffices to consider the case where $\mathfrak{C} = \mathfrak{F}'/\Delta'$ is the canonical $\mathfrak{m}$-representation of an $\mathfrak{m}$-representable $\mathfrak{m}$-algebra. Let g be a $(\mathbf{J}, \mathbf{M})$-homomorphism from $\mathfrak{A}$ into $\mathfrak{C}$. By the same argument as in the proof of 32.6 (see also 22.5) we assert that the homomorphism $h' = g(i^{\Delta^*})^{-1}$ is induced by a point mapping φ which is $(\mathbf{J}, \mathbf{M})$-continuous. In other words,

$$h'([A]_{\Delta^*}) = [\varphi^{-1}(A)]_{\Delta'} \quad \text{for every} \quad A \in \mathfrak{F},$$

and $\varphi^{-1}(B)$ belongs to Δ' for every $(\mathbf{J}, \mathbf{M})$-nowhere dense set B in X, and consequently for all $B \in \Delta^*$. This implies that the formula

$$h([A]_{\Delta^*}) = [\varphi^{-1}(A)]_{\Delta'} \quad \text{for all} \quad A \in \mathfrak{F}_{\mathfrak{m}}$$

defines an $\mathfrak{m}$-homomorphism from $\mathfrak{F}_{\mathfrak{m}}/\Delta^*$ into $\mathfrak{C} = \mathfrak{F}'/\Delta'$ which is an extension of h', i.e. $g = hi^{\Delta^*}$.

Thus we have proved that (8) has the extension property mentioned in 36.9. Consequently every isomorph of (8), i.e. every maximal representable $(\mathbf{J}, \mathbf{M}, \mathfrak{m})$-extension of $\mathfrak{A}$, has this property.

On the other hand, it follows from 23.3 that all elements of K_r with the extension property are isomorphic. Since (8) has this property, all elements of K_r having this property are isomorphic to (8), i.e. are maximal representable.

36.10. *Any maximal representable* $(\mathbf{J}, \mathbf{M}, \mathfrak{m})$-*extension of* $\mathfrak{A}$ *is a greatest element in* K_r.

This follows directly from 36.9.

36.11. *For every* $\{i, \mathfrak{B}\}$ *in* K_r *there exists an ideal* $\Delta \in I_r$ *such that* $\{i, \mathfrak{B}\}$ *is isomorphic to* $\{i^\Delta, \mathfrak{F}_\mathfrak{m}/\Delta\}$.

By 36.9 there exists an $\mathfrak{m}$-homomorphism h from $\mathfrak{F}_\mathfrak{m}/\Delta^*$ into $\mathfrak{B}$ such that $i = h i^{\Delta^*}$, i.e. h is an extension of $i(i^{\Delta^*})^{-1}$. Since $i(\mathfrak{A})$ $\mathfrak{m}$-generates $\mathfrak{B}$ and is contained in $h(\mathfrak{F}_\mathfrak{m}/\Delta^*)$, we have $\mathfrak{B} = h(\mathfrak{F}_\mathfrak{m}/\Delta^*)$. The $\mathfrak{m}$-ideal Δ of all $A \in \mathfrak{F}_\mathfrak{m}$ such that $h([A]_{\Delta^*}) = \wedge_\mathfrak{B}$ belongs to I_r. In fact, it satisfies (e_1) since h is an extension of the isomorphism $i(i^{\Delta^*})^{-1}$. Δ also satisfies (e_2) by definition. The formula

$$h'([A]_\Delta) = h([A]_{\Delta^*}) \quad \text{for} \quad A \in \mathfrak{F}_\mathfrak{m}$$

defines an isomorphism h' of $\mathfrak{F}_\mathfrak{m}/\Delta$ onto $\mathfrak{B}$ such that $i = h' i^\Delta = h i^{\Delta^*}$. Thus $\{i, \mathfrak{B}\}$ is isomorphic to $\{i^\Delta, \mathfrak{F}_\mathfrak{m}/\Delta\}$.

36.12. *For any* Δ', Δ'' *in* I_r,

$$\{i^{\Delta'}, \mathfrak{F}_\mathfrak{m}/\Delta'\} \leq \{i^{\Delta''}, \mathfrak{F}_\mathfrak{m}/\Delta''\} \quad \text{if and only if} \quad \Delta'' \subset \Delta' .$$

Consequently $\{i^{\Delta'}, \mathfrak{F}_\mathfrak{m}/\Delta'\}$ *and* $\{i^{\Delta''}, \mathfrak{F}_\mathfrak{m}/\Delta''\}$ *are isomorphic if and only if* $\Delta' = \Delta''$.

The proof of 36.12 is similar to that of the analogous theorem 36.3.

The following simple remark is a consequence of the fact that every $\mathfrak{m}$-homomorphic image of an $\mathfrak{m}$-representable $\mathfrak{m}$-algebra is an $\mathfrak{m}$-representable $\mathfrak{m}$-algebra.

36.13. *If* $\{i, \mathfrak{B}\} \in K_r$, $\{i', \mathfrak{B}'\} \in K$ *and* $\{i', \mathfrak{B}'\} \leq \{i, \mathfrak{B}\}$ *then* $\{i', \mathfrak{B}'\} \in K_r$.

Theorems 36.8, 36.11 and 36.12 describe precisely the class K_r. Namely $\{i, \mathfrak{B}\}$ is in K_r if and only if it is an isomorph of $\{i^\Delta, \mathfrak{F}_\mathfrak{m}/\Delta\}$ for some $\Delta \in I_r$. After identification of isomorphic elements, the set K_r is isomorphic (in the sense of the theory of partially ordered sets) to the set I_r partially ordered by the converse of the set-theoretical inclusion. Theorems 36.10 and 36.13 explain how is situated K_r in K.

It is not known whether $\mathfrak{m}$-completions of $\mathfrak{A}$ are in K_r. It follows easily from 36.13 that if an $\mathfrak{m}$-completion of $\mathfrak{A}$ is the smallest element in K, then it is in K_r. This is e.g. the case if $\mathfrak{A}$ is $\mathfrak{m}$-distributive, or if $\mathfrak{A}$ is weakly $\mathfrak{m}$-distributive and satisfies the $\mathfrak{m}$-chain condition.

In the case where $\mathfrak{m} = \aleph_0$, we have $K = K_r$ since every Boolean algebra is σ-representable.

Observe that in the case of $\mathfrak{m}$-extensions the ideal $\varDelta^*$ is composed of all sets $A \in \mathfrak{F}_{\mathfrak{m}}$ of the $\mathfrak{m}$-category.

In the case where $\mathfrak{A}$ is a subalgebra of $\mathfrak{B}$, i is the identity mapping and $\{i, \mathfrak{B}\}$ is a $(\mathbf{J}, \mathbf{M}, \mathfrak{m})$-extension (an $\mathfrak{m}$-extensions) of $\mathfrak{A}$, then $\mathfrak{B}$ itself is called a $(\mathbf{J}, \mathbf{M}, \mathfrak{m})$-*extension* (an $\mathfrak{m}$-*extension*) of $\mathfrak{A}$.

Examples. A) If $\mathbf{J}$ and $\mathbf{M}$ are empty, then $\{h_0, \mathfrak{F}_{\mathfrak{m}}\}$ is the maximal representable $(\mathbf{J}, \mathbf{M}, \mathfrak{m})$-extension of $\mathfrak{A}$ (provided $\mathfrak{A}$ is $\mathfrak{m}$-representable)[1].

B) If $\mathbf{J}$ and $\mathbf{M}$ are empty, then $\mathfrak{A}_{\mathfrak{m},\mathfrak{n}}$ is the free (i.e. maximal) $(\mathbf{J}, \mathbf{M}, \mathfrak{m})$-extension of $\mathfrak{A}_{0,\mathfrak{n}}$ (for notation, see p. 136). If $\mathfrak{n} \geq \aleph_0$ and $\mathfrak{m} \geq 2^{\aleph_0}$, then $\mathfrak{A}_{\mathfrak{m},\mathfrak{n}}$ is not representable, by 31.3. Thus the $K = K_{\mathfrak{r}}$ does not hold, in general, for uncountable $\mathfrak{m}$.

C) There exists an atomic Boolean algebra $\mathfrak{A}$ such that the minimal σ-extension of $\mathfrak{A}$ is not isomorphic to the maximal σ-extension of $\mathfrak{A}$.

Let D be the Cantor set of real numbers (see § 9 B)), let B be a Borel subset of D such that $C = D - B$ is dense in D and, for every enumerable set $E \subset D$, the union $B \cup E$ is not an F_{σ}-set. Let ψ be a one-to-one mapping of C onto a set C' disjoint from D and let $X = D \cup C'$. The set X becomes a totally disconnected compact space if the following definition of the topology in X is assumed: every point in C' is isolated; if $x \in D$, the sets of the form $G \cup \psi(G \cap C) - K$, where G is open in D ($x \in G$) and K is a finite subset of C', constitute the basis of neighbourhoods for x.

The Boolean algebra $\mathfrak{A}$ of all open-closed subsets of X is atomic (since the set C' of isolated points of X is dense in the Stone space X of $\mathfrak{A}$). However the algebra $\mathfrak{F}_{\sigma}/\varDelta^*$ (see the notation on p. 170 where $\mathfrak{m} = \sigma$; $\varDelta^*$ is now the σ-ideal of sets of the σ-category) is not atomic and therefore the maximal σ-extension $\{i^{\varDelta^*}, \mathfrak{F}_{\sigma}/\varDelta^*\}$ of $\mathfrak{A}$ is not isomorphic to any σ-completion of $\mathfrak{A}$ (see § 35 I)).

In fact, it is easy to verify that a set $A \subset X$ is nowhere dense and σ-closed in X if and only if it is a closed subset of the space D and the set $A - B$ is at most enumerable. Consequently the ideal $\varDelta^*$ of all sets $A \in \mathfrak{F}_{\sigma}$ of the σ-category is composed of all Borel subsets of F_{σ}-sets $A \subset D$ such that $A - B$ is at most enumerable. Hence it follows that $B \notin \varDelta^*$, i.e. $[B]_{\varDelta^*} \neq \wedge$. The element $[B]_{\varDelta^*}$ contains no atom[2].

§ 37. $\mathfrak{m}$-independent subalgebras. The field $\mathfrak{m}$-product

An indexed set $\{\mathfrak{A}_t\}_{t \in T}$ of subalgebras of a Boolean algebra $\mathfrak{A}$ is said to be $\mathfrak{m}$-*independent* provided

(1) $\bigcap\limits_{t \in T'}^{\mathfrak{A}} A_t \neq \wedge$

---

[1] Recently YAQUB [1] proved that (for empty $\mathbf{J}$ and $\mathbf{M}$) if $\mathfrak{m} \geq 2^{\aleph_0}$ and if $\{h_0, \mathfrak{F}_{\mathfrak{m}}\}$ is the maximal $(\mathbf{J}, \mathbf{M}, \mathfrak{m})$-extension, then $\mathfrak{A}$ is superatomic. DAY [1] proved the converse statement. Thus $\{h_0, \mathfrak{F}_{\mathfrak{m}}\}$ is a maximal $(\mathbf{J}, \mathbf{M}, \mathfrak{m})$-extension of $\mathfrak{A}$ for empty $\mathbf{J}$ and $\mathbf{M}$ and $\mathfrak{m} \geq 2^{\aleph_0}$ if and only if $\mathfrak{A}$ is superatomic. See DAY and YAQUB [1].

[2] Example C) is due to M. KATĚTOV (not published).

for every m-indexed set $\{A_t\}_{t\in T'}$ such that $T'' \subset T$, $A_t \neq \Lambda$ and $A_t \in \mathfrak{A}_t$ for every $t \in T'$ (the last condition implies that elements A_t with different indices belong to subalgebras $\mathfrak{A}_t$ with different indices). The inequality (1) should be read as follows: the meet on the left side exists but is not equal to the zero element [in application we shall consider only the case where $\mathfrak{A}$ is m-complete, therefore the meet (1) always exists].

Every m-independent indexed set of subalgebras is also independent in the sense defined in § 13, therefore it has all the properties of independent subalgebras mentioned in § 13.

Every m-independent indexed set of subalgebras of a Boolean m-algebra has the following important extension property for homomorphisms (see also the analogous theorem 13.1).

37.1. *If $\{\mathfrak{A}_t\}_{t\in T}$ is an m-independent indexed set of m-subalgebras of a Boolean m-algebra $\mathfrak{A}$ and if, for every $t \in T$, h_t is an m-homomorphism of $\mathfrak{A}_t$ into an m-algebra $\mathfrak{A}'$ having the strong m-extension property (in particular, into an m-distributive m-algebra $\mathfrak{A}'$, or into an m-field of sets $\mathfrak{A}'$), then all the homomorphisms h_t can be (uniquely) extended to an m-homomorphism of the m-subalgebra $\mathfrak{A}_0 \subset \mathfrak{A}$ m-generated by the set-theoretical union of all the $\mathfrak{A}_t$ into the m-algebra $\mathfrak{A}'$.*[1]

The proof of theorem 37.1 is the same as that of 13.1, except that one uses theorem 34.3 instead of theorem 12.4.

Analogously to the case of 13.2, we can deduce from 37.1 the following theorem.

37.2. *If $\{\mathfrak{A}_t\}_{t\in T}$ and $\{\mathfrak{A}_t'\}_{t\in T}$ are m-independent indexed sets of m-subalgebras of m-algebras $\mathfrak{A}$, $\mathfrak{A}'$ respectively, having the strong m-extension property (in particular, of m-distributive m-algebras $\mathfrak{A}$, $\mathfrak{A}'$, or of m-fields of sets $\mathfrak{A}$, $\mathfrak{A}'$), the union of all the $\mathfrak{A}_t$ m-generates $\mathfrak{A}$, the union of all the $\mathfrak{A}_t'$ m-generates $\mathfrak{A}'$, and for every $t \in T$, h_t is an isomorphism of $\mathfrak{A}_t$ onto $\mathfrak{A}_t'$, then all the h_t can be extended to an isomorphism h of $\mathfrak{A}$ onto $\mathfrak{A}'$.*[2]

The following operations of forming m-products of fields yields an important example of m-independent subalgebras.

We recall some notation from § 13. For every $t \in T$, let $\mathfrak{F}_t$ be a (non-degenerate) field of subsets of a non-empty space X_t. Let X be the Cartesian product of all the spaces X_t, i.e. the set of all $x = \{x_t\}_{t\in T}$ where $x_t \in X_t$ for every $t \in T$. For every set $A \subset X_t$, let A^* be the set of all points $x \in X$ whose t^{th} co-ordinate x_t belongs to A. Let $\mathfrak{F}_t^*$ be the field (of subsets of X) composed of all sets A^* where $A \in \mathfrak{F}_t$.

The least m-field $\mathfrak{F}$ (of subsets of X) containing all the fields $\mathfrak{F}_t^*$ ($t \in T$) is called the *field m-product of the indexed set* $\{\mathfrak{F}_t\}_{t\in T}$. In other

---

[1] SIKORSKI [11, 14].
[2] SIKORSKI [11].

words, the field $\mathfrak{m}$-product of $\{\mathfrak{F}_t\}_{t\in T}$ is the $\mathfrak{m}$-field $\mathfrak{m}$-generated by the field product of $\{\mathfrak{F}_t\}_{t\in T}$ (see § 13).

$\{\mathfrak{F}_t^*\}_{t\in T}$ is an $\mathfrak{m}$-independent indexed set of subfields of $\mathfrak{F}$. The proof is similar to that of an analogous statement in § 13 (p. 40). $\mathfrak{F}_t^*$ is isomorphic to $\mathfrak{F}_t$ for every $t \in T$.

In the case $\mathfrak{m} = \sigma$ the operation of forming the field σ-product is used very often in measure theory. The following theorem[1] is one of the fundamental theorems in measure theory: if, for each $t \in T$, m_t is a given σ-measure on $\mathfrak{F}_t$ such that $m_t(X_t) = 1$, then there exists exactly one σ-measure m on the field σ-product $\mathfrak{F}$ of $\{\mathfrak{F}_t\}_{t\in T}$ such that

$$(2) \qquad m(A_1^* \cap \cdots \cap A_n^*) = m_{t_1}(A_1) \cdot \cdots \cdot m_{t_n}(A_n)$$

for arbitrary sets $A_j \in \mathfrak{F}_{t_j}$, $t_l \neq t_j$ for $l \neq j$ ($l, j = 1, \ldots, n$). The measure m is said to be the σ-product of the measures m_t.

37.3. *Let $\{\mathfrak{F}_t\}_{t\in T}$ be an indexed set of non-degenerate $\mathfrak{m}$-fields of sets and let $\mathfrak{A}$ be an $\mathfrak{m}$-field of sets. The following conditions are equivalent:*

(i) *$\mathfrak{A}$ is isomorphic to the field $\mathfrak{m}$-product of $\{\mathfrak{F}_t\}_{t\in T}$;*

(ii) *$\mathfrak{A}$ contains an $\mathfrak{m}$-independent indexed set $\{\mathfrak{A}_t\}_{t\in T}$ of $\mathfrak{m}$-subfields isomorphic respectively to $\mathfrak{F}_t$, such that the set-theoretical union of all $\mathfrak{A}_t$ $\mathfrak{m}$-generates $\mathfrak{A}$;*

(iii) *$\mathfrak{A}$ contains an indexed set $\{\mathfrak{A}_t\}_{t\in T}$ of $\mathfrak{m}$-subfields isomorphic respectively to $\mathfrak{F}_t$, such that the set-theoretical union of all $\mathfrak{A}_t$ $\mathfrak{m}$-generates $\mathfrak{A}$ and, for arbitrary $\mathfrak{m}$-homomorphisms h_t of $\mathfrak{A}_t$ into any $\mathfrak{m}$-field of sets $\mathfrak{A}'$, there exists an $\mathfrak{m}$-homomorphism h of $\mathfrak{A}$ into $\mathfrak{A}'$ which is a common extension of all h_t[2].*

(i) implies (ii). This is obvious since the $\mathfrak{m}$-field product $\mathfrak{F}$ of $\{\mathfrak{F}_t\}_{t\in T}$ and the $\mathfrak{m}$-subfields $\mathfrak{F}_t^*$ have the properties mentioned in (ii).

(ii) implies (iii). This follows from **37.1.**

(iii) implies (i). The proof of this implication is the same as that of **37.2** (and the analogous statement **13.2**). In fact, only the extension property of $\mathfrak{m}$-homomorphisms was used in the proof of **37.2.**

Example. A) It follows, in particular, from **37.1** that if $\mathfrak{F}$ is the field $\mathfrak{m}$-product of $\{\mathfrak{F}_t\}_{t\in T}$ and $\mathfrak{F}_t^*$ has the same meaning as above, then arbitrary $\mathfrak{m}$-homomorphisms h_t of $\mathfrak{F}_t^*$ into an $\mathfrak{m}$-algebra $\mathfrak{A}'$ having the strong $\mathfrak{m}$-extension property can be extended to an $\mathfrak{m}$-homomorphism h of $\mathfrak{F}$ into $\mathfrak{A}'$. The hypothesis that $\mathfrak{A}'$ has the strong $\mathfrak{m}$-extension property is essential and cannot be replaced e.g. by the weaker condition that $\mathfrak{A}'$ is an $\mathfrak{m}$-representable $\mathfrak{m}$-algebra, even in the case where T has only two elements and the h_t are isomorphisms. Consequently the hypo-

---

[1] See e.g. HALMOS [1], p. 154—158.

[2] SIKORSKI [11]. The equivalence of (i) and (ii) was also proved simultaneously and independently by SHERMAN [1].

thesis in 37.1 and 37.3 (iii) about $\mathfrak{A}'$ cannot be replaced by the weaker condition that $\mathfrak{A}'$ is an m-representable m-algebra.

For instance, let R be the set of all real numbers, let C be an analytic[1] non-Borel subset of R, and $X_1 = R - C$, $X_2 = R$. Let $\mathfrak{F}_1$ and $\mathfrak{F}_2$ be the σ-fields of all Borel subsets of the spaces X_1 and X_2 respectively. Let X' be the Cartesian product $X' = R \times R$ (i.e. X' is the plane), let $\mathfrak{F}'$ be the σ-field of all Borel subsets of X', and let Δ' be the σ-ideal (of $\mathfrak{F}'$) σ-generated by all sets of the form $B \times R$ where B is a Borel subset of R and $B \cap X_1$ is empty (i.e. $B \subset C$). The formulas

$$h_1((B \cap X_1) \times X_2) = [B \times R]_{\Delta'}, \quad h_2(X_1 \times B) = [R \times B]_{\Delta'}$$

(where B is a Borel subset of R) define σ-isomorphisms h_1 and h_2 of $\mathfrak{F}_1^*$ and $\mathfrak{F}_2^*$ respectively into $\mathfrak{A}' = \mathfrak{F}'/\Delta'$. The σ-isomorphisms h_1 and h_2 cannot be extended to a σ-homomorphism h of the least σ-field $\mathfrak{F}$ containing $\mathfrak{F}_1^*$ and $\mathfrak{F}_2^*$ (i.e. of the field σ-product $\mathfrak{F}$ of $\mathfrak{F}_1$ and $\mathfrak{F}_2$) into $\mathfrak{A}' = \mathfrak{F}'/\Delta'$. For suppose that such an extension h exists. Then it satisfies the equality

$$h(A \cap (X_1 \times X_2)) = [A]_{\Delta'}$$

for every Borel subset A of $X' = R \times R$ (this follows from the fact that the class of all sets $A \subset X'$ which satisfy this equality is a σ-field of sets containing all the sets of the form $B \times R$ and $R \times B$ where B is a Borel subset of R). The analytic set C is the projection of a plane G_δ-set $D \subset X'$, i.e. $x \in C$ if and only if $(x, y) \in D$ for some $y \in R$. Since D is disjoint from $X_1 \times X_2$, we have $h(D \cap (X_1 \times X_2)) = \Lambda$. On the other hand, D does not belong to Δ' (since $D \in \Delta'$ would imply that $D = C \times R$ and C is a Borel subset of R). Hence $[D]_{\Delta'} \neq \Lambda$. Contradiction[2].

By 37.1 the (σ-representable) σ-algebra $\mathfrak{F}'/\Delta'$ does not have the strong σ-extension property.

§ 38. Boolean (m, n)-products

In this section $\{\mathfrak{A}_t\}_{t \in T}$ will be a fixed indexed set of non-degenerate Boolean algebras, and m a fixed infinite cardinal.

$\{\{i_t\}_{t \in T}, \mathfrak{B}\}$ will be called a *Boolean* (m, 0)-*product*[3] (or simply an (m, 0)-*product*) of $\{\mathfrak{A}_t\}_{t \in T}$ provided that

(a) $\mathfrak{B}$ is a Boolean m-algebra;
(b) for every $t \in T$, i_t is an m-isomorphism of $\mathfrak{A}_t$ into $\mathfrak{B}$;
(c) the indexed set $\{i_t(\mathfrak{A}_t)\}_{t \in T}$ of subalgebras of $\mathfrak{B}$ is independent;
(d) the union of all the subalgebras $i_t(\mathfrak{A}_t)$ $(t \in T)$ m-generates $\mathfrak{B}$.

---

[1] For definition and fundamental properties, see e.g. KURATOWSKI [3], §§ 34 and 35.

[2] Example A) was given by SIKORSKI [14].

[3] (m, 0)-products were examined by SIKORSKI [13] (in the case where $m = \aleph_0$) and SIKORSKI [32] (in the case of any $m \geq \aleph_0$).

The class of all $(\mathfrak{m}, 0)$-products of $\{\mathfrak{A}_t\}_{t \in T}$ will be denoted by P. Let

(1) $\{\{i_t\}_{t \in T}, \mathfrak{B}\}$

and

(2) $\{\{i'_t\}_{t \in T}, \mathfrak{B}'\}$

be two elements in P. (2) is said to be an $\mathfrak{m}$-*homomorphic image* of (1) if there exists an $\mathfrak{m}$-homomorphism f from $\mathfrak{B}$ into $\mathfrak{B}'$ such that

(3) $i'_t = h i_t$ for every $t \in T$.

We write then

(4) $\{\{i'_t\}_{t \in T}, \mathfrak{B}'\} \leqq \{\{i_t\}_{t \in T}, \mathfrak{B}\}$.

Observe that condition (3) is equivalent to the following condition

(5) h is a common extension of all the isomorphisms $i''_t i_t^{-1}$ (from $i_t(\mathfrak{A}_t)$ onto $i'_t(\mathfrak{A}_t)$), $t \in T$.

Hence it follows that if the $\mathfrak{m}$-homomorphism h with the required properties exists, then it is unique. Moreover h maps $\mathfrak{B}$ onto $\mathfrak{B}'$. The two last statements are direct consequences of (d).

If the $\mathfrak{m}$-homomorphism h is an isomorphism, then we say that (2) is *isomorphic* to (1). Then also

(6) $\{\{i_t\}_{t \in T}, \mathfrak{B}\} \leqq \{\{i'_t\}_{t \in T}, \mathfrak{B}'\}$

and (1) is isomorphic to (2) since h^{-1} is an isomorphism (and an $\mathfrak{m}$-homomorphism) from $\mathfrak{B}'$ onto $\mathfrak{B}$ such that $i_t = h^{-1} i'_t$ for every $t \in T$. Thus we can then say simply that (1) and (2) are isomorphic.

It is easy to verify that relation (4) is a quasi-ordering in P, i.e. it is reflexive and transitive. Moreover, two elements (1) and (2) in P are isomorphic if and only if (4) and (6) hold simultaneously (see 23.3 and (d)). Sometimes it is convenient to identify isomorphic elements of P. After this identification, P is partially ordered by $\leqq$.

Let

(7) $\{\{i_{0t}\}_{t \in T}, \mathfrak{A}_0\}$

be a fixed Boolean product of $\{\mathfrak{A}_t\}_{t \in T}$ (see § 13). We recall that the Boolean product of $\{\mathfrak{A}_t\}_{t \in T}$ is determined by $\{\mathfrak{A}_t\}_{t \in T}$ uniquely up to an isomorphism. Moreover the i_{0t} are complete isomorphisms from $\mathfrak{A}_t$ into $\mathfrak{A}_0$ (see § 22 J)), i.e. the $i_{0t}(\mathfrak{A}_t)$ are regular subalgebras of $\mathfrak{A}_0$.

Let J be the class of all sets $\mathfrak{S} \subset \mathfrak{A}_0$ such that

(8) $\overline{\overline{\mathfrak{S}}} \leqq \mathfrak{m}$,

(9)   there exists a $t \in T$ such that $\mathfrak{S} \subset i_{0t}(\mathfrak{A}_t)$ and

(10) the join $\bigcup_{A \in \mathfrak{S}}^{\mathfrak{A}_0} A$ exists.

Let $\mathbf{M}$ be the class of all sets $\mathfrak{S} \subset \mathfrak{A}_0$ such that (8) and (9) hold and

(10') the meet $\bigcap_{A \in \mathfrak{S}}^{\mathfrak{A}_0} A$ exists.

The following theorem assures the existence of $(\mathfrak{m}, 0)$-products of $\{\mathfrak{A}_t\}_{t \in T}$.

38.1. *If* $\{i, \mathfrak{B}\}$ *is a* $(\mathbf{J}, \mathbf{M}, \mathfrak{m})$*-extension of the product* (7) *of* $\{\mathfrak{A}_t\}_{t \in T}$, *then*

(11) $\{\{i i_{0 t}\}_{t \in T}, \mathfrak{B}\}$

is an $(\mathfrak{m}, 0)$*-product of* $\{\mathfrak{A}_t\}_{t \in T}$. *If* $\{i', \mathfrak{B}'\}$ *is another* $(\mathbf{J}, \mathbf{M}, \mathfrak{m})$*-extension of* (7), *then*

(12) $\{\{i' i_{0 t}\}_{t \in T}, \mathfrak{B}'\} \leq \{\{i i_{0 t}\}_{t \in T}, \mathfrak{B}\}$ *if and only if* $\{i', \mathfrak{B}'\} \leq \{i, \mathfrak{B}\}$.

In fact, (a) follows from § 36 (a). (b) follows § 36 (b) since $i_{0 t}$ is a complete isomorphism. (c) follows from § 13 (c). (d) follows from § 13 (d) and § 36 (c).

If h is an $\mathfrak{m}$-homomorphism from $\mathfrak{B}$ into $\mathfrak{B}'$ such that $i' = hi$, then $i' i_{0 t} = h i i_{0 t}$ for every $t \in T$.

Conversely, if h is an $\mathfrak{m}$-homomorphism from $\mathfrak{B}$ into $\mathfrak{B}'$ such that $i' i_{0 t} = h i i_{0 t}$ for every $t \in T$, then

$$i'(A) = h i(A)$$

for every A belonging to the union of all the subalgebras $i_{0 t}(\mathfrak{A}_t)$. Since this union generates $\mathfrak{A}_0$ by § 13 (d), this equality holds for all $A \in \mathfrak{A}_0$, i.e. $i' = hi$.

This proves the equivalence (12).

38.2. *For every* $(\mathfrak{m}, 0)$*-product* (1) *of* $\{\mathfrak{A}_t\}_{t \in T}$ *there exists a* $(\mathbf{J}, \mathbf{M}, \mathfrak{m})$*-extension* $\{i, \mathfrak{B}\}$ *of* (7) *such that* (11) *is identical with* (1).

Let $\mathfrak{B}_0$ be the smallest subalgebra of $\mathfrak{B}$ which contains all the subalgebras $i_t(\mathfrak{A}_t)$. It follows from (a)—(d) that

(13) $\{\{i_t\}_{t \in T}, \mathfrak{B}_0\}$

is a Boolean product of $\{\mathfrak{A}_t\}_{t \in T}$. Since all Boolean products of $\{\mathfrak{A}_t\}_{t \in T}$ are isomorphic (see § 13, p. 41), there is an isomorphism i from $\mathfrak{A}_0$ onto $\mathfrak{B}_0$ such that $i_t = i i_{0 t}$ for every $t \in T$. The $(\mathbf{J}, \mathbf{M}, \mathfrak{m})$-extension $(i, \mathfrak{B})$ of $\mathfrak{A}_0$ has the required properties.

By 38.1 and 38.2 the examination of $(\mathfrak{m}, 0)$-products of $\{\mathfrak{A}_t\}_{t \in T}$ can be reduced to an examination of $(\mathbf{J}, \mathbf{M}, \mathfrak{m})$-extensions of (7).

It follows from (12) that if $\{i, \mathfrak{B}\}$ is a maximal (i.e. free) $(\mathbf{J}, \mathbf{M}, \mathfrak{m})$-extension of (7), then (11) is a maximal element of P, i.e. (1) $\leq$ (11) for every $(\mathfrak{m}, 0)$-product (1) of $\{\mathfrak{A}_t\}_{t \in T}$. Consequently this $(\mathfrak{m}, 0)$-product (11) and all its isomorphs will be called *maximal* $\mathfrak{m}$-*products* or *free* $\mathfrak{m}$-*products* of $\{\mathfrak{A}_t\}_{t \in T}$.

38.3. *In order that an* $(\mathfrak{m}, 0)$-*product* (1) *of* $\{\mathfrak{A}_t\}_{t \in T}$ *be a maximal* $\mathfrak{m}$-*product of* $\{\mathfrak{A}_t\}_{t \in T}$ *it is necessary and sufficient that it have the following property:*

(e) *for any collection of* $\mathfrak{m}$-*homomorphisms* h_t *from* $\mathfrak{A}_t$ *into any Boolean* $\mathfrak{m}$-*algebra* $\mathfrak{C}$ *there exists an* $\mathfrak{m}$-*homomorphism* h *from* $\mathfrak{B}$ *into* $\mathfrak{C}$ *such that*

(14) $h_t = h i_t$ *for every* $t \in T$.

Clearly property (e) is equivalent to the following:

(e′) *for any collection of* $\mathfrak{m}$-*homomorphisms* h_t' *from* $i_t(\mathfrak{A}_t)$ *into a Boolean* $\mathfrak{m}$-*algebra* $\mathfrak{C}$ *there exists an* $\mathfrak{m}$-*homomorphism* h *from* $\mathfrak{B}$ *into* $\mathfrak{C}$ *that is a common extension of all* h_t', $t \in T$.

To establish the equivalence of (e) and (e′) it suffices to assume $h_t' = h_t i_t^{-1}$ or $h_t = h_t' i_t$.

To prove the necessity of (e) it suffices to show that if $\{i, \mathfrak{B}\}$ is a maximal extension of (7), then (11) has property (e). Indeed, by **13.3** there exists a homomorphism h_0 from $\mathfrak{A}_0$ into $\mathfrak{C}$ such that $h_t = h_0 i_t$ for every $t \in T$. Hence it follows that h_0 is a $(\mathbf{J}, \mathbf{M})$-homomorphism from $\mathfrak{A}_0$ into $\mathfrak{C}$. By **36.4** there exists an $\mathfrak{m}$-homomorphism h from $\mathfrak{B}$ into $\mathfrak{C}$ such that $h_0 = h i$. Consequently $h_t = h i i_t$ for every $t \in T$.

On the other hand, it follows from **23.3** that all $(\mathfrak{m}, 0)$-products (of $\{\mathfrak{A}_t\}_{t \in T}$) with the property (e) are isomorphic. Since the maximal $\mathfrak{m}$-product (11), where $(i, \mathfrak{B})$ is a maximal $(\mathbf{J}, \mathbf{M}, \mathfrak{m})$-extension of (7), has property (e), every $(\mathfrak{m}, 0)$-product with property (e) is isomorphic to it; i.e. is also a maximal $\mathfrak{m}$-product.

Example. A) If all the algebras $\mathfrak{A}_t$ have four elements, and (1) is a maximal $\mathfrak{m}$-product of $\{\mathfrak{A}_t\}_{t \in T}$, then $\mathfrak{B}$ is a free Boolean $\mathfrak{m}$-algebra with $\overline{\overline{T}}$ free $\mathfrak{m}$-generators. The proof is similar to that of **14.1**.

It follows from (12) that if $(i, \mathfrak{B})$ is a minimal $\mathfrak{m}$-extension of (7), i.e. an $\mathfrak{m}$-completion of (7), then (11) is a minimal element in $\boldsymbol{P}$; i.e. if an $(\mathfrak{m}, 0)$-product (1) satisfies $(1) \leq (11)$, then also $(11) \leq (1)$. In other words, (1) is equal (11) after identification of isomorphic elements in $\boldsymbol{P}$. This $(\mathfrak{m}, 0)$-product and all its isomorphs will be called *minimal* $(\mathfrak{m}, 0)$-*products* of $\{\mathfrak{A}_t\}_{t \in T}$. It is not known whether the minimal $(\mathfrak{m}, 0)$-product (11) is the smallest element in $\boldsymbol{P}$, i.e. $(11) \leq (1)$ for every $(\mathfrak{m}, 0)$-product (1).

38.4. *An* $(\mathfrak{m}, 0)$-*product* (1) *of* $\{\mathfrak{A}_t\}_{t \in T}$ *is minimal if and only if the set of all elements*

(15) $\bigcap_{t \in T'} A_t$ *where* $A_t \in i(\mathfrak{A}_t)$ *for* $t \in T'$,

and where T' *is any finite subset of* T, *is dense in* $\mathfrak{B}$.

It suffices to prove **38.4** in the case where (1) is of the form (11). Since the set of all elements

$$\bigcap_{t \in T'} A_t \quad \text{where} \quad A_t \in i_t(\mathfrak{A}_t) \quad \text{for} \quad t \in T',$$

and T' is finite, is dense in $\mathfrak{A}_0$ (see § 13 (9), p. 41), their isomorphic images (15) compose a set dense in $i(\mathfrak{A}_0)$. Thus the set of all elements (15) is dense in $\mathfrak{B}$ if and only if $i(\mathfrak{A}_0)$ is dense in $\mathfrak{B}$, i.e. if $(i, \mathfrak{B})$ is an $\mathfrak{m}$ completion of $\mathfrak{A}_0$ (see 35.3).

Let $\mathfrak{n}$ be an infinite cardinal $\leq \mathfrak{m}$.

A Boolean $(\mathfrak{m}, 0)$-product (1) of $\{\mathfrak{A}_t\}_{t \in T}$ is said to be an $(\mathfrak{m}, \mathfrak{n})$-*product of* $\{\mathfrak{A}_t\}_{t \in T}$ provided that

$(c_\mathfrak{n})$ the indexed set $\{i_t(\mathfrak{A}_t)\}_{t \in T}$ of subalgebras of $\mathfrak{B}$ is $\mathfrak{n}$-independent.

$(\mathfrak{m}, \mathfrak{m})$-products are called $\mathfrak{m}$-*products*[1].

The class of all $(\mathfrak{m}, \mathfrak{n})$-products of $\{\mathfrak{A}_t\}_{t \in T}$ will be denoted by $P_\mathfrak{n}$. Note that any isomorph of an $(\mathfrak{m}, \mathfrak{n})$-product is also an $(\mathfrak{m}, \mathfrak{n})$-product.

By definition
$$P_{\mathfrak{n}'} \subset P_\mathfrak{n} \quad \text{for} \quad \mathfrak{n} \leq \mathfrak{n}' \leq \mathfrak{m}$$
and
$$P_\mathfrak{m} \subset P_\mathfrak{n} \subset P \quad \text{for} \quad \mathfrak{n} \leq \mathfrak{m}.$$

38.5. *If* (1) *is an* $(\mathfrak{m}, 0)$-*product of* $\{\mathfrak{A}_t\}_{t \in T}$, (2) *is an* $(\mathfrak{m}, \mathfrak{n})$-*product of* $\{\mathfrak{A}_t\}_{t \in T}$ *and* (2) $\leq$ (1), *then* (1) *is an* $(\mathfrak{m}, \mathfrak{n})$-*product of* $\{\mathfrak{A}_t\}_{t \in T}$.

Indeed, let h be an $\mathfrak{m}$-homomorphism of $\mathfrak{B}$ into $\mathfrak{B}'$ such that $i''_t = h i_t$, i.e. h is a common extension of all the isomorphisms $i''_t i_t^{-1}$. If $T' \subset T$, $\overline{\overline{T'}} \leq \mathfrak{n}$ and $A_t \in \mathfrak{A}_t$ for $t \in T'$, then
$$h\left(\bigcap_{t \in T'} i_t(A_t)\right) = \bigcap_{t \in T'} i''_t(A_t) \neq \bigwedge_{\mathfrak{B}'}$$
by $(c_\mathfrak{n})$. Hence $\bigcap_{t \in T'} i(A_t) \neq \bigwedge_\mathfrak{B}$.

For any $\mathfrak{n} \leq \mathfrak{m}$ the class $P_\mathfrak{n}$ is not empty. To construct an example of an $(\mathfrak{m}, \mathfrak{n})$-product let us assume the following notation.

Let g_t be an isomorphism of $\mathfrak{A}_t$ onto the field $\mathfrak{F}_t$ of all open-closed subsets of a Stone space X_t of $\mathfrak{A}_t$, let X be the Cartesian product of all the spaces X_t and (according to notation in § 13 and § 37) let

(16) $g_t^*(A) = g_t(A)^* =$ the set of all $x \in X$ whose t^{th} coordinate is in $g_t(A)$ for $A \in \mathfrak{A}_t$. Let $\mathfrak{F}$ be the field (of subsets of X) generated by all the sets

(17) $\bigcap_{t \in T'} g_t^*(B_t)$ where $B_t \in \mathfrak{A}_t$ for $t \in T'$, $T' \subset T$ and $\overline{\overline{T'}} \leq \mathfrak{n}$,

where $\bigcap$ denotes the set-theoretical intersection. Note that for every $t \in T$, g_t^* is a complete isomorphism from $\mathfrak{A}_t$ into $\mathfrak{F}$.

It is convenient to introduce in X a special topology, called the $\mathfrak{n}$-*topology*, by assuming $\mathfrak{F}$ as an open basis. Thus, in the $\mathfrak{n}$-topology, open sets are any unions of the sets (17). The space X with the $\mathfrak{n}$-topology is not compact and it is not the Stone space of $\mathfrak{F}$ (except when T is finite). However it has the following important property; no non-void open set is of the first category in X.

---

[1] $\mathfrak{m}$-products were investigated by SIKORSKI [13] (in the case of $\mathfrak{m} = \aleph_0$) and SIKORSKI [32] (in the case of any $\mathfrak{m} \geq \aleph_0$).

To prove the last statement, it will be convenient to write sets (17) in the product form

(17') $P_{t \in T} G_t$ where $G_t \in \mathfrak{F}_t$ for $t \in T$, and $G_t = X_t$ for $t \in T - T'$, $T' \leqq \mathfrak{n}$.

Let G be a non-empty set open in the space X (with the $\mathfrak{n}$-topology), and let $\{N_n\}$ be a sequence of closed nowhere dense subsets of X. Since $G - N_1$ is not empty, there exists a non-empty open set

$$G_1 = P_{t \in T} G_{1, t} \subset G - N_1$$

such that the $G_{1, t}$ are non-empty open-closed subsets of X_t. By induction we define a sequence of sets

$$G_n = P_{t \in T} G_{n, t} \subset G_{n-1} - N_n$$

open in X, where the $G_{n, t}$ are non-empty open-closed subsets of X_t. Since $G_{1, t}$, $G_{2, t}$, ... is a decreasing sequence of non-empty closed subsets of the compact space X_t, the intersection $\bigcap_{1 \leqq n < \infty} G_{n, t}$ is not empty and consequently the intersection

$$\bigcap_{1 \leqq n < \infty} G_n = P_{t \in T} \bigcap_{1 \leqq n < \infty} G_{n, t}$$

is a non-empty subset of $G - \bigcup_{1 \leqq n < \infty} N_n$. This proves that G is not a subset of any union of nowhere dense subsets N_n, i.e. G is not of the first category in X.

Now let $\{i, \mathfrak{B}\}$ be any $\mathfrak{m}$-completion of $\mathfrak{F}$. It is easy to verify that

(18) $\{\{i g_t^*\}_{t \in T}, \mathfrak{B}\}$

is an $(\mathfrak{m}, \mathfrak{n})$-product of $\{\mathfrak{A}_t\}_{t \in T}$. The $(\mathfrak{m}, \mathfrak{n})$-product (18) and all its isomorphs will be called *minimal* $(\mathfrak{m}, \mathfrak{n})$-*products* of $\{\mathfrak{A}_t\}_{t \in T}$.

38.6. *In order that an* $(\mathfrak{m}, \mathfrak{n})$-*product* (1) *be a minimal* $(\mathfrak{m}, \mathfrak{n})$-*product it is necessary and sufficient that the set of all the elements*

(19) $\bigcap_{t \in T'}^{\mathfrak{B}} A_t$ *where* $A_t \in i(\mathfrak{A}_t)$ *for* $t \in T'$, $T' \subset T$ *and* $\overline{\overline{T'}} \leqq \mathfrak{n}$

be dense in $\mathfrak{B}$.

To prove the necessity it suffices to show that (18) has the property mentioned in 38.6. First observe that the collection of all sets (17) is dense in $\mathfrak{F}$. Consequently the set of their isomorphic images $\bigcap_{t \in T'} i g_t^* (B_t)$ is dense in $i(\mathfrak{F})$ which is dense in $\mathfrak{B}$ by 35.3. Thus this set is dense in $\mathfrak{B}$.

To prove the sufficiency it suffices to show that if two $(\mathfrak{m}, \mathfrak{n})$-products (1) and (2) have the property mentioned in 38.6, then they are isomorphic.

Let $\mathfrak{S}$ be the set of all the non-zero elements (19) in $\mathfrak{B}$, and let $\mathfrak{S}'$ be the analogous set in $\mathfrak{B}'$.

Any two elements $A, B \in \mathfrak{S}$ can always be represented in the form

$$A = \bigcap_{t \in T'} A_t, \quad B = \bigcap_{t \in T'} B_t \quad (A_t, B_t \in i(\mathfrak{A}_t), A_t \neq \wedge \neq B_t)$$

with the same set $T' \subset T$ of power $\leqq \mathfrak{n}$. Note that

(20) $A \subset B$ if and only if $A_t \subset B_t$ for every $t \in T'$.

It suffices to prove only that if $A_{t_0} \not\subset B_{t_0}$ for some $t_0 \subset T'$, then $A \not\subset B$. Indeed, the non-zero element

$$C = \bigcap_{t \in T'} C_t \quad \text{where} \quad C_{t_0} = A_{t_0} - B_{t_0} \quad \text{and} \quad C_t = A_t \quad \text{for} \quad t \neq t_0$$

is a subelement of A and is disjoint from B.

Since elements in the corresponding subset $\mathfrak{S}'$ of $\mathfrak{B}'$ have the same property (20), the formula

(21) $$f(\bigcap_{t \in T'} A_t) = \bigcap_{t \in T'} i_t' i_t^{-1}(A_t),$$

where $\Lambda \neq A_t \in i(\mathfrak{A}_t)$, $T' \subset T$, $\overline{\overline{T'}} \leq \mathfrak{n}$, defines a mapping f from $\mathfrak{S}$ onto $\mathfrak{S}'$ such that for any $A, B \in \mathfrak{S}$,

$$A \subset B \quad \text{if and only if} \quad f(A) \subset f(B).$$

Since $\mathfrak{S}$ and $\mathfrak{S}'$ are dense in $\mathfrak{B}$ and $\mathfrak{B}'$ respectively, by 12.5 f can be extended to an isomorphism h_0 from the subalgebra $\mathfrak{B}_0$ generated by $\mathfrak{S}$ onto the subalgebra $\mathfrak{B}_0'$ generated by $\mathfrak{S}'$. Since $\mathfrak{S}$ and $\mathfrak{S}'$ are dense in $\mathfrak{B}$ and $\mathfrak{B}'$ respectively, the subalgebras $\mathfrak{B}_0$ and $\mathfrak{B}_0'$ are dense in $\mathfrak{B}$ and $\mathfrak{B}'$, respectively. Moreover $\mathfrak{B}_0$ and $\mathfrak{B}_0'$ $\mathfrak{m}$-generate $\mathfrak{B}$ and $\mathfrak{B}'$, respectively. Thus $\mathfrak{B}$ and $\mathfrak{B}'$ are $\mathfrak{m}$-completions of $\mathfrak{B}_0$ and $\mathfrak{B}_0'$ respectively. Since $\mathfrak{m}$-completions of algebras are determined by these algebras uniquely up to isomorphism, the isomorphism h_0 can be extended to an isomorphism h of $\mathfrak{B}$ onto $\mathfrak{B}'$. Since h is an extension of f, it follows from (21) (the case where $T' = (t)$) that $h(A) = i_t' i_t^{-1}(A)$ for $A \in i(\mathfrak{A}_t)$, i.e. h is a common extension of all the isomorphisms $i_t' i_t^{-1}$. This proves that (1) and (2) are isomorphic.

The following theorem justifies the name "minimal $(\mathfrak{m}, \mathfrak{n})$-product".

38.7. *Any minimal $(\mathfrak{m}, \mathfrak{n})$-product* (1) *of* $\{\mathfrak{A}_t\}_{t \in T}$ *is a minimal element in* $P_{\mathfrak{n}}$, *and the isomorphisms* i_t *are complete isomorphisms from* $\mathfrak{A}_t$ *into* $\mathfrak{B}$.

It is not known whether minimal $(\mathfrak{m}, \mathfrak{n})$-products are smallest elements in $P_{\mathfrak{n}}$.

To prove the first part of 38.7, suppose that (1) is a minimal $(\mathfrak{m}, \mathfrak{n})$-product, (2) is an $(\mathfrak{m}, \mathfrak{n})$-product and h is an $\mathfrak{m}$-homomorphism from $\mathfrak{B}$ onto $\mathfrak{B}'$ such that h is a common extension of all the isomorphisms $i_t' i_t^{-1}$, $t \in T$. Since (2) is an $(\mathfrak{m}, \mathfrak{n})$-product, $h(A) \neq \Lambda$ for every $A \in \mathfrak{S}$, where $\mathfrak{S}$ is the set of all non zero elements (19) in $\mathfrak{B}$. Since $\mathfrak{S}$ is dense in $\mathfrak{B}$, we have $h(A) \neq \Lambda$ for every non-zero element in $\mathfrak{B}$. This proves that h is an isomorphism, i.e. (1) and (2) are isomorphic.

It suffices to prove the second part of 38.7 in the case where (1) is identical with (18). Since g_t^* is a complete isomorphism from $\mathfrak{A}_t$ into $\mathfrak{F}$, and i is a complete isomorphism from $\mathfrak{A}$ into $\mathfrak{B}$, $i g_t^*$ is also complete.

The following theorem yields a simple topological representation for minimal $(\mathfrak{m}, 0)$-products and minimal $(\mathfrak{m}, \mathfrak{n})$-products. X and g_t^* have here the same meaning as on p. 179 (see (16)).

38.8. *Let $\mathfrak{F}'$ be the field of all Borel subsets of the product X of the Stone spaces of $\mathfrak{A}_t$, $t \in T$, with the ordinary product topology (with the $\mathfrak{n}$-topology), and let Δ be the ideal of all sets $A \in \mathfrak{F}'$ of the first category in the topological space X. Let $\mathfrak{B}$ be the smallest $\mathfrak{m}$-subalgebra of the complete Boolean algebra $\mathfrak{F}'/\Delta$ such that all the elements $[g_t^*(A)]_\Delta$ where $t \in T$ and $A \in \mathfrak{A}_t$ are in $\mathfrak{B}$, and let*

$$i_t(A) = [g_t^*(A)]_\Delta \quad \text{for} \quad t \in T, A \in \mathfrak{A}_t.$$

Then

$$\{\{i_t\}_{t \in T}, \mathfrak{B}\}$$

is a minimal $(\mathfrak{m}, 0)$-product (a minimal $(\mathfrak{m}, \mathfrak{n})$-product) of $\{\mathfrak{A}_t\}_{t \in T}$.

i_t is an isomorphism since for any non-zero element $A \in \mathfrak{A}_t$ the set $g_t^*(A)$ is non-void and open in X and, consequently, it is not of the first category in X, which yields $i_t(A) \neq \wedge$.

i_t is a complete isomorphism from $\mathfrak{A}_t$ into $\mathfrak{F}'/\Delta$ for if $A = \bigcup_{s \in S}^{\mathfrak{A}_t} A_s$, then $g_t^*(A) \supset g_t^*(A_s)$ for all $s \in S$ and $g_t^*(A) - \bigcup_{s \in S} g_t^*(A_s)$ is nowhere dense in X, which proves that $i_t(A) = \bigcup_{s \in S}^{\mathfrak{F}'/\Delta} i_t(A_s)$. Since $i_t(\mathfrak{A}_t)$ is contained in the $\mathfrak{m}$-subalgebra $\mathfrak{B}$ of $\mathfrak{F}'/\Delta$, i_t is an $\mathfrak{m}$-isomorphism of $\mathfrak{A}_t$ into $\mathfrak{B}$.

Let A_t be a non-zero element in $\mathfrak{A}_t$ for $t \in T'$ where $T' \subset T$ and T' is finite (T' has power $\leq \mathfrak{n}$). Since the g_t^* are closed sets in X, we have

$$\bigcap_{t \in T'} i_t(A_t) = \bigcap_{t \in T'} [g_t^*(A_t)] = [\bigcap_{t \in T'} g_t^*(A_t)]$$

by § 21 (1''). Since the intersection $\bigcap_{t \in T'} g_t^*(A_t)$ is an open non-void subset of X, it is not of the first category in X which proves that $\bigcap_{t \in T'} i_t(A_t) \neq \wedge$. Thus the subalgebras $i_t(\mathfrak{A}_t)$ are independent (are $\mathfrak{n}$-independent).

By definition of $\mathfrak{B}$, the union of all the subalgebras $i_t(\mathfrak{A}_t)$ $\mathfrak{m}$-generates $\mathfrak{B}$. This proves that (1) is an $(\mathfrak{m}, 0)$-product (an $(\mathfrak{m}, \mathfrak{n})$-product) of $\{\mathfrak{A}_t\}_{t \in T}$. To prove that it is minimal, let us recall that every element A in $\mathfrak{F}'/\Delta$, and consequently every element in $\mathfrak{B}$, is of the form $A = [G]$ where G is open in X (see § 21 C)). If A is a non-zero element, G is not empty and, consequently, by the definition of topology in X, G contains a set $\bigcap_{t \in T'} g_t^*(A_t)$ where A_t is a non-zero element in $\mathfrak{A}_t$ and T' is finite (and T' has power $\leq \mathfrak{n}$). Hence

$$\bigcap_{t \in T'} i_t(A_t) = [\bigcap_{t \in T'} g_t^*(A_t)] \subset A.$$

Since (1) satisfies the condition of theorem 38.4 (of theorem 38.6), (1) is a minimal $(\mathfrak{m}, 0)$-product (is a minimal $(\mathfrak{m}, \mathfrak{n})$-product).

Observe that if $\{i, \mathfrak{B}\}$ is any $\mathfrak{m}$-extension of $\mathfrak{F}$, then (18) is an $(\mathfrak{m}, \mathfrak{n})$-product of $\{\mathfrak{A}_t\}_{t \in T}$. It is not known whether the converse statement is true.

By 38.5, the maximal $\mathfrak{m}$-product is a maximal element in $P_\mathfrak{n}$ for every $\mathfrak{n} \leqq \mathfrak{m}$ (in particular, it belongs to $P_\mathfrak{m}$, i.e. it is an $\mathfrak{m}$-product in the sense defined on p. 179). We do not know any topological representation of maximal $\mathfrak{m}$-products, except the case $\mathfrak{m} = \sigma$ (then the representation in question is a particular case of theorem 38.9 below).

If (1) is an $(\mathfrak{m}, 0)$-product of $\{\mathfrak{A}_t\}_{t \in T}$ and $\mathfrak{B}$ is $\mathfrak{m}$-representable, then all the algebras $\mathfrak{A}_t$ are $\mathfrak{m}$-representable on account of (a) and (b). Conversely, if all the $\mathfrak{A}_t$ are $\mathfrak{m}$-representable, the class P_r of all $(\mathfrak{m}, 0)$-products (1) such that $\mathfrak{B}$ is $\mathfrak{m}$-representable is not empty. An example of an element in P_r is given by the following theorem, where X_t, X, g_t and g_t^* have the same meaning as on p. 179 (the topology in the Cartesian product X of all X_t is ordinary).

38.9. *Let $\mathfrak{F}_\mathfrak{m}$ be the $\mathfrak{m}$-field $\mathfrak{m}$-generated by the field of all open-closed subsets in X, let $\varDelta^*$ be the smallest $\mathfrak{m}$-ideal in $\mathfrak{F}_\mathfrak{m}$ which contains all the sets*

$$(22) \quad g^*(A) - \bigcup\nolimits_{s \in S} g_t^*(A_s) \quad \text{where} \quad A = \bigcup\nolimits_{s \in S}^{\mathfrak{A}_t} A_s \quad \text{and} \quad \bar{\bar{S}} \leqq \mathfrak{m}, t \in T,$$

and let

$$i_t(A) = [g_t^*(A)]_{\varDelta^*} \quad \text{for} \quad A \in \mathfrak{A}_t \quad (t \in T).$$

If all the algebras $\mathfrak{A}_t$ are $\mathfrak{m}$-representable, then

$$(23) \qquad\qquad \{\{i_t\}_{t \in T}, \mathfrak{F}_\mathfrak{m}/\varDelta^*\}$$

is an $\mathfrak{m}$-product in P_r.

In theorem 38.9 and in its proof below the signs $\cup$, $\cap$ without superscripts denote the set-theoretical operations.

According to the notation introduced on p. 40 and p. 173, if B is a subset of X_t, then B^* denotes the set of all points in the Cartesian product X of all X_t whose t^{th} coordinate is in B. Thus the ideal $\varDelta^*$ is the ideal $\mathfrak{m}$-generated by all the sets B^* where B is any subset of X_t of the $\mathfrak{m}$-category, i.e. every set in $\varDelta^*$ is a subset of a set $\bigcup_{t \in T'} B_t^*$ where $T' \subset T$, $\bar{\bar{T'}} \leqq \mathfrak{m}$ and, for every $t \in T'$, B_t is a subset of X_t of the $\mathfrak{m}$-category.

First observe that if $\wedge \neq A_t \in \mathfrak{A}_t$ and $B_t \subset X_t$ is of the $\mathfrak{m}$-category in X_t for $t \in T'$, $T' \subset T$, $\bar{\bar{T'}} \leqq \mathfrak{m}$, then the inclusion

$$(24) \qquad\qquad \bigcap\nolimits_{t \in T'} g_t^*(A_t) \subset \bigcup\nolimits_{t \in T'} B_t^*$$

never holds. In fact, the set $g_t(A_t) - B_t$ is not empty since $\mathfrak{A}_t$ is $\mathfrak{m}$-representable (see 29.3 (r_4)). Thus the set $\bigcap_{t \in T'}(g_t^*(A_t) - B_t^*)$ is a non-empty subset of the set on the left side of (24), and it is disjoint from the set on the right side of (24).

Since (24) never holds, the mappings i_t are isomorphisms and the subalgebras $i_t(\mathfrak{A}_t)$ are $\mathfrak{m}$-independent. It follows immediately from the definition of i_t and $\varDelta^*$ that the i_t are $\mathfrak{m}$-homomorphisms from $\mathfrak{A}_t$ into $\mathfrak{B}$.

Since $\mathfrak{F}_m$ is m-generated by the union of all the fields $g_t^*\,(\mathfrak{A}_t)$, the union of all $i_t\,(\mathfrak{A}_t)$ m-generates the m-representable m-algebra $\mathfrak{F}_m/\varDelta^*$. Thus (23) is an m-product in $\boldsymbol{P}_r$.

The m-product (23) and all its isomorphs are called *maximal represent-able* m-*products*. This name is justified by the second part of the following theorem.

38.10. *In order that an* $(m, 0)$-*product* (1) *of* m-*representable algebras* $\mathfrak{A}_t\ (t \in T)$ *be a maximal representable* m-*product of* $\{\mathfrak{A}_t\}_{t \in T}$ *it is necessary and sufficient that it have the following property:*

(e_r) *for any collection of* m-*homomorphisms* h_t *from* $\mathfrak{A}_t$ *into any* m-*representable* m-*algebra* $\mathfrak{C}$ *there exists an* m-*homomorphism* h *from* $\mathfrak{B}$ *into* $\mathfrak{C}$ *such that*

$$h_t = h i_t \quad \text{for every} \quad t \in T .$$

Consequently any maximal representable product of $\{\mathfrak{A}_t\}_{t \in T}$ *is a greatest element in* $\boldsymbol{P}_r$.

Clearly property (e_r) is equivalent to the following:

(e_r') *for any collection of* m-*homomorphisms* h_t' *from* $i\,(\mathfrak{A}_t)$ *into an* m-*representable* m-*algebra* $\mathfrak{C}$ *there exists an* m-*homomorphism* h *from* $\mathfrak{B}$ *into* $\mathfrak{C}$ *that is a common extension of all* h_t', $t \in T$.

To prove the necessity of (e_r) it suffices to show the existence of h in the case where (1) is the m-product (23), and $\mathfrak{C} = \mathfrak{F}_0/\varDelta_0$ is the canonical m-representation of an m-representable m-algebra. The m-homomorphism h_t is induced by an m-continuous mapping φ_t (see 22.5). Since φ_t is m-continuous, the mapping

$$\varphi(x) = \{\varphi_t(x)\} \in X$$

satisfies

$$\varphi^{-1}(A) \in \varDelta_0 \quad \text{for every set} \quad A \in \varDelta^*$$

(verify first the case where $A = B^*$ and B is of the m-category in X_t!). Hence it follows that the formula

$$h([A]_{\varDelta^\bullet}) = [\varphi^{-1}(A)]_{\varDelta_0} \quad \text{for} \quad A \in \mathfrak{F}_m$$

defines an m-homomorphism from $\mathfrak{F}_m/\varDelta^*$ into $\mathfrak{F}_0/\varDelta_0$. Moreover, for any $A \in \mathfrak{A}_t$

$$h_t(A) = [\varphi_t^{-1}(g_t(A))]_{\varDelta_0} = [\varphi^{-1}(g_t^*(A))]_{\varDelta_0} = h(i_t(A)) ,$$

i.e. $h_t = h i_t$ for every $t \in T$.

To prove the sufficiency let us observe that all $(m, 0)$-products in $\boldsymbol{P}_r$ with property (e_r) are isomorphic by 23.3. Since (23) has property (e_r), all $(m, 0)$-products in $\boldsymbol{P}_r$ with this property are isomorphic to (23), i.e. they are maximal representable m-products.

The second part of 38.10 follows directly from the first; if (2) is any $(m, 0)$-product in $\boldsymbol{P}_r$, then taking $\mathfrak{C} = \mathfrak{B}'$ and $h_t = i_t'$ we infer from (e_r) that $(2) \leqq (1)$.

Let $\mathfrak{F}_\mathfrak{m}$ and $\varDelta^*$ have the same meaning as in theorem 38.9. Let $I_\mathfrak{r}$ be the class of all $\mathfrak{m}$-ideals $\varDelta$ in $\mathfrak{F}_\mathfrak{m}$ such that

$(\mathfrak{r}_1)$ $$\varDelta^* \subset \varDelta ,$$

$(\mathfrak{r}_2)$ no set $\bigcap_{t \in T'} g_t^*(A_t)$, where $\varLambda \neq A_t \in \mathfrak{A}_t,\ T' \subset T,\ \overline{\overline{T'}} \leq \mathfrak{n}$, belongs to $\varDelta$.

For any $\varDelta$ in $I_\mathfrak{r}$, let

$$i_t^\varDelta(A) = [g_t^*(A)]_\varDelta \quad \text{for} \quad A \in \mathfrak{A}_t .$$

38.11. *For any $\varDelta$ in $I_\mathfrak{r}$,*

$$\{\{i_t^\varDelta\}_{t \in T},\ \mathfrak{F}_\mathfrak{m}/\varDelta\}$$

is an element in $P_\mathfrak{r} \cap P_\mathfrak{n}$. Conversely, every element in $P_\mathfrak{r} \cap P_\mathfrak{n}$ is isomorphic to an element of the above form for some $\varDelta$ in $I_\mathfrak{r}$.

For any $\varDelta'$, $\varDelta''$ in $I_\mathfrak{r}$,

$$\{\{i_t^{\varDelta'}\}_{t \in T},\ \mathfrak{F}_\mathfrak{m}/\varDelta'\} \leq \{\{i_t^{\varDelta''}\}_{t \in T},\ \mathfrak{F}_\mathfrak{m}/\varDelta''\} \quad \text{if and only if} \quad \varDelta'' \subset \varDelta'.$$

The proof of 38.11 is similar to that of 36.8, 36.11 and 36.12.

Theorem 38.11 yields a topological representation for any representable $(\mathfrak{m}, \mathfrak{n})$-product. Identifying isomorphic elements, we also infer from the last part of 38.11 that the partially ordered set $P_\mathfrak{r} \cap P_\mathfrak{n}$ of all representable $(\mathfrak{m}, \mathfrak{n})$-products is isomorphic (in the sense of the theory of partially ordered sets) to the set $I_\mathfrak{r}$ partially ordered by the converse of the set-theoretical inclusion.

It is not known whether minimal $(\mathfrak{m}, \mathfrak{n})$-products of $\mathfrak{m}$-representable algebras $\mathfrak{A}_t$ are in $P_\mathfrak{r}$.

If (1) is an $(\mathfrak{m}, 0)$-product of $\mathfrak{m}$-algebras $\mathfrak{A}_t$ and $\mathfrak{B}$ is isomorphic to an $\mathfrak{m}$-field of sets, then every $\mathfrak{A}_t$ is isomorphic to an $\mathfrak{m}$-field of sets, by (b). The converse statement is also true; if each of the algebras $\mathfrak{A}_t$ is isomorphic to an $\mathfrak{m}$-field of sets, then there is an $\mathfrak{m}$-product (1) such that $\mathfrak{B}$ is isomorphic to an $\mathfrak{m}$-field of sets.

It suffices to prove this statement in the case where, for every $t \in T$, $\mathfrak{A}_t$ is an $\mathfrak{m}$-field of subsets of a space X_t. Let X be the Cartesian product all the spaces X_t. For every $A \in \mathfrak{A}_t$, let

$i_t(A) = A^* =$ the set of all $x \in X$ whose t^{th} coordinate is in A, and let $\mathfrak{F}$ be the field $\mathfrak{m}$-product of all $\mathfrak{A}_t$, i.e. the $\mathfrak{m}$-field (of subsets of X) $\mathfrak{m}$-generated by the union of all $i_t(\mathfrak{A}_t),\ t \in T$. Then

(25) $$\{\{i_t\}_{t \in T},\ \mathfrak{F}\}$$

is an $\mathfrak{m}$-product of $\{\mathfrak{A}_t\}_{t \in T}$. We shall call (25) the *field $\mathfrak{m}$-product* of $\{\mathfrak{A}_t\}_{t \in T}$, according to § 37, p. 173.

38.12. *If all the algebras $\mathfrak{A}_t$ are $\mathfrak{m}$-fields of sets, (1) is an $\mathfrak{m}$-product of $\{\mathfrak{A}_t\}_{t \in T}$ and $\mathfrak{B}$ is isomorphic to an $\mathfrak{m}$-field of sets, then (1) is a smallest* *element in P_m and, consequently, a minimal m-product, and* (1) *is isomorphic to* (25).

The inequality (1) $\leq$ (2) for every m-product (2) follows directly from theorem 34.3. Thus (1) is a smallest element in P_m and consequently a minimal element in P_m. By the same argument, (25) is the smallest element in P_m and a minimal element in P_m. Since all minimal elements in P_m are isomorphic, (1) is isomorphic to (24).

Thus, after identification of isomorphic m-products, we can say that if all $\mathfrak{A}_t$ are m-fields of sets, then there exists exactly one m-product (1) such that $\mathfrak{B}$ is an m-field, viz. the field m-product (25).

If an $(m, 0)$-product (1) of m-algebras $\mathfrak{A}_t$ is m-distributive, then every algebra $\mathfrak{A}_t$ is m-distributive by (b). The converse statement is also true: if all the $\mathfrak{A}_t$ are m-distributive, then an m-product of $\{\mathfrak{A}_t\}_{t \in T}$ is m-distributive. Namely,

38.13. *Let $\mathfrak{A}_t$ be an m-distributive m-algebra for every $t \in T$, and let* (1) *be an m-product of $\{\mathfrak{A}_t\}_{t \in T}$. The m-algebra $\mathfrak{B}$ is m-distributive if and only if* (1) *is a minimal m-product. Then* (1) *is a smallest element in P_m.*[1]

Suppose (1) is a minimal m-product. To prove that $\mathfrak{B}$ is m-distributive, we have to show (see 24.6) that every m-subalgebra of $\mathfrak{B}$ m-generated by at most m elements is isomorphic to an m-field of sets. It suffices to show that every m-subalgebra $\mathfrak{B}_0$ m-generated by the union of m-subalgebras $\mathfrak{B}_t \subset i_t(\mathfrak{A}_t)$ where $t \in T'$ ($T' \subset T$) is atomic if $\overline{\overline{T'}} \leq m$ and every $\mathfrak{B}_t$ is m-generated by at most m elements. In fact, every subalgebra of $\mathfrak{B}$ m-generated by at most m elements is an m-subalgebra of an m-subalgebra $\mathfrak{B}_0$ of the above form.

Since $\mathfrak{A}_t$ is m-distributive, so is $\mathfrak{B}_t$. Let $\{A_{t,s}\}_{s \in S_t}$ be an indexed set of all atoms of $\mathfrak{B}_t$,

(26) $$A_{t,s} \cap A_{t,s'} = \wedge \quad \text{for} \quad s \neq s'.$$

By 24.6 (i), (ii),

$$\bigcup_{s \in S_t}^{i(\mathfrak{A}_t)} A_{t,s} = V.$$

By the last part of 38.7, $i(\mathfrak{A}_t)$ is a regular subalgebra of $\mathfrak{B}$. Thus

(27) $$\bigcup_{s \in S_t}^{\mathfrak{B}} A_{t,s} = V.$$

Let S be the set of all functions f defined on T' such that $f(t) \in S_t$ for every $t \in T'$. For every $f \in S$, let

$$A_f = \bigcap_{t \in T'}^{\mathfrak{B}} A_{t,f(t)} \in \mathfrak{B}_0.$$

---

[1] Theorem 38.13 was proved by CHRISTENSEN and PIERCE [1]. The proof given above is due to SIKORSKI and TRACZYK [2].

It follows from (26) that

(28) $$A_f \cap A_{f'} = \wedge \quad \text{for} \quad f \neq f' .$$

We shall prove that

(29) $$\bigcup\nolimits_{f \in S}^{\mathfrak{B}} A_f = V .$$

Since the class of all elements

(29') $A = \bigcap\nolimits_{t \in T''}^{\mathfrak{B}} A_t$ where $\wedge \neq A_t \in i(\mathfrak{A}_t)$ for $t \in T''$, $T'' \subset T$, $\overline{\overline{T}}'' \leq \mathfrak{m}$

is dense in $\mathfrak{B}$ (see 38.6), it suffices to prove that for any element A of the form (29') there exists an $f \in S$ such that $A \cap A_f \neq \wedge$. By (27), for any $t \in T' \cap T''$ there is an $s_t \in S_t$ such that $A_t \cap A_{t, s_t} \neq \wedge$. Let f be any mapping in S such that $f(t) = s_t$ for $t \in T' \cap T''$. We have

$$A \cap A_f = \bigcap\nolimits_{t \in T' \cap T''}^{\mathfrak{B}} (A_t \cap A_{t, s_t}) \cap \bigcap\nolimits_{t \in T'' - T'}^{\mathfrak{B}} A_t \cap \bigcap\nolimits_{t \in T' - T''}^{\mathfrak{B}} A_{t, f(t)} \neq \wedge$$

since the subalgebras $i_t(\mathfrak{A}_t)$ are $\mathfrak{m}$-independent.

It follows from (28) and (29) that the class of all elements $A \in \mathfrak{B}$ such that both A and $-A$ are joins (in $\mathfrak{B}$) of some elements A_f is an $\mathfrak{m}$-subalgebra of $\mathfrak{B}$. Multiplying both sides of (29) by any element $A \in \mathfrak{B}_t$, we infer that this $\mathfrak{m}$-subalgebra contains all the subalgebras $\mathfrak{B}_t$, $t \in T'$. Thus it contains $\mathfrak{B}_0$. This proves that every non-zero element in $\mathfrak{B}_0$ contains at least an A_f as its subelement. Thus $\mathfrak{B}_0$ is atomic and the elements A_f $(f \in S)$ are all the atoms of $\mathfrak{B}_0$.

Suppose now that (1) is an $\mathfrak{m}$-product of $\{\mathfrak{A}_t\}_{t \in T}$ and $\mathfrak{B}$ is $\mathfrak{m}$-distributive. It follows directly from theorem 34.3 that (1) $\leq$ (2) for every $\mathfrak{m}$-product (2) of $\{\mathfrak{A}_t\}_{t \in T}$. Thus (1) is a smallest element in $P_\mathfrak{m}$ and, consequently, a minimal $\mathfrak{m}$-product of $\{\mathfrak{A}_t\}_{t \in T}$.

One can prove, by means of the characterization 38.3, that the operation of forming a maximal $\mathfrak{m}$-product is associative, i.e. if $\{\{i_{s, t}\}_{t \in T_s}, \mathfrak{B}_s\}$ is a maximal $\mathfrak{m}$-product of $\{\mathfrak{A}_{t, s}\}_{t \in T_s}$ and $\{\{i_s\}_{s \in S}, \mathfrak{B}\}$ is a maximal $\mathfrak{m}$-product of $\{\mathfrak{B}_s\}_{s \in S}$, then $\{\{i_s i_{s, t}\}_{t \in T_s, s \in S}, \mathfrak{B}\}$ is a maximal $\mathfrak{m}$-product of $\{\mathfrak{A}_{t, s}\}_{t \in T_s, s \in S}$. An analogous remark is true for maximal representable $\mathfrak{m}$-products of $\mathfrak{m}$-representable algebras. Similarly one can prove by means of the characterizations 38.4 and 38.6 that the operation of forming a minimal $(\mathfrak{m}, 0)$-product or a minimal $(\mathfrak{m}, \mathfrak{n})$-product is associative[1].

Now we shall examine the case where $\mathfrak{m} = \sigma$. Then $P_r = P$ since all Boolean algebras are σ-representable. Theorem 38.11 yields a topological representation of all σ-products. Since maximal representable σ-products coincide with maximal σ-products, theorem 38.9 yields a topological representation of maximal σ-products.

---

[1] For detailed proofs of these statements see SIKORSKI [13].

The following theorem is a Boolean analogue of the theorem on forming products of σ-measures on σ-fields of sets [see § 37 (2)].

38.14. *For every $t \in T$, let m_t be a σ-measure defined on $\mathfrak{A}_t$ such that $m_t(V) = 1$ and let (1) be a maximal σ-product of $\{\mathfrak{A}_t\}_{t \in T}$. Then there exists a σ-measure m defined on $\mathfrak{B}$ such that*

(30) $$m\left(\bigcap_{1 \leq n \leq k} i_{t_n}(A_n)\right) = \prod_{1 \leq n \leq k} m_{t_n}(A_n)$$

for every sequence $A_n \in \mathfrak{A}_{t_n}$, $t_l \neq t_j$ for $l \neq j$.

Since maximal σ-products coincide with representable maximal σ-products, it suffices to prove 38.14 in the case where (1) is the σ-product (23). We assume below the notation from theorem 38.9 (where $\mathfrak{m} = \sigma$) and its proof.

The formula

$$\mu_{0,t}(g_t(A)) = m_t(A) \quad \text{for} \quad A \in \mathfrak{B}_t$$

defines a measure $\mu_{0,t}$ on the field of all open-closed subsets of the bicompact space X_t. By Carathéodory's outer measure method the measure $\mu_{0,t}$ can be extended to a σ-measure μ_t on the σ-field $\mathfrak{F}_t$.[1] If $B \in \mathfrak{F}_t$ is a nowhere dense σ-closed subset of X_t, i.e. if $B = \bigcap_{1 \leq n < \infty} g_t(A_n)$ where $\bigcap_{1 \leq n < \infty}^{\mathfrak{A}_t} A_n = \Lambda$ ($A_{n+1} \subset A_n$ for $n = 1, 2, \ldots$), we have

$$0 \leq \mu_t(B) = \lim_{n \to \infty} \mu_t(g_t(A_n)) = \lim_{n \to \infty} m_t(A_n) = m_t(\Lambda) = 0$$

since μ_t and m_t are σ-measures[2]. This proves that

$$\mu_t(B) = 0$$

for every set B nowhere dense and σ-closed in X_t.

Let μ be the σ-product of all the measures μ_t. By definition, μ is a σ-measure on $\mathfrak{F}_\sigma$,

$$\mu\left(\bigcap_{1 \leq n \leq k} g_t^*(A_n)\right) = \prod_{1 \leq n \leq k} m_{t_n}(A_n)$$

for every sequence $A_n \in \mathfrak{B}_{t_n}$ ($t_l \neq t_j$ for $l \neq j$), and

$$\mu(B^*) = 0$$

for every set $B \in \mathfrak{F}_t$ nowhere dense and σ-closed in X_t. Consequently

$$\mu(B) = 0$$

for every set $B \in \Delta^*$. Hence it follows that the formula

$$m([B]_{\Delta^*}) = \mu(B)$$

defines a σ-measure m on $\mathfrak{F}_\sigma/\Delta^*$ and (30) holds.

It is easy to see that there exists only one σ-measure m on $\mathfrak{A}$ such that (30) holds. We call m the *Boolean σ-product* of $\{m_t\}_{t \in T}$.

---

[1] For more detailed explanation, see § 42, p. 203.
[2] See e.g. HALMOS [1], p. 38.

Examples. B) Let $\mathfrak{F}_1$ and $\mathfrak{F}_2$ be the σ-fields of sets defined in § 37 A). It follows from § 37 A) that the field σ-product (25) of $\mathfrak{F}_1$ and $\mathfrak{F}_2$ does not have the extension property (e) in 38.3 [by 37.3 (iii), it has this property only in the case where the σ-homomorphisms in question map into a σ-field of sets], therefore it is not the maximal σ-product of $\mathfrak{F}_1$ and $\mathfrak{F}_2$. By 38.11 the field σ-product of $\mathfrak{F}_1$ and $\mathfrak{F}_2$ is the minimal σ-product of $\mathfrak{F}_1$ and $\mathfrak{F}_2$. Hence it follows that the maximal and minimal σ-products of $\mathfrak{F}_1$ and $\mathfrak{F}_2$ are not isomorphic.

C) Theorem 38.14 is not true, in general, if the maximal σ-product is replaced by the minimal σ-product (however this modification of 38.14 is true if all the $\mathfrak{A}_t$ are σ-fields of sets — see 38.12).

In fact, let $\mathfrak{F}'$ be the σ-field of all Borel subsets of the unit interval U, let μ' be the Lebesgue measure on $\mathfrak{F}'$ and let Δ' be the σ-ideal of all sets (in $\mathfrak{F}'$) of μ' measure zero. Let $T = (1, 2)$, $\mathfrak{A}_1 = \mathfrak{A}_2 = \mathfrak{F}'/\Delta'$, let $\{\{i_t\}_{t \in T}, \mathfrak{B}_0\}$ be a Boolean product of $\mathfrak{A}_1$ and $\mathfrak{A}_2$ (see § 13) and let $\mathfrak{B}$ be a σ-completion of $\mathfrak{B}_0$. By definition, $\{\{i_t\}_{t \in T}, \mathfrak{B}\}$ is a minimal σ-product of $\mathfrak{A}_1$ and $\mathfrak{A}_2$. The formulas

$$m_t([A]) = \mu'(A) \quad \text{for} \quad A \in \mathfrak{F}' \qquad (t = 1, 2)$$

define σ-measures m_t on $\mathfrak{A}_t$. By the remark at the end of § 13, there exists a product m_0 of the measures m_1, m_2 on $\mathfrak{B}_0$. However there exists no σ-product m of m_1, m_2 on $\mathfrak{B}$. For suppose that such a σ-product m exists. Then m is an extension of m_0. Since m is a σ-measure, we have[1] $\lim_{n \to \infty} m(A_n)$ = 0 for every sequence $A_n \in \mathfrak{B}$ such that $A_{n+1} \subset A_n$ $(n = 1, 2, \ldots)$ and

$$\bigcap_{1 \leq n < \infty}^{\mathfrak{B}} A_n = \Lambda .$$

On the other hand, there exists a sequence $B_n \in \mathfrak{B}_0$ such that $B_{n+1} \subset B_n$ $(n = 1, 2, \ldots)$, $\bigcap_{1 \leq n < \infty}^{\mathfrak{B}} B_n = \bigcap_{1 \leq n < \infty}^{\mathfrak{B}_0} B_n = \Lambda$ but

$$\lim_{n \to \infty} m_0(B_n) = \lim_{n \to \infty} m(B_n) > 0 .$$

Contradiction.

To define this sequence $\{B_n\}$ we may suppose that $\mathfrak{B}_0$ is of the form $\mathfrak{F}/\Delta$ where $\mathfrak{F}$ is the least field (of Borel subsets of the unit square $U \times U$) containing all the sets

$$A \times U, \ U \times A \quad \text{where} \quad A \in \mathfrak{F}' ,$$

and Δ is the ideal of all sets (in $\mathfrak{F}$) of Lebesgue plane measure zero. Let this measure be denoted by μ. In fact, the subalgebras of elements of the form $[A \times U]_\Delta$ and $[U \times A]_\Delta$ respectively are independent, isomorphic to $\mathfrak{B}_1$ and $\mathfrak{B}_2$, and generate $\mathfrak{F}/\Delta$. Therefore $\mathfrak{F}/\Delta$ is isomorphic to $\mathfrak{B}_0$ (see § 13).

The measure m_0 on $\mathfrak{B}_0$ is defined by the formula

(31) $$m_0([A]) = \mu(A) \quad \text{for} \quad A \in \mathfrak{F} .$$

---

[1] See e.g. HALMOS [1], p. 38.

Now let N be a nowhere dense closed subset of U with $\mu'(N) > 0$, and let S be the set of all points $(x_1, x_2) \in U \times U$ such that $|x_1 - x_2| \in N$. The hypothesis $\mu'(N) > 0$ implies that

(32) $$\mu(S) > 0 .$$

Now we shall prove that

(33)   if $A_1' \times A_2' \in \mathfrak{F}$ and $\mu(A_1' \times A_2') > 0$, then $A_1' \times A_2' - S \neq \wedge$.

For suppose that $A_1' \times A_2' \subset S$ for some sets A_1', $A_2' \in \mathfrak{F}$ of positive μ' measure. By a known theorem[1], the set N_0 of numbers $|x_1 - x_2|$ where $x_t \in A_t'$ $(t = 1, 2)$ would then contain an interval. This is impossible because N_0 is a subset of the nowhere dense set N.

Property (33) can be reinforced as follows:

(34)   if $A_1 \times A_2 \in \mathfrak{F}$ and $\mu(A_1 \times A_2) > 0$, then $\mu(A_1 \times A_2 - S) > 0$.

In fact, A_t contains a subset A_t' of positive μ' measure such that $A_t' \cap G \neq \wedge$ implies $\mu'(A_t' \cap G) > 0$ for every set G open in U $(t = 1, 2)$. Consequently, if H is open in $U \times U$, then $H \cap (A_1' \times A_2') \neq \wedge$ implies $\mu(H \cap (A_1' \times A_2')) > 0$. The set $H = U \times U - S$ is open in $U \times U$ and $H \cap (A_1' \times A_2') \neq \wedge$ on account of (33). Hence

$$0 < \mu(A_1' \times A_2' - S) \leq \mu(A_1 \times A_2 - S)$$

which proves (34).

Since S is closed, there exists a decreasing sequence $S_n \in \mathfrak{F}$ such that S is the intersection of all S_n. It follows from (31), (32) and (34) that the sequence $B_n = [S_n]_A$ has the required properties[2].

D) Let (1) be a Boolean σ-product of Boolean algebras $\{\mathfrak{A}_t\}_{t \in T}$, and let $\mathfrak{B}_0$ be the least subalgebra containing all the $i(\mathfrak{A}_t)$ $(t \in T)$. $\mathfrak{B}_0$ is a Boolean product of $\{\mathfrak{B}_t\}_{t \in T}$. If (1) is not a minimal σ-product of $\{\mathfrak{A}_t\}_{t \in T}$, then the subalgebra $\mathfrak{B}_0$ is not, in general, a σ-regular subalgebra of $\mathfrak{A}$.

In fact, let $\mathfrak{A}_1$ and $\mathfrak{A}_2$ be Boolean algebras defined in example C), $T = (1, 2)$ and let $\{\{i_t\}_{t \in T}, \mathfrak{B}\}$ be a maximal σ-product of $\mathfrak{A}_1, \mathfrak{A}_2$. Then the subalgebra $\mathfrak{B}_0$ generated by $i_1(\mathfrak{A}_1)$ and $i_2(\mathfrak{A}_2)$ is not a σ-regular subalgebra of $\mathfrak{B}$. For suppose that $\mathfrak{B}_0$ is σ-regular. The Boolean σ-product m of the σ-measures m_1, m_2 defined in example C) is an extension of the product m_0 of m_1, m_2. Let $B_n \in \mathfrak{B}_0$ be the sequence defined in example C). By definition, $\bigcap_{1 \leq n < \infty}^{\mathfrak{B}_0} B_n = \wedge$, $B_{n+1} \subset B_n$, and $\lim_{n \to \infty} m_0(B_n) > 0$. On the other hand, $\lim_{n \to \infty} m_0(B_n) = \lim_{n \to \infty} m(B_n) = 0$ since $\bigcap_{1 \leq n < \infty}^{\mathfrak{B}} B_n = \wedge$ and m is a σ-measure on $\mathfrak{B}$. Contradiction.

---

[1] STEINHAUS [1].

[2] Example C) was given by SIKORSKI [15]. A part of the proof is a slight modification of an argument in HELSON [2].

Appendix
§ 39. Relation to other algebras

Boolean algebras are a special case of *universal algebras* (with a finite number of finite operations). Many notions introduced in Chapter I belong to the general theory of *universal* algebras. We quote here such notions as homomorphism, isomorphism, subalgebra, generator, free algebra etc. Boolean m-algebras investigated in Chapter II can also be interpreted as a special case of *universal* algebras but with some infinite operations, viz. the complementation $-A$, the infinite join $\bigcup_{t \in T} A_t$ and the infinite meet $\bigcap_{t \in T} A_t$ where T is a fixed set of cardinality m. Thus such notions as m-subalgebra, m-homomorphism (between two m-algebras), m-generator, free Boolean m-algebra etc. also belong to the general theory of universal algebras. Many remarks in Chapter I and II and also some theorems belong to the theory of universal algebras. We mention here, for instance, theorems 12.1 and 23.3 which are particular cases of a general theorem on universal algebras. The notions of field product and m-product, and of Boolean product and maximal m-product, and maximal representable m-product are particular cases of a general notion of product of universal algebras[1].

Also some other general algebraic notions not examined in Chapters I and II can be applied to the theory of Boolean algebras. As an example we quote here the notion of inverse and direct systems[2] and the notion of projectivity and injectivity[3] from the general theory of categories.

It follows immediately from the set of axioms assumed in § 1 that Boolean algebras are distributive *lattices*. More exactly, the notion of Boolean algebra coincides with the notion of distributive complemented lattice[4]. Some theorems proved here for Boolean algebras can be generalized to distributive lattices. In particular, the fundamental representation theorem 8.2 can be considered as a particular case of a representation theorem for distributive lattices[5].

As we have observed in § 17, the theory of Boolean algebras coincides (if only finite joins and meets are taken into consideration) with a part of the theory of algebraic rings. The term "ideal" is taken from this

---

[1] SIKORSKI [19]. See also CHRISTENSEN and PIERCE [1]. From the point of view of the general theory of categories Boolean products and maximal m-products are free joins in the category of Boolean algebras or of Boolean m-algebras (see e.g. KUROŠ, LIVŠIC and ŠULGEIFER [1], SEMADENI [4, 5]).

[2] See ENGELKING and KURATOWSKI [1], HAIMO [1], WALLACE [1].

[3] For an investigation of projective and injective Boolean algebras, see HALMOS [8, 10].

[4] For other characterizations of Boolean algebras among lattices, see BALACHANDRAN [1, 2, 3], BIRKHOFF and WARD [1], DILWORTH [2], MICHIURA [1], NACHBIN [1].

[5] Proved by BIRKHOFF [1] and STONE [9]. See also RIEGER [2].

theory. In fact, it is easy to verify that a set $\varDelta$ of elements of a Boolean algebra $\mathfrak{A}$ is an ideal in the sense defined in § 3 if and only if it is an ideal of the Boolean ring $\mathfrak{A}$ in the sense assumed in the general theory of algebraic rings. The forming of quotient algebras described in § 10 is a particular case of forming algebraic quotient rings. It is also a particular case of forming quotient algebras modulo a congruence relation in the general theory of universal algebras.

It is worth noticing that the method of two-valued homomorphisms used in the proof of the fundamental representation theorem[1] 8.2 is a particular case of a general method of investigation of more complicated algebras by means of homomorphisms into certain special simple algebras with the same operations. An example of the use of this method is given by the notion of the dual space to a given Banach space. This space is the set of all continuous linear functionals, i.e. homomorphisms (of the theory of Banach spaces) into the simplest Banach space, viz. the field of scalars (real or complex).

Another example of this method is given by the theory of normed rings in functional analysis. Here the dual space $\mathfrak{A}^*$ of a given normed ring $\mathfrak{A}$ is the set of all continuous multiplicative linear functionals, i.e. homomorphisms χ into the simplest normed ring, viz. the ring of scalars. If $A \in \mathfrak{A}$ is fixed, then the formula

$$\varPhi_A(\chi) = \chi(A) \qquad\qquad (\chi \in \mathfrak{A}^*)$$

defines a continuous mapping from $\mathfrak{A}^*$ to the scalars. Under some additional conditions, the mapping h defined by the formula

(1) $h(A) = \varPhi_A$

is one-to-one. In this case $\mathfrak{A}$ can be represented as a ring of continuous functions on the dual space $\mathfrak{A}^*$.

The last example has an analogue in the theory of Boolean algebras. Every Boolean algebra $\mathfrak{A}$ is not only an algebraic ring but also a linear ring over the two-element algebraic field $\mathfrak{R}$ (see § 17, p. 52). According to the remark on p. 18 and the definition in the proof of 8.2, the Stone space X of a given Boolean algebra $\mathfrak{A}$ can be interpreted as the set of all two-valued homomorphisms of $\mathfrak{A}$ into $\mathfrak{R}$, i.e. the set of all multiplicative linear functionals with values in the field $\mathfrak{R}$ of scalars. The class of all continuous mappings from the Stone space X into the two-element Hausdorff space $\mathfrak{R}$ can be identified with the class of all open-closed subsets of X since every mapping of this type is uniquely determined by the (open-closed) set of points where it assumes the unit as its value. Thus the isomorphism h in the proof of the fundamental representation

---

[1] In the proof of 8.2 we used the notion of maximal filter instead of two-valued homomorphisms but these notions are equivalent (see p. 18).

theorem 8.2 can be considered as a particular case of (1). Consequently the Stone space of $\mathfrak{A}$ is sometimes called *dual space* of $\mathfrak{A}$[1].

The analogy between the representation theory of normed rings and Boolean algebras is so deep that it is possible to develop a general theory of maximal ideals which gives, as particular cases, the two representation theorems[2].

The notion of Boolean σ-algebras is a particular case of the notion of *cardinal algebras*[3]. Cardinal algebras are abstract algebras with an infinite operation which is a common generalization of the Boolean join of an enumerable sequence of elements and of the sum of an enumerable sequence of cardinal numbers. The axioms characterizing this operation are such that from them one may deduce a large part of the additive arithmetic of cardinal numbers. However they are also satisfied by Boolean σ-algebras, types of isomorphisms for Boolean σ-algebras, the set of non-negative integers or reals (with $+\infty$), the set of all non-negative functions on a set, and other systems. Theorems on cardinal algebras are also theorems on Boolean algebras but they were not quoted in this book. It is worth noticing that some of these theorems on Boolean algebras are generalizations of some well-known fundamental theorems on cardinals. For instance, the fundamental Cantor-Bernstein theorem on cardinals

$$\text{if } \; \mathfrak{n}_1 \leqq \mathfrak{n}_2 \; \text{ and } \; \mathfrak{n}_2 \leqq \mathfrak{n}_1 \; \text{ then } \; \mathfrak{n}_1 = \mathfrak{n}_2$$

is a particular case of the following theorem on Boolean σ-algebras $\mathfrak{A}$, $\mathfrak{B}$ which is another formulation of theorem 22.4:

if $\mathfrak{B}$ is isomorphic to $\mathfrak{A} \mid A$ $(A \in \mathfrak{A})$ and $\mathfrak{A}$ is isomorphic to $\mathfrak{B} \mid B$ $(B \in \mathfrak{B})$, then $\mathfrak{A}$ and $\mathfrak{B}$ are isomorphic[4].

To obtain the Cantor-Bernstein theorem it suffices to assume that $\mathfrak{A}$ and $\mathfrak{B}$ are fields of all subsets of some sets X, Y with cardinality $\mathfrak{n}_1$, $\mathfrak{n}_2$ respectively.

By the same method we infer that the Bernstein theorem

$$\text{if } \; 2\mathfrak{n}_1 = 2\mathfrak{n}_2, \; \text{ then } \; \mathfrak{n}_1 = \mathfrak{n}_2$$

is a particular case of the following theorem valid for every Boolean σ-algebra $\mathfrak{A}$:

if A, $B \in \mathfrak{A}$, $\mathfrak{A} \mid A$ is isomorphic to $\mathfrak{A} \mid -A$, and $\mathfrak{A} \mid B$ is isomorphic to $\mathfrak{A} \mid -B$, then $\mathfrak{A} \mid A$ and $\mathfrak{A} \mid B$ are isomorphic[5].

---

[1] See HALMOS [4], [8].

[2] SŁOWIKOWSKI and ZAWADOWSKI [1].

[3] The notion and the theory of cardinal algebras are due to TARSKI [8]. See also JÓNSSON and TARSKI [2].

[4] SIKORSKI [1] and TARSKI [8]. See also BRUNS and SCHMIDT [2].

[5] TARSKI [8]. The hypothesis that $\mathfrak{A}$ is σ-complete is essential. See HANF [1].

Many papers are devoted to the study of Boolean algebras with some additional operations and of various generalizations of Boolean algebras and rings[1].

§ 40. Applications to mathematical logic. Classical calculi

The most important applications of the theory of Boolean algebras are those to mathematical logic. That is not surprising because the notion of a Boolean algebra was created as the result of Boole's investigation of the algebraic structure of the "laws of thought"[2]. In the first part of its development, the theory of Boolean algebras was also called the *algebra of logic*.

Consider first the case of the (two-valued) propositional calculus. Denote the propositional connectives "or", "and", "not", "if ... , then ..." by $\cup$, $\cap$, $-$, $\rightarrow$ respectively.

The set of all formulas of the propositional calculus becomes a Boolean algebra after identification of equivalent formulas (see an analogous remark in § 1 D)). We recall that formulas α, β are said to be equivalent if both the implications $\alpha \rightarrow \beta$ and $\beta \rightarrow \alpha$ are derivable. The Boolean algebra $\mathfrak{A}$ so obtained will be called the *Lindenbaum-Tarski algebra* of the propositional calculus in question. Let $|\alpha|$ denote the element of $\mathfrak{A}$ determined by a formula α. We have the fundamental identities.

$$
\begin{aligned}
|\alpha| \cup |\beta| &= |\alpha \cup \beta| \\
|\alpha| \cap |\beta| &= |\alpha \cap \beta| \\
-|\alpha| &= |-\alpha| \\
|\alpha| \rightarrow |\beta| &= |\alpha \rightarrow \beta| .
\end{aligned}
$$

(1)

The first three identities are definitions of the Boolean operations in the Lindenbaum-Tarski algebra $\mathfrak{A}$. The Boolean operation $\rightarrow$ on the left-hand side of the fourth identity is that defined on p. 10. We see that the Boolean operations $\cup, \cap, -, \rightarrow$ are Boolean analogues of the propositional connectives $\cup, \cap, -, \rightarrow$ respectively.

The fundamental completeness theorem on the propositional calculus states that the class of all derivable formulas (i.e. formulas obtained

---

[1] BIAŁYNICKI-BIRULA [1], BIAŁYNICKI-BIRULA and RASIOWA [1], CHANG and HORN [1], CHOUDHURY [1], COPELAND [1], COPELAND and HARARY [1, 2], CRAWLEY [1], C. DAVIS [1], EPSTEIN [1], EVERETT and ULAM [1], FELL and TARSKI [1], FOSTER [1], HARARY [1], JÓNSSON and TARSKI [1], L'ABLE [1], McCOY [1], McCOY and MONTGOMERY [1], McKINSEY [5], MOISIL [1], RIBEIRO [1, 2], SŁOWIKOWSKI and ZAWADOWSKI [1], SUSSMAN [1], TRACZYK [3, 4, 6], WOOYENAKA [1]. See also algebras mentioned in §§ 40 and 41.

[2] BOOLE [1, 2].

from the assumed set of axioms of the propositional calculus by means
of the rules of inference) coincides with the class of all tautologies, i.e.
intuitively true formulas. Using the notion of the Lindenbaum-Tarski
algebra, this theorem can be easily translated into the language of the
theory of Boolean algebras. Its equivalent Boolean formulation is that
every non-zero element of the Lindenbaum-Tarski algebra $\mathfrak{A}$ belongs to
a maximal filter. Therefore the completeness theorem can be easily
obtained from the theorem on the existence of maximal filters in Boolean
algebras (see 6.1) or from the fundamental representation theorem
stating that every Boolean algebra is isomorphic to a field of sets (see 8.2).
The role played by the set of axioms of the propositional calculus in
that proof of the completeness theorem reduces to showing that the
Lindenbaum-Tarski algebra $\mathfrak{A}$ is a Boolean algebra.

Conversely, the fundamental representation theorem for Boolean
algebras can also be deduced directly from the completeness theorem
formulated in a little stronger form. Thus both theorems express the
same mathematical content formulated in different languages[1].

Consider now the case of the (two-valued) lower predicate calculus.
Just as previously, the set of all formulas becomes a Boolean algebra $\mathfrak{A}$
(called the *Lindenbaum-Tarski algebra* of the predicate calculus) after
identification of equivalent formulas. We also have identities (1), the
first three being definitions of the Boolean operations in $\mathfrak{A}$. Assume that
$\bigcup_\tau$ and $\bigcap_\tau$ denote the quantifiers "there exists a τ such that . . ." and
"for every τ . . .". Then, for every formula α, we have also the identities
(see § 18 F))

(2)
$$|\bigcup_\tau \alpha(\tau)| = \bigcup{}^{\mathfrak{A}}_{t \in T} |\alpha(t)| \, ,$$
$$|\bigcap_\tau \alpha(\tau)| = \bigcap{}^{\mathfrak{A}}_{t \in T} |\alpha(t)| \, ,$$

where T denotes the set of all terms of the assumed formalized language
of the predicate calculus in question.

A simple analysis shows that the analogous completeness theorem for
the predicate calculus coincides with the theorem stating that there
exists an isomorphism h of the Lindenbaum-Tarski algebra $\mathfrak{A}$ onto a
field of sets such that h transforms all the joins and meets (2) onto the
corresponding set-theoretical unions and intersections respectively. Thus
the completeness theorem for the predicate calculus can be deduced
easily (under the hypothesis that the set of all signs of the predicate
calculus is countable) from the representation theorem 24.10 or from the
theorem on the existence of maximal filters preserving a given countable
set of infinite joins and meets (the last theorem is, in fact, the basis of the
proof of 24.10).

---

[1] See Henkin [2] and Łoś [1, 3].

The representation theorem 24.10 also can be applied to prove easily the theorem on the existence of enumerable or finite semantic models for any consistent enumerable set of formulas[1].

Also other fundamental theorems on the predicate calculus and on formalized elementary theories can be obtained easily by Boolean methods by investigating certain appropriate Lindenbaum-Tarski algebras. We mention here the theorem on the existence of models for any (not necessarily enumerable) consistent set of formulas, the first and the second ε-theorems[2] and the Herbrand theorem[3]. The previously known proofs of the theorems mentioned above were complicated. Boolean methods permit one to obtain simple proofs for them. Boolean methods also enable one to explain better the mathematical content of metamathematical theorems and to discover new theorems[4]. By Boolean methods we understand here a systematic translation of logical problems into the language of Boolean algebras, and the investigation of Lindenbaum-Tarski algebras instead of sets of formulas. From this point of view, the examination of the predicate calculus coincides with the study of Boolean algebras with a distinguished set of infinite joins and meets corresponding to the logical quantifiers [see (2)]. The notion of such Boolean algebras lies between Boolean algebras examined in Chapter I and Boolean $\mathfrak{m}$-algebras examined in Chapter II. The Lindenbaum-Tarski algebras of predicate calculi play a special part in the investigation of this class of Boolean algebras because, from the point of view of the general theory of universal algebras, they are the free algebras in this class[5]. Observe that Lindenbaum-Tarski algebras of propositional calculi coincide with free Boolean algebras (see § 14) with a corresponding number of free generators (viz. the free generators in these Lindenbaum-Tarski algebras are elements of the form $|p|$ where p is any propositional variable).

Lindenbaum-Tarski algebras of predicate calculi and of formalized theories (see § 1 D)) are particular cases of *polyadic Boolean algebras*[6]

---

[1] This method of proof of the completeness theorem and the theorem on existence of semantic models for consistent enumerable sets of formulas is due to RASIOWA and SIKORSKI [1, 2]. See also BETH [1], HASENJAEGER [1], HENKIN [1, 3, 4], LOŚ [1,3], REICHBACH [1], RIEGER [4, 5, 7]. For a detailed study of connection between the existence of semantic models and the representation of Lindenbaum-Tarski algebras, see RASIOWA and SIKORSKI [7].

[2] RASIOWA [5, 6].

[3] LOŚ, MOSTOWSKI and RASIOWA [1, 2], SIKORSKI [23].

[4] See e.g. a topological characterization of open theories given by SIKORSKI [28, 29, 30].

[5] This remark is due to RIEGER [5]. See also RASIOWA and SIKORSKI [7].

[6] The notion and the theory of polyadic Boolean algebras are due to HALMOS [3, 4, 5, 6, 7, 9]. See also BASS [1], COPELAND [2], DAIGNEAULT and MONK [1], GALLER [1], LEBLANC [4, 5], VARSAVSKY [1], WRIGHT [1]. Another algebraization of predicate calculi (by means of *cylindric algebras*) is due to HENKIN and TARSKI [1, 2]. See also COPELAND [2], GALLER [1], KASNER [1], MONK [1].

whose theory has been developed in recent years. A polyadic algebra is, by definition, a Boolean algebra $\mathfrak{A}$ with an additional set of operations, each of which is, roughly speaking, an abstract formulation of the operation which, with every element $|\alpha|$ of a Lindenbaum-Tarski algebra, associates the element $|\bigcup_\tau \alpha|$, τ being a fixed individual variable. A set of endomorphisms (i.e. of homomorphisms of $\mathfrak{A}$ into $\mathfrak{A}$) is distinguished. These endomorphisms are abstract analogues of the operation of substitution in mathematical logic. More exactly, they are abstract analogues of the endomorphisms h in a Lindenbaum-Tarski algebra $\mathfrak{A}$ given by

$$h(|\alpha|) = |\alpha^*|$$

where α^* denotes the formula obtained from α by a simultaneous substitution of terms $t_1, t_2, \ldots$ for individual variables $\tau_1, \tau_2, \ldots$, the sequences $t_1, t_2, \ldots$ and $\tau_1, \tau_2, \ldots$ being fixed. The additional operations, analogous to logical quantifiers, and the set of endomorphisms corresponding to performance of substitutions are characterized by a suitable set of axioms.

An important example of a polyadic algebra (called a *functional polyadic algebra*) is given by the Boolean algebra $\mathfrak{A}$ of all mappings from X^V into a complete Boolean algebra $\mathfrak{B}$. The symbol X^V denotes here the set of all mappings $x = \{x_v\}$ of a non-empty set V into a non-empty set X. Every element $v' \in V$ determines an operation (corresponding to logical quantifier $\bigcup_\tau$) which with every $\beta \in \mathfrak{A}$ associates the element $\beta' \in \mathfrak{A}$ defined by the formula

$$\beta'(x) = \bigcup \beta(x')$$

where the join $\bigcup$ is extended over all points $x' = \{x'_v\} \in X^V$ such that $x'_v = x_v$ for all $v \neq v'$. The set of distinguished endomorphisms corresponding to substitutions is determined by a set of mappings φ of V into itself. Viz. the endomorphism corresponding to a fixed φ transforms any $\beta \in A$ into $\beta' \in A$ defined by the formula

$$\beta'(\{x_v\}) = \beta(\{x_{\varphi(v)}\}) .$$

Polyadic algebras are, roughly speaking, an algebraization of the lower predicate calculus without any reference to the notion of formula. The representation problem for polyadic algebras consists in establishing a connection between general polyadic algebras and functional polyadic algebras defined above. One of the fundamental representation theorems directly implies the completeness theorem for the lower predicate calculus and the theorem on existence of semantic models for consistent sets of formulas.

The notion of Boolean algebras and some Boolean methods are also helpful in the algebraization of other parts of mathematical logic. We mention here only *relation algebras*. Consider the set of all binary relations $R, R', \ldots$ between elements of a given space X. This set forms a Boolean algebra $\mathfrak{A}$ with respect to the logical operations "R' or R''" (join), "R' and R''" (meet), "not R" (complement). If R is a relation, then

$R^\cup$ will denote its *converse*, i.e. the relation defined by the condition: $xR^\cup y$ holds if and only if yRx holds. The symbol $R;R'$ will denote a new operation, called *relative product*, defined as follows: $R;R'$ is a binary relation in X such that $xR;R'y$ holds if and only if there exists a $z \in X$ such that xRz and $zR'y$ hold. Relation algebras are a generalization of the algebras $\mathfrak{A}$ described above. By definition, they are Boolean algebras with two additional operations "$\cup$" and ";" characterized by a set of axioms so that they are Boolean analogues of the corresponding operations on binary relations[1].

§ 41. Topology in Boolean algebras. Applications to non-classical logic

A *closure algebra* is, by definition, a Boolean algebra $\mathfrak{A}$ with an operation which, with every element $A \in \mathfrak{A}$, associates an element $\mathbf{C}A \in \mathfrak{A}$ called the *closure* of A, in such a way that the following axioms[2] are satisfied

$$\mathbf{C}(A \cup B) = \mathbf{C}A \cup \mathbf{C}B, \quad A \subset \mathbf{C}A$$
$$\mathbf{C}\mathbf{C}A = \mathbf{C}A, \quad \mathbf{C}\wedge = \wedge.$$

The notion of closure algebra is a generalization of topological spaces. In fact, if X is a topological space, then the field of all subsets of X is a closure algebra.

Closure algebras have been examined from a topological point of view by many writers[3]. Many topological notions can be extended to closure algebras. For instance, the interior $\mathbf{I}A$ of $A \in \mathfrak{A}$ is defined as the complement of the closure of its complement

$$\mathbf{I}A = -\mathbf{C}-A$$

as in set-theoretical topology. An element A is called closed (open) if $A = \mathbf{C}A$ (if $A = \mathbf{I}A$). It is called a boundary element if $\mathbf{I}A = \wedge$. It is nowhere dense if $\mathbf{I}\mathbf{C}A = \wedge$ etc. Many topological theorems are valid also for closure algebras.

Infinite set-theoretical operations play an essential role in topology. To extend this part of topology to closure algebras it is necessary to suppose σ-completeness. The part of topology which can be generalized to closure σ-algebras is very large[4]. Just as in set-theoretical topology, restrictions to special types of closure algebras are sometimes necessary.

---

[1] For the theory of relation algebras, see Bernays [1], Chin and Tarski [1], Jónsson and Tarski [1], Kamel [1], Keedy [1], Lyndon [1, 2], Moisil [2], Tarski [10, 13].

[2] Due to Kuratowski [1].

[3] C. Davis [1], Hofmann [1], Monteiro and Ribeiro [1], Nöbeling [1], Ridder [2, 3], Rieger [9], Rubin [1], Ruprecht [1], Sikorski [2, 3, 7, 16, 20, 24], Terasaka [1, 2], Zarickij [1]. The book of Nöbeling [2] contains a systematic exposition of the theory of closure algebras but ignores the names of authors of earlier publications on closure algebras.

[4] Sikorski [7].

For instance, in order to reproduce the theory of separable metric spaces it is necessary to restrict the investigation to the class of closure algebras satisfying the following axiom which is a combination of the known axioms of regularity, normality and separability:

(a) there exists a sequence $\{G_n\}$ of open elements (the open basis) such that every open element G is the join of all G_n such that $\mathbf{C}G_n \subset G$.[1]

Some rather non-elementary parts of set-theoretical topology, e.g. the theory of dimension[2], the theory of Baire functions[3] etc., can be extended to the class of closure σ-algebras satisfying axiom (a). This class is an essential generalization of the class of all metric spaces. For instance, if $\mathfrak{B}$ is the closure algebra of all subsets of separable metric space, and Δ is a σ-ideal in $\mathfrak{B}$, then the closure operation in $\mathfrak{B}$ induces, in a natural way, a closure operation in $\mathfrak{A} = \mathfrak{B}/\Delta$[4]. The closure algebra $\mathfrak{A}$ satisfies axiom (a) but, if Δ is not principal, is essentially distinct from topological spaces.

The notion of closure algebras has important applications to the theory of certain non-classical propositional and predicate calculi in mathematical logic.

Consider first the case of Lewis' modal propositional calculus S4 called in the sequel the Lewis propositional calculus, for brevity. This calculus, besides the usual logical connectives $\cup, \cap, \rightarrow, -$ contains also a unary connective $\mathbf{C}$. If α is a formula, then the formula $\mathbf{C}\alpha$ should be read: it is possible that α. The connective $\mathbf{C}$ has properties of the closure operation. More precisely, if we form the Lindenbaum-Tarski algebra $\mathfrak{A}$ for the Lewis propositional calculus by the method described in § 40, we obtain a Boolean algebra with a closure operation defined by

$$(1) \qquad\qquad \mathbf{C}|\alpha| = |\mathbf{C}\alpha| \ .$$

Thus the examination of the Lewis propositional calculus can be reduced to the examination of closure algebras. Indeed, the Lindenbaum-Tarski algebra of the Lewis propositional calculus is a free closure algebra, the elements $|p|$ (where p is any sentential variable) being the free generators.

This method of investigation of the Lewis propositional calculus is a convenient tool in this part of logic. For instance, it permits one to prove easily the decidability for the Lewis propositional calculus[5].

This method also can be applied to the examination of the Lewis predicate calculus[6].

---

[1] Sikorski [7].

[2] Hofmann [1], Sikorski [16].

[3] Sikorski [7].

[4] Sikorski [7].

[5] The discovery and the development of this method is due to McKinsey and Tarski. See McKinsey [4], McKinsey and Tarski [1, 3]. See also C. Davis [1].

[6] Rasiowa [1], Rasiowa and Sikorski [3, 4, 5].

The notion of closure algebra is very helpful in the examination of the propositional and predicate intuitionistic calculi. In this case the Lindenbaum-Tarski algebras are not Boolean algebras. Algebras with operations corresponding to the intuitionistic connectives (disjunction $\cup$, conjunction $\cap$, implication $\longrightarrow$ and negation $\neg$) are distributive lattices of a special type. However every lattice of this type can be represented as the algebra of all open elements of a closure algebra $\mathfrak{A}$ with the same join and meet and with the following definition of operations $\longrightarrow$, $\neg$ corresponding to the intuitionistic implication $\longrightarrow$ and negation $\neg$

$$A \longrightarrow B = \mathbf{I}(-A \cup B) = \mathbf{I}(A \to B) \, , \quad \neg A = \mathbf{I} - A \, .$$

Consequently the investigation of the intuitionistic logic can be reduced to the investigation of algebras of open elements in closure algebras. Thus also in this domain, the notion of closure algebra is an adequate and powerful tool. For instance, the theorem on decidability for the intuitionistic propositional calculus can be obtained easily by this method[1].

The same remark as in the case of the intuitionistic logic can be applied to the positive logic.

Observe that the notion of polyadic Lewis or intuitionistic algebras can be also introduced.

In the theory of closure algebras, the elements of Boolean algebras play a role analogous to subsets of topological spaces. However another point of view is also possible; a Boolean algebra $\mathfrak{A}$ can be interpreted as a topological space, and the elements of $\mathfrak{A}$ as points of this space. The topology introduced into $\mathfrak{A}$ can be e.g. the (sequential or neighbourhood) topology determined by the partial Boolean ordering $\subset$, or in another way in connection with the Boolean operations in A. Some theorems of this type are known[2].

---

[1] The discovery of the connection between the intuitionistic logic and lattices is due to STONE [9] and TARSKI [4]. The development of the method mentioned for the intuistionistic propositional calculus is due to McKINSEY [4], McKINSEY and TARSKI [2,3]. See also RIEGER [1, 3]. This method was applied to a problem of the intuitionistic predicate caluculus for the first time by MOSTOWSKI [3], and systematically developed by RASIOWA [1, 3, 4], RASIOWA and SIKORSKI [3, 4, 5, 8], SIKORSKI [22, 24, 26]. For a similar investigation of the intuitionistic and Lewis propositional calculus with quantifiers, see RASIOWA and SIKORSKI [6].

The systematic development of Boolean methods in mathematical logic is the subject of the monograph of RASIOWA and SIKORSKI [9].

[2] See e.g. AMEMIYA and MORI [1], ANTONOWSKIJ, BOLTJANSKIJ and SARYMSAKOV [1], FLOYD [1], NOVÁK and NOVOTNÝ [1], WARD [1]. See also metric spaces of measures discussed in § 42.

For applications of Boolean notions to topology, see STONE [6], SPECKER [1].

A theorem analogous to topological fixed-point theorems was recently proved for lattices by TARSKI [9]. See also A. C. DAVIS [1], WOLK [1].

§ 42. Applications to measure theory

Let μ be a σ-measure on a σ-field $\mathfrak{F}$ of subsets of a space X. Sets in $\mathfrak{F}$ will be called *measurable* according to the ordinary terminology in measure theory. Let Δ be the σ-ideal of sets of measure zero.

Two sets $A, B \in \mathfrak{F}$ differing only in a set of measure zero (i.e. such that $(A - B) \cup (B - A) \in \Delta$) have the same properties from the point of view of measure theory and practically they are identified each with another. Thus, in practice, we examine the Boolean algebra

(1) $\mathfrak{A} = \mathfrak{F}/\Delta$

and the σ-measure $\bar{\mu}$ defined on $\mathfrak{A}$ by the equality

(2) $\bar{\mu}([B]_\Delta) = \mu(B)$ $(B \in \mathfrak{F})$.

The measure $\bar{\mu}$ so defined is strictly positive, i.e. it vanishes only on the zero element. A Boolean σ-algebra with a strictly positive σ-measure will be called a *measure algebra*. The algebra (1) with the measure (2) will be called the *measure algebra of μ*.

The fundamental notion of measure theory is that of a σ-measure on a σ-field of sets. The point of view mentioned above leads to a more general definition of a σ-measure on a Boolean σ-algebra. This extended definition was assumed in this book[1] (see § 20 M)). The theory of σ-measures on Boolean σ-algebras can be developed as the theory of σ-measures on σ-fields of sets without any essential change. The extended notion of σ-measure on a Boolean σ-algebra $\mathfrak{A}$ is not any essential generalization of the notion of σ-measure on a σ-field of sets. In fact, it follows from the fundamental representation theorem for Boolean σ-algebras 29.1 that $\mathfrak{A}$ can be represented in the form (1) where $\mathfrak{F}$ is a σ-field of sets and Δ is a σ-ideal of $\mathfrak{F}$. If $\bar{\mu}$ is a σ-measure on $\mathfrak{A}$, then equality (2) defines a σ-measure μ on $\mathfrak{F}$, and the examination of $\bar{\mu}$ can be reduced to the study of μ.

However, in some measure-theoretical questions, the possibility of elimination of sets of measure zero and the passage to the corresponding strictly positive σ-measure seems to be convenient and adequate. Here the introduction of the notion of Boolean algebras is essential because no σ-measures on σ-fields of sets are strictly positive except in some trivial cases.

---

[1] Measures and σ-measures on Boolean algebras have been examined by many writers. See e.g. AUMANN [3], BAUER [1, 2], CARATHÉODORY [1, 2, 3, 4], DUBINS [1], HAUPT and PAUC [1, 2, 3, 4], HEIDER [2], HEWITT [1], HODGES and HORN [1], HORN and TARSKI [1], KAPPOS [2, 3, 4, 5], KAWADA [1], KELLEY [2], KRICKEBERG [1, 2, 3, 4], KOLMOGOROFF [2], MACKEY [1], MAHARAM [1, 2, 3, 4, 5], MARCZEWSKI [4], MARCZEWSKI and SIKORSKI [3], MIBU [1], NIKODYM [2, 3, 4, 5, 7], NOVAK and NOVOTNY [1], OGASAWARA [1], OLMSTED [1], ONICESCU [1], PAUC [1, 4], PETTIS [1], RIDDER [1, 3], RIVKIND [1], SEGAL [1], TOMITA [1], VINOKUROV [1], VLADIMIROV [1], WECKEN [1].

An application of such a procedure is given by forming the metric space of measurable sets. By the metric space of a σ-measure μ on a σ-field $\mathfrak{F}$ we understand the measure algebra $\mathfrak{A}$ defined by (1) with the following definition of the distance

(3) $\varrho(A, B) = \bar{\mu}(A - B) + \bar{\mu}(B - A)$ for $A, B \in \mathfrak{A}$.

It can be proved that this metric space is complete[1]. Sometimes we consider only the subspace of all elements of finite measure. This subspace is a closed subset of the whole space and therefore it is also complete. The completeness of the metric space $\mathfrak{A}$ and its subspace of elements of finite measure permits one to introduce into measure theory a topological method based on the Baire theorem on sets of the first category (the so-called *category method*). As an example we mention here certain proofs of the Hahn-Vitali theorem[2].

Another application is given by the notion of isomorphism of σ-measures. This notion can be introduced in various ways. For instance, two σ-measures μ_1 and μ_2 (on fields $\mathfrak{F}_1$ and $\mathfrak{F}_2$ of subsets of spaces X_1 and X_2 respectively) are called isomorphic if there exists a measure preserving isomorphism h of $\mathfrak{F}_1$ onto $\mathfrak{F}_2$, i.e. a Boolean isomorphism such that

$$\mu_2(h(A)) = \mu_1(A)$$

for every $A \in \mathfrak{F}_1$. Another definition is as follows: μ_1 and μ_2 are called isomorphic if there exists sets $X_{1,0} \in \mathfrak{F}_1$ and $X_{2,0} \in \mathfrak{F}_2$ with $\mu_1(X_{1,0}) = 0 = \mu_2(X_{2,0})$, such that there exists a measure preserving isomorphism of $\mathfrak{F}_1 | X_1 - X_{1,0}$ onto $\mathfrak{F}_2 | X_2 - X_{2,0}$. However, both definitions are inadequate because, by a non-essential addition of a class of sets of measure zero, we can disturb the existing isomorphism (for instance, the Lebesgue measure on all Borel sets and the Lebesgue measure on all sets measurable in the sense of Lebesgue are not isomorphic if one of the two above definitions of isomorphism is assumed). The structural properties of the fields $\mathfrak{F}_1$ and $\mathfrak{F}_2$ play too great a part in both definitions.

The following definition avoids this difficulty. The σ-measures μ_1 and μ_2 are said to be *isomorphic* if there exists a measure-preserving isomorphism between their measure algebras [see (1) and (2)]. This definition permits one to give a complete classification of types of isomorphisms for finite σ-measures[3].

The advantage of considering σ-measures on Boolean σ-algebras lies in the possibility of passing to the corresponding measure algebras. The problem arises as to which Boolean algebras are measure algebras (up to isomorphism) of finite measures. If $\mathfrak{A}$ is the measure algebra of a finite σ-measure, then $\mathfrak{A}$ is a complete Boolean algebra (see § 21 D)). However

---

[1] See e.g. NIKODYM [1].
[2] Due to SAKS [1].
[3] MAHARAM [1]. See also ZINK [1].

the completeness is not a sufficient condition because there are complete Boolean algebras such that all σ-measures on them vanish identically (see § 21 F)). A necessary and sufficient condition for the existence of a finite strictly positive σ-measure will be quoted at the end of this section.

An important auxiliary notion in measure theory is that of measure defined in § 3 C). As is known, not every measure μ defined on a given field $\mathfrak{F}$ of subsets of a space X can be extended to a σ-measure μ' on the σ-field $\mathfrak{F}'$ σ-generated by $\mathfrak{F}$. The necessary and sufficient condition for existence of such an extension is as follows:

(c) for every sequence $\{A_n\}$ of disjoint sets in $\mathfrak{F}$, if the set-theoretical union $\bigcup_{1 \le n < \infty} A_n$ belongs to $\mathfrak{F}$, then $\mu(\bigcup_{1 \le n < \infty} A_n) = \Sigma_{1 \le n < \infty} \mu(A_n)$.

This condition is not always satisfied. Boolean methods show that, roughly speaking, the reason for this difficulty lies in the fact that the space X under consideration may have too few points. If $\mathfrak{F}$ is perfect, then every measure on $\mathfrak{F}$ can be extended to a σ-measure. In fact, the union of any sequence of disjoint non-empty sets in a perfect field $\mathfrak{F}$ never belongs to $\mathfrak{F}$ (see § 7 J)) which implies that condition (c) is satisfied. On the other hand, every field of sets is isomorphic to a perfect field of sets (see 8.2) and, indeed, this isomorphism can be realized by adding some points to X (see § 8 G)).

Thus, replacing the field $\mathfrak{F}$ under consideration by an isomorphic perfect field, we infer that every measure μ has a natural extension μ' which is a σ-measure. The isomorphism type of μ' is uniquely determined by μ and can also be obtained by another method, without passing to the perfect isomorph, provided the measure μ is finite. In fact, let Δ be the ideal of all sets of μ measure zero. Then the formula (2) defines a strictly positive measure $\bar\mu$ on $\mathfrak{A} = \mathfrak{F}/\Delta$. Just as in the case of σ-measures, $\mathfrak{A}$ can be considered as a metric space. This space is not complete, in general. By the known method of completing metric spaces, we can define a complete metric space $\mathfrak{A}'$ such that $\mathfrak{A}$ is dense in $\mathfrak{A}'$. The measure $\bar\mu$ satisfies the condition

$$|\bar\mu(A) - \bar\mu(B)| \le \varrho(A, B) .$$

Therefore it can be extended to a real continuous function $\bar\mu'$ on $\mathfrak{A}'$. It can be proved that the Boolean operations in $\mathfrak{A}$ can be also extended over $\mathfrak{A}'$ so that $\mathfrak{A}'$ becomes a Boolean σ-algebra. $\bar\mu'$ is a strictly positive σ-measure on $\mathfrak{A}'$ and is isomorphic to the previously defined extension μ' of μ.

The following question arises: under what condition does there exist a strictly positive finite measure on a given Boolean algebra $\mathfrak{A}$[1]? The σ-chain condition is, of course, a necessary condition. However this condition is not sufficient. There exists a Boolean algebra $\mathfrak{A}$ with no strictly positive measure, such that $\mathfrak{A}$ is the union of an enumerable sequence of sets, the nth set containing no more than n disjoint elements[2]. The last property implies the σ-chain condition.

To formulate a necessary and sufficient condition let us assume the following definition. For every finite sequence $\mathfrak{s} = \{A_1, \ldots, A_n\}$ of elements of a Boolean algebra $\mathfrak{A}$, let $i(\mathfrak{s}) = m/n$ where m is the greatest integer with the property that there exists a sequence $1 \leq k_1 < k_2 < \cdots \cdots < k_m \leq n$ such that $A_{k_1} \cap A_{k_2} \cap \cdots \cap A_{k_m} \neq \wedge$. For every set $\mathfrak{S}$ of elements in $\mathfrak{A}$, let the *intersection number of* $\mathfrak{S}$ be the greatest lower bound of the numbers $i(\mathfrak{s})$ where $\mathfrak{s}$ is any sequence of elements in $\mathfrak{S}$ (sequences with repetitions of elements are admitted).

The following condition is necessary and sufficient for the existence of a finite strictly positive measure on a Boolean algebra $\mathfrak{A}$:

(k) the set $\mathfrak{A} - (\wedge)$ is the union of an enumerable sequence of sets each of which has a positive intersection number[3].

In order that there exist a finite strictly positive σ-measure on a complete Boolean algebra $\mathfrak{A}$ it is necessary and sufficient that $\mathfrak{A}$ be weakly σ-distributive (see § 30 A)) and satisfy condition (k)[4].

Another necessary and sufficient condition for the existence of a finite strictly positive σ-measure on a Boolean σ-algebra $\mathfrak{A}$ is as follows:

(k') the set $\mathfrak{A} - (\wedge)$ is the union of an enumerable sequence of sets $\{\mathfrak{S}_m\}$ each of which has positive intersection number and possesses the following property: if $A_1 \subset A_2 \subset \cdots$ and $\bigcup_{1 \leq i < \infty} A_i \in \mathfrak{S}_m$, then there exists an integer n such that $A_n \in \mathfrak{S}_m$[5].

Note that on every Boolean algebra there exists a strictly positive measure whose values belong to a non-archimedean ordered algebraic field[6].

§ 43. Measurable functions and real homomorphisms

Let $\mathfrak{F}$ be a σ-field of subsets of a space X. We recall that a real function φ (assuming, possibly, the values $\pm \infty$) defined on X is said to

---

[1] For a detailed investigation of this question, see HORN and TARSKI [1]. The positive answer for separable Boolean algebras was given by MARCZEWSKI. See TARSKI [14], footnote [16], and HORN and TARSKI [1]. The answer is negative for the algebra of all subsets of an enumerable set modulo the ideal of finite sets. This follows e.g. from a theorem of SIERPIŃSKI [4] (see HORN and TARSKI [1]). The problem of the existence of strictly positive measure is closely related the so-called Souslin hypothesis. See HORN and TARSKI [1], KELLEY [2], MAHARAM [2].

[2] This result is due to GAIFMAN [1, 2].

[3] Condition (k) and the whole theorem are due to KELLEY [2].

[4] This theorem is due to KELLEY [2] (another proof of this theorem was given by C. RYLL-NARDZEWSKI). Another more complicated necessary and sufficient condition was given earlier by MAHARAM [2]. A simplification of MAHARAM's condition was given by HODGES and HORN [1].

[5] This modification of KELLEY's theorem is due to C. RYLL-NARDZEWSKI (not published).

[6] NIKODYM [7, 8]. See also LUXEMBURG [1, 2].

be $\mathfrak{F}$-*measurable*, or briefly *measurable*, if for every real number a the set of all $x \in X$ satisfying $\varphi(x) < a$ belongs to $\mathfrak{F}$.

In some investigations it is necessary to identify measurable functions modulo a σ-ideal $\varDelta$ of $\mathfrak{F}$. More precisely, we identify two measurable functions φ_1 and φ_2 if and only if the set of all $x \in X$ such that $\varphi_1(x) \neq$ $\neq \varphi_2(x)$ belongs to $\varDelta$. We say then that φ_1, φ_2 are $\varDelta$-*equivalent*, or briefly, *equivalent*. This procedure is, for instance, followed in measure and integration theory, $\varDelta$ being the σ-ideal of sets of measure zero.

Instead of considering abstract classes of $\varDelta$-equivalence, we can also perform the identification in the following way.

Let $\mathfrak{B}$ denote the σ-field of all Borel sets of real numbers (including $\pm \infty$). Every σ-homomorphism h of $\mathfrak{B}$ into any Boolean algebra $\mathfrak{A}$ will be called *real homomorphism* into $\mathfrak{A}$. A real homomorphism h is said to be *finite* provided $h((\infty)) = \varLambda = h((-\infty))$. It is called *bounded* provided there exists a finite interval B such that $h(B) = V$. By 29.1 we can always represent $\mathfrak{A}$ as a quotient algebra of a σ-field by a σ-ideal. Then, by 32.5, h is induced by a real function φ. It is easy to verify that h is finite (bounded) if and only if h is induced by a finite (bounded) function φ.

Let $\mathfrak{F}$ and $\varDelta$ have the same meaning as before. Every $\mathfrak{F}$-measurable function induces a real homomorphism into $\mathfrak{F}/\varDelta$. Two $\mathfrak{F}$-measurable functions induce the same real homomorphism if and only if they are $\varDelta$-equivalent. Conversely, every real homomorphism h into $\mathfrak{F}/\varDelta$ is induced by an $\mathfrak{F}$-measurable function, and every function inducing h is $\mathfrak{F}$-measurable.

Hence we infer that instead of considering $\mathfrak{F}$-measurable functions modulo the σ-ideal $\varDelta$, we can consider real homomorphisms into $\mathfrak{F}/\varDelta$. These notions have the same mathematical content. Real homomorphisms are the Boolean analogue of the set-theoretical notion of measurable real functions.

The usual operations performed on measurable functions can also be performed on real homomorphisms into any Boolean σ-algebra $\mathfrak{A}$.

For instance, if h, h_1, h_2 are finite real homomorphisms into $\mathfrak{A}$ and c is a finite real number, then $ch, h_1 + h_2, h_1 - h_2, h_1 \cdot h_2$ can easily be defined so that they are Boolean analogues of the corresponding operations on measurable functions. In fact, we may assume (see 29.1) that $\mathfrak{A} = \mathfrak{F}/\varDelta$ where $\mathfrak{F}$ and $\varDelta$ satisfy conditions mentioned above. We may assume (see 32.5) that h, h_1, h_2 are induced by some finite measurable functions $\varphi, \varphi_1, \varphi_2$ respectively. Then, by definition, $ch, h_1 + h_2, h_1 - h_2, h_1 \cdot h_2$ are real homomorphisms (into $\mathfrak{A}$) induced by the measurable functions $c\varphi, \varphi_1 + \varphi_2, \varphi_1 - \varphi_2, \varphi_1 \cdot \varphi_2$ respectively. It is easy to verify that the real homomorphisms so defined depend neither on the representation of $\mathfrak{A}$ in the form $\mathfrak{F}/\varDelta$, nor on the choice of the inducing functions φ,

φ_1, φ_2. In a similar way we can define the quotient h_1/h_2 under the hypothesis that $h_2((0)) = \varLambda$.

We write $h_1 \leqq h_2$ if there exist functions φ_1, φ_2 inducing h_1 and h_2 respectively, such that $\varphi_1 \leqq \varphi_2$.

A sequence $\{h_n\}$ of real homomorphisms into $\mathfrak{A}$ is said to *converge* (to *converge uniformly*) to a real homomorphism h (into $\mathfrak{A}$) provided there exist functions φ, φ_n inducing h and h_n ($n = 1, 2, \ldots$) respectively such that $\{\varphi_n\}$ converges (converges uniformly) to φ.

The definitions of the above operations on real homomorphisms can be also formulated without any reference to inducing functions. We can, for instance, make use of the fact that every real homomorphism is uniquely determined by the values it assumes on the infinite intervals B_a: $-\infty \leqq x < a$. Then the sum $h_1 + h_2$ is the real homomorphism h such that

$$h(B_a) = \bigcup_{w \in W} (h_1(B_w) \cap h_2(B_{a-w})) \quad \text{for every real } a,$$

where W denotes the set of all rational numbers. In a similar way we can formulate other definitions. However these definitions require more difficult proofs.

The fundamental operations on real homomorphisms having been established, we can operate on them in the same way as on measurable functions without any difficulty[1].

We may also perform some more advanced operations on real homomorphisms. For instance, suppose that $\bar{\mu}$ is a σ-measure on $\mathfrak{A}$ and that h is a real homomorphism into $\mathfrak{A}$. Then the integral $\int h \, d\bar{\mu}$ can be defined by

$$\int h \, d\bar{\mu} = \int \varphi \, d\mu$$

(provided the integral on the right side exists) where φ is a function inducing h, and μ is a measure (on $\mathfrak{F}$) determined by $\bar{\mu}$ [see § 42 (1) and (2)][2].

§ 44. Measurable functions. Reduction to continuous functions

Another model for the space of all measurable functions modulo a σ-ideal can be obtained as follows.

Just as in § 43, let $\mathfrak{F}$ be a σ-field of subsets of a space X and let $\varDelta$ be a σ-ideal of $\mathfrak{F}$. Let X' be the Stone space of the quotient algebra $\mathfrak{F}/\varDelta$ and let h_0 be a fixed isomorphism of $\mathfrak{F}/\varDelta$ onto the field of all open-closed subsets of X'.

---

[1] For other definitions of Boolean analogues of point-mappings, see BERSTEIN [1], CARATHÉODORY [1, 4], GÖTZ [1], KAPPOS [1], NIKODYM [3], OLMSTED [1], POSPIŠIL [1, 2, 4], RIDDER [1], WECKEN [1].

[2] SIKORSKI [9]. For other definitions based on other generalizations of point-functions, see BISCHOF [1], CARATHÉODORY [1, 2, 4], FORADORI [1], KAPPOS [2], OLMSTED [1], RIDDER [1], WECKEN [1].

For every $\mathfrak{F}$-measurable function φ on X, let φ' denote the function defined uniquely on X' by the following condition:

$$\varphi'^{-1}(B_a) = \bigcup_{b < a} h_0(\varphi^{-1}(B_b)) \quad \text{for every real } a,$$

where B_a is the infinite interval defined at the end of § 43. In other words,

$$\varphi'(x') = \sup_{x' \notin h_0(\varphi^{-1}(B_a))} a \quad \text{for every } x' \in X'.$$

The function φ' is always continuous (it may assume infinite values). The canonical transformation, mapping φ onto φ', is one-to-one provided we identify Δ-equivalent functions. More exactly, $\varphi'_1 = \varphi'_2$ if and only if φ_1, φ_2 are Δ-equivalent.

The canonical transformation preserves the natural partial ordering of functions and the algebraic operations. More precisely, $\varphi'_1(x') \leq \varphi'_2(x')$ for every $x' \in X'$ if and only if $\varphi_1(x) \leq \varphi_2(x)$ for every $x \in X$ except for a set belonging to Δ. If φ is finite, then $\varphi'(x')$ is finite for all $x' \in X'$ except for a nowhere dense set where φ' assumes the infinite values. If $\varphi = \varphi_1 + \varphi_2$ and φ_1, φ_2 are finite, then $\varphi'(x') = \varphi'_1(x') + \varphi'_2(x')$ for all $x' \in X'$ except for a nowhere dense set where the addition on the right side is not feasible. Analogous statements hold for difference, etc.

Thus the space of all $\mathfrak{F}$-measurable functions modulo Δ can be identified with the space of all continuous functions (including functions assuming infinite values) on the Stone space X' of $\mathfrak{F}/\Delta$.

This representation is especially valuable for bounded measurable functions φ[1]. Then the class of all corresponding functions φ' coincides with the class of all finite continuous functions on X'. Moreover

(1)
$$\sup_{x' \in X'} \varphi'(x') = \inf_{A \in \Delta} \sup_{x \in X - A} \varphi(x),$$

(2)
$$\inf_{x' \in X'} \varphi'(x') = \sup_{A \in \Delta} \inf_{x \in X - A} \varphi(x).$$

Observe that the canonical transformation preserves uniform convergence. The analogous statement on pointwise convergence is not true, in general.

§ 45. Applications to functional analysis

Let μ be a σ-finite σ-measure on a σ-field $\mathfrak{F}$ of subsets of a space X, and let Δ be the σ-ideal of all sets of μ measure zero. Consider the Banach space $\boldsymbol{L}$ of all integrable functions (identified modulo Δ) on X with the ordinary norm. It is known that the space adjoint to $\boldsymbol{L}$ (i.e. the space of all continuous linear functionals on $\boldsymbol{L}$) coincides with the space $\boldsymbol{M}$ of all bounded $\mathfrak{F}$-measurable functions (identified modulo Δ) on X with the usual norm [given by the right side of equality § 44 (1) where φ is replaced by its absolute value]. Text books in functional analysis often omit

---

[1] For applications, see e.g. DIEUDONNÉ [1], SEMADENI [2, 3, 7]. See also § 45.

the question as to what is the space adjoint to M. An answer to this question can be given easily by the Boolean investigation in § 44. In fact, M can be considered as the space of all finite continuous functions on the Stone space X' of $\mathfrak{F}/\varDelta$ with the ordinary norm [see § 44 (1)]. By a fundamental theorem in functional analysis[1], the space adjoint to the space of all finite continuous functions on a compact topological space X' coincides with the set of all finite signed σ-measures on the σ-field of all Baire subsets of X'. Thus the space adjoint to M coincides with the space of all signed σ-measures (with the ordinary norm) on the σ-field generated by all open-closed subsets of the Stone space of $\mathfrak{F}/\varDelta$.

Boolean notions sometimes appear in other parts of functional analysis. We mention here, for instance, the problem of whether a given Banach space E' has the following property:

(p) for every Banach space E, every continuous linear operation from a subspace $E_0 \subset E$ into E' can be extended to a continuous linear operation from E into E' with the same norm.

The answer is given by the following theorem: E' has property (p) if and only if E' is norm-isomorphic to the space of all finite continuous functions on an extremally disconnected compact topological space, i.e. to the space of all bounded real homomorphisms into a complete Boolean algebra[2].

The notion of complete Boolean algebras also figures, in an essential way, in the theory of partially ordered linear spaces[3]. In particular, extremally disconnected spaces appear in problems connected with the lattice-completeness of spaces of continuous functions[4]. Stone spaces also appear in the proof of the fundamental representation theorem for abstract (L)-spaces[5].

§ 46. Applications to foundations of the theory of probability

The fundamental notions of the theory of probability are *event, probability* of an event, and *random variable*.

As we have observed in § 1 C), the set of all events is always supposed to form a Boolean algebra $\mathfrak{A}_0$. The probability is a *normed* measure μ_0 on $\mathfrak{A}_0$, i.e. a measure assuming the value 1 at the unit element of $\mathfrak{A}_0$.

---

[1] KAKUTANI [3].

[2] See KELLEY [1]. Some partial solutions were given earlier by GOODNER [1] and NACHBIN [3].

[3] See KANTOROVIČ, VULICH and PINSKER [1].

[4] See NAKANO [1], DILWORTH [3], STONE [13].

[5] See KAKUTANI [2]. For other applications to functional analysis see also AMEMIYA and MORI [1], FOGUEL [1], MAEDA [2], MAHARAM [9], MCCARTHY [1].

Observe also that STONE's representation theorem for Boolean algebras can be obtained as a corollary to KAKUTANI's [3] representation theorem for abstract (M)-spaces.

By theorem 8.2, we may always assume that $\mathfrak{A}_0$ is a field of sets. Thus the investigation of events and their probabilities can be reduced to an examination of normed measures on fields of sets. However, for purely technical reasons, it is more convenient to reduce the examination to the case of σ-measures on σ-fields because the hypothesis of σ-additivity has important mathematical consequences. This can always be realized. For instance, by 8.2 we may assume that $\mathfrak{A}_0$ is the perfect reduced field of subsets of all open-closed subsets of its Stone space X. Then μ_0 can be uniquely extended to a σ-measure μ on the σ-field $\mathfrak{A}$ generated by $\mathfrak{A}_0$ (see § 42). Observe that points in X have a simple probabilistic interpretation. In fact, from the intuitive point of view, an event is something which can occur or not; this can be verified, e.g., by a suitable trial. Suppose that such a trial was performed. Then the class of all events which occurred in this trial is a maximal filter in $\mathfrak{A}_0$, i.e. a point in X (see the remarks on p. 21). Thus points in X can be interpreted as all theoretically possible results of trials[1].

By the above consideration, the previously given set $\mathfrak{A}_0$ of events can always be extended to a σ-field $\mathfrak{A}$, the elements of which (i.e. sets $A \in \mathfrak{A}$) also can be interpreted as events. The probability function μ_0 on $\mathfrak{A}_0$ can be extended to a probability function μ on $\mathfrak{A}$ so that μ is a σ-measure. Therefore the examination can be reduced to the case of normed σ-measures on σ-fields of sets.

According to the old-fashioned non-precise definition, "a random variable is a variable ξ such that, for every real number a, the probability of the event $\xi < a$ is defined". Thus $\xi < a$ must be interpreted as an event, i.e. a set $A_a \in \mathfrak{A}$. It is natural to assume that the event $\xi < a$ is the union of all events $\xi < b$ where $b < a$. Thus $\{A_a\}$ is an indexed set of sets in $\mathfrak{A}$, such that

(1) $$A_a = \bigcup_{b < a} A_b \quad \text{for every real } a \, .$$

On the other hand, for every indexed set of sets $A_a \in \mathfrak{A}$ satisfying (1) there exists exactly one real $\mathfrak{A}$-measurable function φ on X such that

(2) $$A_a \text{ is the set of all } x \in X \text{ such that } \varphi(x) < a \, .$$

Conversely, if φ is an $\mathfrak{A}$-measurable function on X, then (2) defines an indexed set $\{A_a\}$ of sets in $\mathfrak{A}$, such that (1) holds. Hence we infer that the random variables coincide with $\mathfrak{A}$-measurable functions on X. So we get, in a rather natural way, the known set-theoretical model for the theory of probability[2].

The question arises whether it would be better or more natural to consider, more generally, normed σ-measures on Boolean σ-algebras

---

[1] This interpretation is due to Łoś [2, 4]. The quoted papers contain an exhaustive exposition of the problems discussed in § 46.

[2] Due to KOLMOGOROFF [1].

instead of σ-fields of sets[1]. Then random variables should be interpreted as real homomorphisms, according to § 43. Such a generalization is in fact possible. In particular, the notion of Boolean σ-products (see § 38) gives the mathematical basis for the investigation of the independence of events, which in the set-theoretical model is based on forming products of σ-measures in Cartesian products. However, on the other hand, the fundamental representation theorem 29.1 for Boolean σ-algebras shows that this generalization is not essential and, by the method described in § 42 and § 43, the Boolean model for the theory of probability can always be replaced by a set-theoretical one.

§ 47. Problems of effectivity

As the reader has observed, the fundamental representation theorem 8.2 is the basic theorem for the whole theory of Boolean algebras. This theorem was obtained as a consequence of theorem 6.1 on the existence of maximal ideals and filters. The proof of 6.1 is not effective because it is based on the well-ordering principle:

(a_1) every set can be well ordered

which is, as it is well known, equivalent to the axiom of choice quoted below in full generality:

(a_2) for every indexed set $\{A_t\}_{t \in T}$ of non-empty sets, there exists an indexed set $\{a_t\}_{t \in T}$ such that $a_t \in A_t$ for every $t \in T$.

The problem arises as to whether it is possible to prove the fundamental representation theorem 8.2 in an effective way, i.e. without any use of the axiom of choice. Some of the results quoted below show that the answer seems to be negative.

It is easy to verify that the following statements on Boolean algebras are effectively equivalent (i.e. each of them can be deduced from every other one without any use of the axiom of choice):

(b_1) every proper ideal (filter) can be extended to a maximal ideal (filter);

(b_2) every non-degenerate Boolean algebra has at least one maximal ideal (filter);

(b_3) every two-valued measure on a subalgebra can be extended to a two-valued measure on the whole Boolean algebra;

(b_4) every Boolean algebra is isomorphic to a field of sets;

---

[1] Such an attitude towards probability has been suggested by HALMOS [2], KOLMOGOROFF [2], SEGAL [1]. It permits one to consider probability as strictly positive σ-measure and to omit the difficulty connected with the existence of non-empty measurable sets of measure zero, which have no probabilistic interpretation. For applications of Boolean notions to the theory of probability, see DUBINS [1], KRICKEBERG [2, 3], SEGAL [1], THEODORESCU [1]. For another approach to foundations of the theory of probability, see ŁOŚ [6].

(b_5) every Boolean algebra is isomorphic to a field of open-closed subsets of a totally disconnected compact space;

(b_6) every proper ideal (filter) is the intersection of a class of maximal ideals (filters).

It can be proved that each of statements (b_1)—(b_6) is effectively equivalent to each of the following statements:

(b_7) every finite measure m_0 on a subalgebra can be extended to a measure m on the whole Boolean algebra so that the set of all values of m is contained in the closure of the set of all values of m_0;

(b_8) the Cartesian product of any number of non-empty compact Hausdorff spaces is a non-empty compact space.

Each of the statements (b_1)—(b_8) implies effectively the following statement:

(c) every partial ordering can be extended to a linear ordering.

It is evident that (c) implies effectively the ordering principle, i.e. the statement

(d) every set can be linearly ordered,

and that (d) implies effectively the axiom of choice for finite sets, i.e. the statement

(e) for every indexed set $\{A_t\}_{t \in T}$ of non-empty finite sets there exists an indexed set $\{a_t\}_{t \in T}$ such that $a_t \in A_t$ for every $t \in T$.

Thus we have effective implications

$$(a) \to (b) \to (c) \to (d) \to (e)$$

where (a) and (b) denote respectively one of the statements (a_1)—(a_2) or (b_1)—(b_8)[1]. The implication (a) → (b) cannot be replaced by an effective equivalence[2]. However the statement (b_1) formulated for arbitrary lattices is effectively equivalent to the axiom of choice[3]. The implication (d) → (e) also cannot be replaced by an effective equivalence[4].

The field $\mathfrak{F}$ of all subsets of a countable set X is one of the simplest infinite Boolean algebras. It follows from (b) that there exists a two-valued measure defined on $\mathfrak{F}$ and vanishing on all finite subsets of X.

[1] All results mentioned are due to Łoś and Ryll-Nardzewski [1]. Some of these results and many other effective equivalences and implications have been announced simultaneously by Henkin [5], Rudin and Scott [1], Scott [2], Tarski [15, 16, 17]. The implication (b) → (e) was observed earlier by A. Davis.

[2] This result was recently obtained by Halpern [1, 2, 3]. Earlier Mostowski [2] proved that (d) does not imply effectively (a). See also Specker [2]. All the proofs of non-equivalence concern a Zermelo-Fraenkel axiomatic set theory that does not exclude the existence of objects which are not sets or does not contain the axiom of regularity.

[3] This result was announced by Scott [2]. See also Mrówka [1, 2]. Mrówka's papers contain also another similar statement (for distributive lattices) equivalent to the axiom of choice.

[4] Läuchli [1].

Bibliography

However we do not know any effective proof of this statement and it is doubtful whether we would be able to find such a proof in the present state of mathematical knowledge because the existence of the required measure implies effectively the existence of a non-measurable set of real numbers. In fact, suppose that X is the set of all positive integers and that m is a two-valued measure on all subsets of X, $m(A) = 0$ for every finite set $A \subset X$. Every real number x can be uniquely represented in the form

$$x = k + \sum_{j=1}^{\infty} \left(\frac{1}{2}\right)^{n_j}$$

where k is an integer and $\{n_j\}$ is an infinite increasing sequence of positive integers. Let A_x be the set of all the integers n_j. The set B of all x such that $m(A_x) = 1$ is not measurable (in the sense of LEBESGUE)[1].

The Hahn-Banach theorem on extensions of linear functionals, one of the most important theorems in functional analysis, is proved in every text book by means of transfinite induction or another statement equivalent to the axiom of choice. It is interesting to know that (b) implies effectively the Hahn-Banach theorem[2].

Note that the axiom of choice (a) implies theorem 33.1 on extension of homomorphisms and this theorem implies effectively (b) (see remarks on p. 142). It is not known whether any of these implications can be replaced by an effective equivalence[3].

Bibliography

ALEXANDROF, P. S., u. H. HOPF: [1] Topologie I. Berlin 1935.
— and P. S. URYSOHN: [1] Mémoire sur les espaces topologiques compacts. Verhand. Kon. Akad. Wetenschappen (Amsterdam) 14, 1—96 (1929).
ANDREOLI, G.: [1] Struttura delle algebre di Boole e loro estensione quale calcolo delle classi (in senso ordinario oppure probabilistico). Giorn. Mat. Battaglini 5, 141—171 (1957). [2] Proprietà delle funzioni simmetriche elementari nelle algebre di Boole e nelle algebre dei livelli. Ricerca (Napoli)10, 1—10 (1959). [3] Algebre di Boole-algebre di insieme-algebre di liveli. Giorn. Mat. Battaglini 7, 3—22 (1959). [4]. Matrici; reticoli booleani di sottomatrici e loro valutazione per caratteristiche. Ricerca, Rivista Math. pur. appl. II Ser. 11, 6—14 (1960).
AMEMIYA, I., and T. MORI: [1] Topological structures in ordered linear spaces. J. math. Soc. Japan 9, 131—142 (1957).

---

[1] SIERPIŃSKI [1]. See also MARCZEWSKI [13].

[2] This result is due to ŁOŚ and RYLL-NARDZEWSKI [1, 2]. Another proof was given by LUXEMBURG [1, 2] who also proved that (b) implies effectively NIKODYM's theorem on existence of strictly positive measures with values in non-archimedean ordered algebraic fields (see p. 204). LUXEMBURG's results make use of a method of forming products of models for mathematical theories, due to ŁOŚ [5].

[3] For an investigation of this problem, see LUXEMBURG [3].

ANTONOVSKIJ, M. YA., V. G. BOLTJANSKIJ and T. A. SARYMSAKOV: [1] Topological Boolean algebras, Taškent 1963, 1—132 (Russian).

AUBERT, K. E.: [1] A generalization of the ideal theory of commutative rings without finiteness assumptions. Math. Scand. 4, 209—230 (1956).

AUMANN, G.: [1] Ein Beweis des Loomisschen Darstellungssatzes für σ-Somenringe. Arch. Math. 2, 321—324 (1950). [2] Alternative-Zerlegungen in Booleschen Verbänden. Math. Z. 55, 109—113 (1951). [3] Reelle Funktionen. Berlin-Göttingen-Heidelberg 1954.

BALACHANDRAN, V. K.: [1] A characterization for complete Boolean algebras. J. Madras Univ. Sect. B 24, 273—278 (1954); J. Osaka Inst. Sci. Tech. 4, 39—44 (1952). [2] On certain BS-representations and a characterization of complete Boolean algebras. Proc. Indian Acad. Sci. Sect. A. 34, 35—46 (1957). [3] On isomorphic BS-representations preserving arbitrary joins. Math. Japon. 4, 55—61 (1956).

BANACH, S.: [1] Über additive Maßfunktionen in abstrakten Mengen. Fund. Math. 15, 97—101 (1930). [2] On measures in independent fields of sets (edited by S. HARTMAN). Stud. Math. 10, 159—177 (1948). [3] Théorème sur les ensembles de première catégorie. Fund. Math. 16, 395—398 (1930).

— et C. KURATOWSKI: [1] Sur une généralisation du problème de la mesure. Fund. Math. 14, 127—131 (1929).

BASS, H.: [1] Finite monadic algebras. Proc. Am. math. Soc. 9, 258—268 (1958).

BAUER, H.: [1] Reguläre und singuläre Abbildungen eines distributiven Verbandes in einem Vektorverband, welche der Funktionalgleichung $f(x \cup y) + f(x \cap y) = f(x) + f(y)$ genügen. J. reine angew. Math. 194, 141—179 (1955). [2] Darstellung additiver Funktionen auf Booleschen Algebren als Mengenfunktionen. Arch. Math. 6, 215—222 (1955).

BELL, C. B.: [1] On the structure of algebras and homomorphisms. Proc. Am. math. Soc. 7, 483—492 (1956).

BENADO, M.: [1] Sur une caractérisation abstraite des algèbres de Boole (I) Compt. rend. 251, 622—623 (1960); (II) Compt. rend. 251, 835—836 (1960).

BENNETT, A. A.: [1] Solution of Huntington's "unsolved problem in Boolean algebra". Bull. Am. math. Soc. 39, 289—295 (1933).

BERNAYS, P.: [1] Über eine natürliche Erweiterung des Relationenkalküls, Constructivity in Mathematics. Proc. of Coll. in Amsterdam, 1957, 1—14.

BERNSTEIN, B. A.: [1] Simplification of the set of four postulates for Boolean algebras in terms of rejection. Bull. Am. math. Soc. 39, 783—787 (1933). [2] A set of four postulates for Boolean algebra in terms of the "implicative" operation. Trans. Am. math. Soc. 36, 876—884 (1934). [3] On finite Boolean algebras. Am. J. Math. 57, 733—742 (1935). [4] Postulates for Boolean algebras involving the operation of complete disjunction. Ann. Math. 37, 317—325 (1936). [5] Sets of postulates for Boolean groups. Ann. Math. 40, 420—422 (1939). [6] Postulate-sets for Boolean rings. Trans. Am. math. Soc. 55, 393—400 (1944). [7] A dual-symmetric definition of Boolean algebra free from postulated special elements. Scripta Math. 16, 157—160 (1950). [8] Operations with respect to which the elements of a Boolean algebra form a group. Trans. Am. math. Soc. 26, 171—175 (1924). [9] On the existence of fields in Boolean algebras. Trans. Am. math. Soc. 28, 645—657 (1926).

BERSTEIN, I.: [1] Sur les classes des fonctions équivalentes et sur les fonctions définies dans une algèbre de Boole. Acad. R. P. Romine 7, 565—581 (1955) (Rumanian).

BETH, E. W.: [1] A topological proof of the theorem of LÖWENHEIM-SKOLEM-GÖDEL. Ind. Math. 13, 436—444 (1951).

Białynicki-Birula, A.: [1] Remarks on quasi-Boolean algebras. Bull. Acad. Polon. Sci. Cl. III **5**, 615—619 (1957).
— and H. Rasiowa: [1] On the representation of quasi-Boolean algebras. Bull. Acad. Pol. Sci. Cl. III, **5**, 259—261 (1957).
Birkhoff, G.: [1] Rings of sets. Duke math. J. **3**, 443—454 (1937). [2] Lattice Theory. New York, 1940; second edition 1948, 1961. [3] Order and the inclusion relation. Proc. Oslo Congress 1936, vol. 2, p. 37.
— and G. D. Birkhoff: [1] Distributive postulates for systems like Boolean algebras. Trans. Am. math. Soc. **60**, 3—11 (1946).
— and M. Ward: [1] A characterization of Boolean algebras. Ann. Math. **40**, 609—610 (1939).
Bischof, A.: [1] Beiträge zur Carathéodoryschen Algebraisierung des Integralbegriffs. Schr. Math. Inst. u. Inst. angew. Math. Univ. Berlin **5**, 237—262 (1941).
Blake, A.: [1] Canonical expressions in Boolean algebra, Chicago Diss. 1938; correction J. Symb. Logic **3**, 112—113 (1938).
Blumenthal, L. M.: [1] Boolean geometry (I). Rend. Coirc. Math. Palermo **1**, 1—18 (1952); (II) Proc. Intern. Congress of Math. **2**, 205 (1954). [2] Theory and applications of distance geometry, Oxford 1953.
Boole, G.: [1] The mathematical analysis of logic. Cambridge 1847. [2] An investigation of the laws of thought. Cambridge 1854.
Brainerd, B., and J. Lambek: [1] On the ring of quotients of a Boolean ring. Canad. Math. Bull. **2**, 25—29 (1959).
Braithwaite, R. B.: [1] Characterization of finite Boolean lattices and related algebras. J. London math. Soc. **17**, 180—192 (1942).
Bruns, G.: [1] On the representation of Boolean algebras. Canad. Math. Bull. **5**, 37—41 (1962). [2] Darstellungen und Erweiterungen geordneter Mengen. J. reine angew. Math. (I) **209**, 167—200 (1962); (II) **210**, 1—23 (1962).
— and J. Schmidt: [1] Ein Zerlegungssatz für gewisse Boolesche Verbände. Abh. math. Sem. Univ. Hamburg **22**, 191—200 (1958). [2] Eine Verschärfung des Bernsteinschen Äquivalenzsatzes. Math. Annalen **135**, 257—262 (1958).
Büchi, J. R.: [1] Die Boolesche Partialordnung und die Paarung von Gefügen. Portugal. Math. **7**, 119—190 (1948).
Byrne, L.: [1] Two brief formulations of Boolean algebra. Bull. Am. math. Soc. **52**, 269—272 (1946). [2] Boolean algebra in terms of inclusion. Am. J. Math. **70**, 139—143 (1948). [3] Short formulations of Boolean algebra using ring operations. Canad. J. Math. **3**, 31—33 (1951).
Carathéodory, C.: [1] Entwurf für eine Algebraisierung des Integralbegriffs. S.-B. bayer. Akad. Wiss. **1938**, 27—69. [2] Maßtheorie und Integral. Reale Acad. Ital. Atti Convegni **9**, 195—208 (1939), Rome 1943. [3] Bemerkungen zum Ergodensatz von G. Birkhoff. S.-B. math. nat. Abt. bayer. Akad. Wiss. **1944**, 189—208. [4]. Über die Differentiation von Maßfunktionen. Math. Z. **46**, 181—189 (1940).
Čech, E.: [1] On bicompact spaces. Ann. Math. **38**, 823—844 (1937).
Chang, C. C.: [1] On the representation of α-complete Boolean algebras. Trans. Am. math. Soc. **85**, 208—218 (1957).
— and A. Horn: [1] Prime ideal characterization of generalized Post algebras. Proc. Symp. pure Math. **II**, 43—48 (1961).
Chin, L. H., and A. Tarski: [1] Distributive and modular laws in the arithmetic of relation algebras. Univ. California Publ. Math., N. S. **1**, 341—384 (1951).
Choudhury, A. C.: [1] On Boolean narings. Bull. Calcutta math. Soc. **46**, 41—46 (1954).

CHRISTENSEN, D. J., and R. S. PIERCE: [1] Free products of α-distributive Boolean algebras. Math. Scand. 7, 81—105 (1959).
CONSTANTINESCU, P.: [1] Sur la classification des fonctions booléennes symétriques. Acad. R. P. Romine. Stud. Cerc. Mat. 11, 193—206 (1960) (Rumanian).
COPELAND, A. H. sr.: [1] Implicative Boolean algebra. Math. Z. 53, 285—290 (1950). [2] Note on cylindric algebras and polyadic algebras. Michigan math. J. 3, 155—157 (1955—1956).
— and F. HARARY: [1] A characterization of implicative Boolean rings. Canad. J. Math. 5, 465—469 (1953). [2] The extension of an arbitrary Boolean algebra to an implicative Boolean algebra. Proc. Am. math. Soc. 4, 751—758 (1953).
CRAWLEY, P.: [1] Lattices whose congruences form a Boolean algebra. Pacific J. Math. 10, 787—795 (1960).
CROISOT, R.: [1] Axiomatique des lattices distributives. Canad. J. Math. 3, 24—27 (1951).
CUESTA, N.: [1] On an article of MACNEILLE. Rev. Mat. Hisp.-Am. 18, 3—9 (1958).
CUNKLE, C. H.: [1] A note on Boolean operations. Portugal. Math. 18, 177—179 (1959).
DAIGNEAULT, A., and D. MONK: [1] Representation theory for polyadic algebras. Fund. Math. 52, 151—176 (1963).
DAVIS, A. C.: [1] A characterization of complete lattices. Pacific J. Math. 5, 311—319 (1955).
DAVIS, C.: [1] Modal operators, equivalence relations, and projective algebras. Amer. J. Math. 76, 747—762 (1954).
DAY, G. W.: [1] Superatomic Boolean algebras. Notices Am. math. Soc. 8, 518 (1961); 8, 602 (1961).
— and F. M. YAQUB: [1] On free α-extensions of Boolean algebras. Notices Am. math. Soc. 8, 255 (1961).
DEKKER, J. C. E.: [1] The constructivity of maximal dual ideals in certain Boolean algebras. Pacific J. Math. 3, 73—101 (1953).
DIAMOND, A. H.: [1] The complete existential theory of the Whitehead-Huntington set of postulates for the algebra of logic. Trans. Am. math. Soc. 35, 940—948 (1933). [2] Simplification of the Whitehead-Huntington set of postulates for the algebra of logic. Bull. Am. math. Soc. 40, 599—601 (1934).
— and J. C. C. McKINSEY: [1] Algebras and their subalgebras. Bull. Am. math. Soc. 53, 959—962 (1947).
DIEUDONNÉ, J.: [1] Sur le théorème de Lebesgue-Nikodym, III. Ann. Univ. Grenoble Sect. Sci. Math. Phys. (N. S.) 23, 25—53 (1948).
DILWORTH, R. P.: [1] Ideals in Birkhoff lattices. Trans. Am. math. Soc. 49, 325—353 (1941). [2] Lattices with unique complements. Trans. Am. math. Soc. 57, 123—154 (1945). [3] The normal completion of the lattice of continuous functions. Trans. Am. math. Soc. 68, 427—438 (1950).
DUBINS, L. E.: [1] Generalized random variables. Trans. Am. math. Soc. 84, 273—309 (1957).
DUNFORD, N., and J. T. SCHWARTZ: [1] Linear operators, Part I: General theory. New York 1958.
— and M. H. STONE: [1] On the representation theorem for Boolean algebras. Rev. Ci. Lima 43, 743—749 (1941).
DURST, L. K.: [1] On certain subsets of finite Boolean algebras. Proc. Am. math. Soc. 6, 695—697 (1955).
DWINGER, PH.: [1] On the completeness of the quotient algebras of a complete Boolean algebra. I. Indag. Math. 20, 448—456 (1958); II. Indag. Math. 21, 26—35 (1959). [2] Remarks on the field representation of Boolean algebras.

Indag. Math. **22**, 213—217 (1960). [3] A note on the normal β-completion of a Boolean algebra. Nieuw Arch. voor Wiskunde **8**, 83—88 (1960). [4] Introduction to Boolean algebras, Würzburg 1961. [5] Retracts in Boolean algebras, Proc. Symp. Pure Math. II, Am. math. Soc. 1961. [6] A note on the completeness of factor algebras of α-complete Boolean algebras. Indag. Math. **21**, 376—383 (1959).

ELLIOT, J. C.: [1] Autometrization and the symmetric difference. Canad. J. Math. **5**, 324—331 (1953).

ENGELKING, R., and K. KURATOWSKI: [1] Quelques théorèmes de l'Algèbre de Boole et leurs applications topologiques. Fund. Math. **50**, 519—535 (1962).

— and A. PEŁCZYŃSKI: [1] Remarks on dyadic spaces. Coll. Math. **11**, 55—63 (1963).

ENOMOTO, S.: [1] Boolean algebras and field of sets. Osaka Math. J. **5**, 99—115 (1953).

EPSTEIN, G.: [1] The lattice theory of Post algebras. Trans. Am. math. Soc. **95**, 300—317 (1960).

ERDÖS, P., and A. TARSKI: [1] On families of mutually exclusive sets. Ann. Math. **44**, 315—329 (1953). [2] On some problems involving inaccessible cardinals. Essays on foundations of mathematics. Jerusalem 1961, 50—82.

EVERETT, C. J., and S. ULAM: [1] Projective algebra. Am. J. Math. **68**, 77—88 (1946).

FADINI, A.: [1] Algoritmi tra operazione unarie e binarie in un'algebra di Boole. Ricerca (Napoli) **11**, 16—34 (1960).

FELL, J. M. G., and A. TARSKI: [1] On algebras whose factor algebras are Boolean. Pacific J. Math. **2**, 297—318 (1952).

FICHTENHOLZ, G., et L. KANTOROVITCH: [1] Sur les opérations dans l'espace des fonctions bornées. Studia Math. **5**, 69—98 (1934).

FLOYD, E. E.: [1] Boolean algebras with pathological order topologies. Pacific J. Math. **5**, 687—689 (1955).

FOGUEL, S. R.: [1] Boolean algebras of projections of finite multiplicity. Pacific J. Math. **9**, 681—693 (1959).

FORADORI, E.: [1] Zur Theorie des Carathéodoryschen Integrals. Dtsch. Math. **4**, 578—582 (1939).

FORTET, R.: [1] L'Algèbre de Boole et ses applications en recherche opérationnelle. Cahiers Centre Etudes Rech. Oper. **4**, 5—36 (1959).

FOSTER, A. L.: [1] The theory of Boolean-like rings. Trans. Am. math. Soc. **59**, 166—187 (1946). [2] p-rings and their Boolean vector representation. Acta Math. **84**, 231—261 (1951). [3] The idempotent elements of a commutative ring. Duke Math. J. **12**, 143—152 (1945).

FRINK, O. jr.: [1] Representations of Boolean algebras. Bull. Am. math. Soc. **47**, 755—756 (1941). [2] On the existence of linear algebras in Boolean algebras. Bull. Am. math. Soc. **34**, 329—333 (1933).

FUNAYAMA, N.: [1] Imbedding infinitely distributive lattices completely isomorphically into Boolean algebras. Nagoya Math. J. **15**, 71—81 (1959).

GAIFMAN, H.: [1] Two contributions to the theory of Boolean algebras (doctoral dissertation), University of California, Berkeley 1962. [2] Concering measures in Boolean algebras. Pacific J. Math. **14**, 61—73 (1964). [3] Infinite Boolean polynomials I. Fund. Math. **54**, 229—250 (1964); errata Fund. Math. **57**, 117 (1965)

GALLER, B. A.: [1] Cylindric and polyadic algebras. Proc. Am. math. Soc. **8**, 176—183 (1957).

GILLMAN, L., and M. JERISON: [1] Rings of continuous functions. Toronto-London-New York 1960.

GINGSBURG, S.: [1] A class of everywhere branching sets. Duke Math. J. 20, 521—526 (1953). [2] On the existence of complete Boolean algebras whose principal ideals are isomorphic to each other. Proc. Am. math. Soc. 9, 130—132 (1958).

GLEASON, A. M.: [1] Projective topological spaces. Ill. J. Math. 2, 482+489 (1958).

GLIVENKO, V.: [1] Sur quelques points de la logique de Brouwer. Bull. Acad. Sci. Belg. 1929, 183—188.

GÖDEL, K.: [1] The consistency of the Continuum Hypothesis. Princeton 1940.

GOODNER, D. B.: [1] Projections in normed linear spaces. Trans. Am. math. Soc. 69, 89—108 (1950).

GÖTZ, A.: [1] Analogues to the notion of point functions in Boolean algebras. Prace Mat. 1, 145—161 (1954) (Polish).

GRAU, A. A.: [1] A ternary operation related to the complete disjunction of Boolean algebra. Univ. Nac. Tucuman Rev. A. 8, 121—126 (1951). [2] Ternary Boolean algebra. Bull. Am. math. Soc. 53, 567—572 (1947).

GRÄTZER, G., and E. T. SCHMIDT: [1] Two notes on lattice congruences. Ann. Univ. Sci. Budapest Eötvös. Sect. Math. 1, 83—87 (1958). [2] Characterizations of relatively complemented distributive lattices. Publ. math. Debrecen 5, 275—287 (1958). [3] On the generalized Boolean algebras generated by a distributive lattice. Nederl. Akad. Wet. Proc. Ser. A. 61, 547—553 (1958).

GUILLAUME, M.: [1] Calculs de conséquences et tableaux d'épreuve pour les classes algébriques générales d'anneaux booléins à opérateurs. Compt. rend. 247, 1542—1544 (1958).

HAIMO, F.: [1] Some limits of Boolean algebras. Proc. Am. math. Soc. 2, 566—576 (1951). [2] A representation for Boolean algebras. Am. J. Math. 73, 725—740 (1951).

HALES, A. W.: [1] On the nonexistence of free complete Boolean algebras, doctoral dissertation. California Institute of Technology, Pasadena 1962. [2] On the nonexistence of free complete Boolean algebras. Fund. Math 54, 45—66 (1964).

HALMOS, P. R.: [1] Measure Theory. New York 1950. [2] The foundations of probability. Am. Math. Monthly 51, 497—510 (1944). [3] Polyadic Boolean algebras. Proc. nat. Acad. Sci. (Wash.) 40, 296—301 (1954). [4] Algebraic logic, I. Monadic Boolean algebras. Comp. Math. 12, 217—249 (1955); II. Homogeneous locally finite polyadic Boolean algebras of infinite degree. Fund. Math. 53, 255—325 (1956); III. Predicates, terms and operations in polyadic algebras. Trans. Am. math. Soc. 83, 430—470 (1956); IV. Equality in polyadic algebras. Trans. Am. math. Soc. 86, 1—27 (1957). [5] The basic concepts of algebraic logic. Am. Math. Monthly 63, 365—387 (1956). [6] The representation of monadic Boolean algebras. Duke Math. J. 26, 447—454 (1959). [7] Free monadic algebras. Proc. Am. math. Soc. 10, 219—227 (1959). [8] Lectures on Boolean algebras. Toronto-New York-London 1963. [9] Algebraic logic. New York 1962. [10] Injective and projective Boolean algebras. Proc. Sympos. Pure Math. II, 114—122 (1961).

HALPERN, J. D.: [1] The independence of the axiom of choice from the Boolean prime ideal theorem. Notices Am. math. Soc. 8, 641 (1961). [2] Contributions to the study of the independence of the axiom of choice, doctoral dissertation, University of California 1962. [3] The independence of the axiom of choice from the Boolean prime ideal theorem. Fund. Math. 55 (to appear).

HAMMER, L. P.: [1] Shefferian algebras. Bull. Math. Soc. Sci. Math. Phys. R. P. R. 3 (1959).

HANF, W.: [1] On some fundamental problems concerning isomorphism of Boolean algebras. Math. scand. 5, 205—217 (1957). [2] Models of languages with infinitely long expressions. International Congress for Logic, Methodology and Philosophy of Sciences; Abstract of contributed Papers, Stanford University 1960 (mimeographed), p. 24. [3] Incompactness in languages with infinitely long expressions. Fund. Math. **53**, 309—324 (1964). [4] On a problem of ERDÖS and TARSKI. Fund. Math. **53**, 325—334 (1964).

HARARY, F.: [1] Atomic Boolean-like rings with finite radical. Duke Math. J. **17**, 273—276 (1950).

HASENJAEGER, G.: [1] Eine Bemerkung zu HENKIN's Beweis für die Vollständigkeit des Prädikatenkalküls der ersten Stufe. Journal of Symbolic Logic **18**, 42—48 (1953).

HAUPT, O., u. CH. Y. PAUC.: [1] Über die Erweiterung eines Inhaltes zu einem Maße. S.-B. math.-nat. Kl. bayer. Akad. Wiss. **1948**, 247—253. [2] Vitalische Systeme in Booleschen σ-Verbänden. S.-B. math.-nat. Kl. bayer. Akad. Wiss. **1950**, 187—207. [3] Holabedingungen und Vitalische Eigenschaft von Somensystemen. Arch. Math. **4**, 107—114 (1953). [4] Über Adjunktion von Idealen in Booleschen Verbänden. Akad. Wiss. Mainz. Abh. Math.-Nat. Kl. **1957**, 177—193.

HAUSDORFF, F.: [1] Über zwei Sätze von G. FICHTENHOLZ und L. KANTOROVITCH. Studia Math. **6**, 18—19 (1936).

HEIDER, L. J.: [1] Prime dual ideals in Boolean algebras. Canad. J. Math. **11**, 397—408 (1959). [2] A representation theorem for measures on Boolean algebras. Mich. Math. J. **5**, 213—221 (1958).

HELSON, H.: [1] On the symmetric difference of sets as a group operation. Coll. Math. **1**, 203—205 (1948). [2] Remark on measures in almost independent fields. Studia math. **10**, 182—183 (1948).

HENKIN, L.: [1] The completeness of the first-order functional calculus. J. Symb. Logic **14**, 159—166 (1949). [2] Boolean representation through propositional calculus. Fund. Math. **41**, 89—96 (1955). [3] The algebraic structure of mathematical theories. Bull. Soc. Math. Belg. **7**, 131—136 (1955). [4] La structure algébrique des théories mathématiques. Paris 1956. [5] Metamathematical theorems equivalent to the prime ideals theorems for Boolean algebras. Bull. Am. math. Soc. **60**, 387—388 (1954).

— and A. TARSKI: [1] Cylindric algebras. Proc. of symposia in pure math. II, 83—113 (1960). [2] Cylindric algebras. Summaries of talks presented at the Summer Institute of Symbolic Logic 3, 332—340 (1957).

HERMES, H.: [1] Einführung in die Verbandstheorie. Berlin-Göttingen-Heidelberg **1955**.

HEWITT, E.: [1] A note on measures in Boolean algebras. Duke Math. J. **20**, 253—256 (1953). [2] A remark on density characters. Bull. Am. math. Soc. **52**, 641—643 (1946).

HOBERMAN, S., and J. C. C. MCKINSEY: [1] A set of postulates for Boolean algebra. Bull. Am. math. Soc. **43**, 588—592 (1937).

HODGES, J. L., and A. HORN: [1] On MAHARAM's conditions for measure. Trans. Am. math. Soc. **64**, 594—595 (1948).

HOFMANN, H.: [1] Über eine Dimensionstheorie in topologischen Verbänden. Fund. Math. **42**, 289—311 (1955).

HORN, A., and A. TARSKI: [1] Measures in Boolean algebras. Trans. Am. math. Soc. **64**, 467—497 (1948).

HUNTINGTON, E. V.: [1] Sets of independent postulates for the algebra of logic. Trans. Am. math. Soc. **5**, 288—309 (1904). [2] A new set of independent postulates for the algebra of logic with special reference to WHITEHEAD and RUSSELL's Principia Mathematica. Proc. nat. Acad. Sci. (Wash.) **18**, 179—180

(1932). [3] New sets of independent postulates for the algebra of logic, with special reference to WHITEHEAD and RUSSELL's Principia Mathematica. Trans. Am. math. Soc. 35, 274—304 (1933); correction p. 557—558 and 971.

ISBELL, J. R., and Z. SEMADENI: [1] Projection constants and spaces of continuous functions. Trans. Am. math. Soc. 107, 38—48 (1963).

ISEKI, K.: [1] A construction of two-valued measure on Boolean algebra. J. Osaka Inst. Sci. Techn. Part I, 2, 43—45 (1950).

ITOH, M.: [1] On Boolean equation with many known elements and generalized PORETZKY's formula. Univ. Nac. Tucumán. Rev. Ser. A 12, 107—112 (1959).

JAKUBIK, J.: [1] Remarks on the JORDAN-DEDEKIND condition in Boolean algebras. Časopis. Pest. Mat. 82, 44—46 (1957) (Slovak). [2] Über Ketten in Booleschen Verbänden. Mat.-Fyz. Časopis. Slovensk. Akad. Vied 8, 193—202 (Russian).

JÓNSSON, B.: [1] A Boolean algebra without proper automorphisms. Proc. Am. math. Soc. 2, 766—770 (1951).

— and A. TARSKI: [1] Boolean algebras with operators. I. Amer. J. Math. 73, 891—939 (1951); II. Am. J. Math. 74, 127—162 (1952). [2] Cardinal products of isomorphism types. Appendix to TARSKI [8].

KAKUTANI, S.: [1] Weak topology, bicompact set and the principle of duality. Proc. Imp. Acad. Jap. 16, 63—67 (1940). [2] Concrete representation of abstract (L)-spaces and the mean ergodic theorem. Ann. Math. 42, 523—537 (1941). [3] Concrete representation of abstract (M)-spaces. Ann. Math. 42, 994—1024 (1941).

KALICKI, J.: [1] On the axioms of GRAU's ternary algebra. Proc. Leeds Phil. Lit. Soc. Sci. Sect. 6, 12—13 (1952).

KAMEL, H.: [1] Relational algebra and uniform spaces. J. London math. Soc. 29, 342—344 (1954).

KANTOROVIČ, L. V., B. Z. VULICH and A. G. PINSKER: [1] Functional analysis in partially ordered spaces. Moscow-Leningrad 1950 (Russian).

KAPPOS, D. A.: [1] Ein Beitrag zur Carathéodoryschen Definition der Ortsfunktion in Booleschen Algebren. Math. Z. 51, 616—634 (1949). [2] Die Cartesischen Produkte und die Multiplikation von Maßfunktionen in Booleschen Algebren. I. Math. Ann. 120, 43—74 (1947); II. Math. Ann. 121, 223—333 (1949). [3] Baire and Borel theory for the Carathéodory „Ortsfunktionen". Bull. Soc. Math. Grèce 25, 130—152 (1951). [4] Erweiterung von Maßverbänden. J. reine angew. Math. 191, 97—109 (1953). [5] Strukturtheorie der Wahrscheinlichkeitsfelder und -räume. Berlin-Göttingen-Heidelberg. Springer Verlag 1960.

KARP, C. R.: [1] A note on the representation of α-complete Boolean algebras. Proc. Am. math. Soc. 14, 705—707 (1963).

KASNER, M.: [1] Les algèbres cylindriques. Bull. Soc. Math. France 86, 315—319 (1958).

KATĚTOV, M.: [1] Remarks on Boolean algebras. Coll. Math. 2, 229—235 (1951). [2] Measures in fully normal spaces. Fund. Math. 38, 73—84 (1951).

KAWADA, Y.: [1] Über die Existenz der invarianten Integrale. Jap. J. Math. 19, 81—95 (1949).

KEEDY, M. L.: [1] On a theorem of JÓNSSON and TARSKI. Portugal. Math. 16, 11—14 (1957).

KEISLER, H. J.: [1] Some applications of the theory of models to set theory; Logic, Methodology and Philosophy of Science. Proceedings of the 1960 International Congress, Stanford 1962, 80—86.

— and A. TARSKI: [1] From accessible to inaccessible cardinals. Results holding for whole accessible cardinal numbers and the problem of their extension for inaccessible ones. Fund. Math. 53, 225—308 (1964).

KELLEY, J. L.: [1] Banach spaces with the extension property. Trans. Am. math. Soc. **72**, 323—326 (1952). [2] Measures in Boolean algebras. Pacific J. Math. **9**, 1165—1177 (1959).

KERSTAN, J.: [1] Tensorielle Erweiterungen distributiver Verbände. Math. Nachr. **22**, 1—20 (1960). [2] Zur topologischen Invarianz der Hausdorffschen $Q^{(\alpha)}$-Mengen. Z. math. Logik u. Grundl. Math. **7**, 259—277 (1961).

KINOSHITA, S.: [1] A solution of a problem of R. SIKORSKI. Fund. Math. **40**, 39—41 (1953).

KLEIN, F.: [1] Boole-Schrödersche Verbände. Dtsch. Math. **1**, 528—537 (1936).

KOLMOGOROFF, A. N.: [1] Grundbegriffe der Wahrscheinlichkeitsrechnung. Berlin 1933. [2] Algèbres de Boole métriques complètes. VI. Zjazd Matematyków Polskich 1948. Appendix to Ann. Soc. Pol. Math. **20**, 21—30 (1948).

KOWALSKY, H. J.: [1] Distributivität in atomaren Booleschen Verbänden. Arch. Math. **6**, 9—12 (1954).

KÖTHE, G.: [1] Die Theorie der Verbände, ein neuer Versuch zur Grundlegung der Algebra und der projektiven Geometrie. Jber. Dtsch. Math. Ver. **47**, 125—144 (1937).

KRICKEBERG, KL.: [1] Extreme Derivierte von Zellenfunktionen in Booleschen σ-Algebren und ihre Integration. Bayer. Akad. Wiss., math.-nat. Kl. 1955, 217—279. [2] Convergence of martingales with a directed index set. Trans. Am. math. Soc. **83**, 313—337 (1956). [3] Stochastische Konvergenz von Semimartingalen. Math. Z. **66**, 470—486 (1957). [4] Stochastische Derivierte. Math. Nachr. **18**, 203—217 (1958).

KURATOWSKI, C.: [1] Sur l'opération $\overline{A}$ de l'Analysis Situs. Fund. Math. **3**, 181—199 (1922). [2] Quelques problèmes concernant les espaces métriques non-séparables, Fund. Math. **25**, 534—535 (1935). [3] Topologie I (second edition). Warszawa-Wrocław 1948. [4] Topologie II, Warszawa-Wrocław 1950.

— et T. POSAMENT: [1] Sur l'isomorphie algébro-logique et les ensembles relativement boréliens. Fund. Math. **22**, 281—286 (1934).

KUROŠ, A. G., A. H. LIVŠIC and E. G. ŠULGEIFER: [1] Foundations of the theory of categories. Uspehi Mat. Nauk (6) **15**, 1—46 (1960) (Russian).

L'ABBÉ, M.: [1] Structures algébriques suggérées par la logique mathématique. Bull. Soc. Math. France **86**, 299—314 (1958).

LALAN, V.: [1] Définition de deux structures d'anneaux dans une algèbre de Boole. Compt. rend. **223**, 1086—1087 (1946). [2] Equations fonctionnelles dans un anneau booléien. Compt. rend. **230**, 603—605 (1950).

LAMPERTI, J.: [1] A note on autometrized Boolean algebras. Am. Math. Monthly **64**, 188—189 (1957).

LÄUCHLI, H.: [1] The independence of the ordering principle from a restricted axiom of choise, to appear.

LEBLANC, L.: [1] Dualité pour égalités booliennes. Compt. rend. **250**, 3552—3553 (1960). [2] Les algèbres boolénnes topologiques bornées. Compt. rend. **250**, 3766—3768 (1960). [3] Les algèbres de transformation. Compt. rend. **250**, 3928—3930 (1960). [4] Représentation des algèbres polyadiques pour anneau. Compt. rend. **250**, 4092—4094 (1960). [5] Nonhomogenous polyadic algebras. Proc. Am. math. Soc. **13**, 59—65 (1962). [6] Duality for Boolean equalities. Proc. Am. math. Soc. **13**, 74—79 (1962).

LIVENSON, E.: [1] On the realization of Boolean algebras by algebras of sets. Rec. math. Moscou, N. S. F. **7**, 309—312 (1940).

LJUBČENKO, G. G.: [1] Representing Boolean functions by formulae (Ukrainian). Dopovidi Akad. Nauk Ukraïn. RSR 1960, 1011—1015. [2] Logical synthesis of

schemes of arrangement realizing Boolean functions of a definite class (Ukrainian).
Dopovidi Akad. Nauk Ukraïn. RSR **1960**, 1331—1333.
Loomis, L. H.: [1] On the representation of σ-complete Boolean algebras. Bull.
Am. math. Soc. **53**, 757—760 (1947).
Łoś, J.: [1] Sur le théorème de Gödel pour les théories indénombrables. Bull. Acad.
Polon. Sci. Cl. III, **2**, 319—320 (1954). [2] On the axiomatic treatment of
probability. Coll. Math. **3**, 125—137 (1955). [3] Remarks on Henkin's paper:
Boolean representation through propositional calculus. Fund. Math. **44**, 82—83
(1957). [4] O ciałach zdarzeń i ich definicji w aksjomatycznej teorii prawdopodobieństwa (On fields of events and their definition in the axiomatic Theory of
Probability). Studia Logica **9**, 95—115 (1960) (Polish; English summary
125—132). [5] Quelques remarques, théorèmes et problèmes sur les classes
définissables d'algèbres. Mathematical Interpretations of Formal Systems,
Amsterdam 1955, p. 98. [6] Remarks on foundations of probability. Semantic
interpretation of the probability of formulas. Proc. Internat. Congres Math.
1962, 225—229.
— and E. Marczewski: [1] Extensions of measure. Fund. Math. **36**, 267—276
(1949).
— A. Mostowski and H. Rasiowa: [1] A proof of Herbrand's theorem. J. Math.
Pur. Appl. **35**, 19—24 (1956). [2] Addition au travail "A proof of Herbrand's
theorem". J. Math. Pur. Appl. **40**, 129—134 (1961).
— and C.Ryll-Nardzewski: [1] Effectiveness of the representation theory for
Boolean algebras. Fund. Math. **41**, 49—56 (1954). [2] On the application of
Tychonoff's theorem in mathematical proofs. Fund. Math. **38**, 233—237 (1951).
Löwig, H.: [1] On transistive Boolean relations. Czechoslovak math. J. **1**, 199—201
(1951). [2] Instrisic topology and completion of Boolean rings. Ann. Math. **42**,
1138—1196 (1941).
Luxemburg, W. A. J.: [1] Non-standard analysis. California Institute of Technology,
Pasadena 1962. [2] Two applications of the Method of Construction by Ultra-
powers to Analysis. Bull. Am. math. Soc. **68**, 416—419 (1962). [3] A remark on
Sikorski's extension theorem for homomorphisms in the theory of Boolean
algebras. Fund. Math. **55**, 239—247 (1963—1964).
Lyndon, R. C.: [1] The representation of relational algebras. I. Ann. Math. **51**,
707—729 (1950); II. Ann. Math. **63**, 294—307 (1956). [2] Relation algebras
and projective geometries. Michigan Math. J. **8**, 21—28 (1961).
Mackey, G. W.: [1] Point realizations of transformation groups. Illinois J. Math.
6, 327—335 (1962).
MacNeille, H.: [1] Partially ordered sets. Trans. Am. math. Soc. **42**, 416—460
(1937). [2] Extension of a distributive lattice to a Boolean ring. Bull. Am.
math. Soc. **45**, 452—455 (1939).
Maeda, F.: [1] Ideals in a Boolean algebra with transfinite chain condition. J. Sci.
Hiroshima Univ. A **10**, 7—36 (1940). [2] Partially ordered linear spaces. J. Sci.
Hiroshima Univ. A **10**, 137—150 (1940).
Maharam, D.: [1] On homogeneous measure algebras. Proc. nat. Acad. Sci. (Wash.)
28, 108—111 (1942). [2] An algebraic characterization of measure algebras.
Ann. Math. **48**, 154—167 (1947). [3] Set functions and Souslin's hypothesis.
Bull. Am. math. Soc. **54**, 587—590 (1948). [4] The representation of abstract
measure functions. Trans. Am. math. Soc. **65**, 279—330 (1949). [5] Decomposition of measure algebras and spaces. Trans. Am. math. Soc. **69**, 142—160
(1950). [6] The representation of abstract integrals. Trans. Am. math. Soc. **75**,
154—184 (1953). [7] Automorphisms of product of measure spaces. Proc.
Am. math. Soc. **9**, 702—707 (1958). [8] On a theorem of von Neumann. Proc.

Am. math. Soc. **9**, 987—994 (1958). [9] Homogenous extensions of positive linear operators. Trans. Am. math. Soc. **99**, 62—82 (1961).

MARCZEWSKI, E. (SZPILRAJN): [1] Remarques sur les fonctions complètement additives d'ensemble et sur les ensembles jouissant de la propriété de Baire. Fund. Math. **22**, 303—311 (1934). [2] The characteristic function of a sequence of sets and some of its applications. Fund. Math. **31**, 207—223 (1938). [3] On the isomorphism and the equivalence of classes and sequences of sets. Fund. Math. **32**, 133—148 (1939). [4] Mesures dans les corps de Boole. Ann. Soc. Pol. Math. **19**, 243—244 (1946). [5] Indépendance d'ensembles et prolongement de mesures. Coll. Math. **1**, 122—132 (1948). [6] Ensembles indépendants, et leurs applications à la théorie de la mesure. Fund. Math. **35**, 1—28 (1948). [7] Concerning the symmetric difference in the theory of sets and in Boolean algebras. Coll. Math. **1**, 199—202 (1948). [8] Measures in almost independent fields. Fund. Math. **38**, 217—229 (1951). [9] Séparabilité et multiplication cartésienne des espaces topologiques. Fund. Math. **34**, 127—143 (1947). [10] Sur l'équivalence des suites d'ensembles et l'équivalence des fonctions. Fund. Math. **26**, 302—326 (1936). [11] On the equivalence of some classes of sets. Fund. Math. **30**, 235—241 (1938). [12] Sur les mesures à deux valeurs et les idéaux premiers dans les corps d'ensembles. Ann. Soc. Pol. Math. **19**, 232—233 (1947). [13] Two-valued measures and prime ideals in fields of sets. Compt. rend. Soc. Sc. Lett. Varsovie Cl. III **1947**, 11—17. [14] Ensembles indépendants et mesures non-séparables. Compt. rend. **207**, 768—770 (1938). [15] Sur les ensembles et les fonctions absolument mesurables. Compt. rend. Soc. Sci. Varsovie Cl. III **30**, 1—30 (1937) (Polish). [16] Independence in algebras of sets and Boolean algebras. Fund. Math. **48**, 135—145 (1959/60).

— et W. SIERPIŃSKI: [1] Remarque sur le problème de la mesure. Fund. Math. **26**, 256—261 (1936).

— and R. SIKORSKI: [1] Measures in non-separable metric spaces. Coll. Math. **1**, 133—139 (1948). [2] Remarks on measures and category. Coll. Math. **2**, 13—19 (1949). [3] On isomorphism types of measure algebras. Fund. Math. **38**, 92—98 (1951).

— and T. TRACZYK: [1] On developable sets and almost-limit points. Coll. Math. **8**, 55—56 (1961).

MARKOV, A. A.: [1] On the inversion complexity of function system. Dokl. Akad. Nauk SSSR **116**, 917—919 (1957).

MATSUSHITA, S.: [1] The algebra of topological operations. I. Math. Jap. **1**, 28—35 (1948); II. J. Osaka Inst. Sci. Techn. Part I, **1**, 77—80 (1949).

MATTHES, K.: [1] Über eine Schar von Regularitätsbedingungen für Verbände. Math. Nachr. **22**, 93—128 (1960). [2] Über die Ausdehnung von ℵ-Homomorphismen Boolescher Algebren. Z. math. Logik u. Grundl. Math. **6**, 97—105 (1960); (II) — **7**, 16—19 (1961).

MAYER, R. D., and R. S. PIERCE: [1] Boolean algebras with ordered bases. Pacific J. Math. **10**, 925—942 (1960).

MAZUR, S.: [1] On continuous mappings on Cartesian products. Fund. Math. **39**, 229—238 (1952).

MAZURKIEWICZ, S.: [1] Podstawy rachunku prawdopodobieństwa (Foundation of the calculus of probability), prepared for print from the late author's manuscript by J. Łoś, Warszawa 1956 (Polish).

— et W. SIERPIŃSKI: [1] Contribution à la topologie des ensembles dénombrables. Fund. Math. **1**, 17—27 (1920).

McCARTHY, C. A.: [1] Commuting Boolean algebras of projections. Pacific J. Math. **2**, 295—307 (1961).

McCoy, N. H.: [1] Subrings of direct sums. Am. J. Math. **60**, 374—382 (1938)
— and D. Montgomery: [1] A representation of generalized Boolean rings. Duke math. J. **3**, 455—459 (1937).
McKinsey, J. C. C.: [1] Boolean functions and points. Duke math. J. **2**, 465—471 (1936). [2] On Boolean functions of many variables. Trans. Am. math. Soc. **40**, 343—362 (1936). [3] A condition that a first Boolean function vanish whenever a second does not. Bull. Am. math. Soc. **43**, 694—696 (1937). [4] A solution of the decision problem for the Lewis system S2 and S4 with an application to topology. J. Symb. Logic **4**, 115—158 (1939). [5] On the representation of projective algebras. Amer. J. Math. **70**, 375—384 (1948).
— and A. Tarski: [1] The algebra of topology. Ann. Math. **45**, 141—191 (1944). [2] On closed elements in closure algebras. Ann. Math. **47**, 122—162 (1946). [3] Some theorems about the sentential calculi of Lewis and Heyting. J. Symb. Logic **13**, 1—15 (1948).
Mibu, Y.: [1] Relations between measures and topology in some Boolean spaces. Proc. Imp. Acad. Tokyo **20**, 454—458 (1944).
Michiura, T.: [1] On characteristic properties of Boolean algebras. J. Osaka Inst. Sci. Tech. Part I. **1**, 129—133 (1949).
Miheev, V. M.: [1] On sets containing the greatest number of pairwise non-comparable Boolean vectors (Russian). Probl. Kibernet. **2**, 69—71 (1959).
Miller, D. G.: [1] Postulates for Boolean algebras. Amer. Math. Monthly **59**, 93—96 (1952).
Moisil, G. C.: [1] Sur les anneaux de caractéristique 2 ou 3 et leurs applications. Bull. Ecole Polytech. Bucurest **12**, 66—90 (1941). [2] Sur l'algèbre des relations binaires (I). Com. Acad. R. P. Romine **8**, 1251—1254 (1958).
Monk, D.: [1] On the representation theory for cylindric algebras. Pacific J. Math. **11**, 1447—1457 (1961).
Montague, R., and J. Tarski: [1] On Bernstein's selfdual set of postulates for Boolean algebras. Proc. Am. math. Soc. **5**, 310—311 (1954).
Monteiro, A.: [1] Propriedades características de los filtros de un algebra de Boole (Characteristic properties of the filters of a Boolean algebra). Acta Cuyana Ingen. **1**, no. 5 (1954) (Spanish). [2] L'arithmétique des filtres et les espaces topologiques. De Segundo Symposium de Matematicas. Buenos Aires 1954.
— et H. Ribeiro: [1] L'opération de fermeture et ses invariants dans les systèmes partiellement ordonnés. Portugal. Math. **3**, 171—184 (1942).
Montgomery, D.: [1] Non-separable metric spaces. Fund. Math. **25**, 527—533 (1935).
Mori, S.: [1] Prime ideals in Boolean rings. J. Sci. Hiroshima Univ. A **9**, 67—71 (1939). [2] On the group structure of Boolean lattices. Proc. Japan Acad. **32**, 423—425 (1956).
Mostowski, A.: [1] Abzählbare Boolesche Körper und ihre Anwendungen auf die allgemeine Metamathematik. Fund. Math. **29**, 34—53 (1937). [2] Über die Unabhängigkeit des Wohlordnungssatzes vom Ordnungsprinzip. Fund. Math. **32**, 201—252 (1939). [3] Proofs of nondeducibility in intuitionistic functional calculus. J. Symb. Logic **13**, 204—207 (1948).
— u. A. Tarski: [1] Boolesche Ringe mit geordneter Basis. Fund. Math. **32**, 69—86 (1939).
Mrówka, S.: [1] On the ideal's extension theorem and its equivalence to the axiom of choice. Fund. Math. **43**, 46—49 (1956). [2] Two remarks to my paper: "On the ideal's extension theorem and its equivalence to the axiom of choice". Fund. Math. **46**, 165—166 (1958).

NACHBIN, L.: [1] Une propriéte caractéristique des algèbres booléiennes. Portugal. Math. 6, 115—188 (1947). [2] On a characterization of the lattice of all ideals of a Boolean ring. Fund. Math. 36, 137—142 (1949). [3] A theorem of the HAHN-BANACH type for linear transformations. Trans. Am. math. Soc. 68, 28—46 (1950).

NAKANO, H.: [1] Über das System aller stetigen Funktionen auf einem topologischen Raume. Proc. Imp. Acad. Tokyo 17, 308—310 (1941).

NEUMANN, J. VON: [1] Einige Sätze über meßbare Abbildungen. Ann. Math. 33, 574—586 (1932). [2] Lectures on continuous geometry. Princeton 1937.

— and M. H. STONE: [1] The determination of representative elements in the residual classes of a Boolean algebra. Fund. Math. 25, 353—376 (1935).

NEWMAN, M. H. A.: [1] A characterization of Boolean lattices and rings. J. London math. Soc. 16, 256—272 (1941).

NIKODYM, O.: [1] Sur une généralisation des intégrales de M. J. Radon. Fund. Math. 15, 131—179 (1930). [2] Sur la mesure vectorielle parfaitement additive dans un corps abstrait de Boole. Acad. roy. Belg. Cl. Sci. Mem. 17, 1—40 (1938). [3] Sur les êtres fonctionnoïdes. Compt. rend. 226, 375—377, 458—460, 541—543 (1948). [4] Tribus de Boole et fonctions mesurables. Compt. rend. 228, 37—38, 150—151 (1949). [5] Remarks on the Lebesgue's measure extension device for finitely additive Boolean lattices. Proc. nat. Acad. Sci. (Wash.) 37, 533—537 (1951). [6] Critical remarks on some basic notions in Boolean lattices. (I) Ann. Acad. Brasil. Ci. 24, 113—136 (1952); (II) Rend. Sem. Mat. d'Univ. Padova 27, 193—217 (1957). [7] Sur l'extension d'une mesure non Archimédienne sur une tribu de Boole simplement additive, à une autre tribu plus extendue. Compt. rend. 241, 1439—1440, 1544—1545, 1695—1696 (1955); 242, 864—866 (1956). [8] On extension of a given finitely additive field-valued non-negative measure on a finitely additive Boolean tribe to another tribe more ample. Rend. Sem. Mat. Univ. de Padova 26, 232—327 (1956). [9] Sur le mesure non-archimédienne effective sur une tribu de Boole arbitraire. Compt. rend. 251, 2113—2115 (1960).

NÖBELING, G.: [1] Topologie der Vereine und Verbände. Arch. Mat. 1, 154—159 (1949). [2] Grundlagen der analytischen Topologie. Berlin-Göttingen-Heidelberg 1954.

NOLIN, L.: [1] Sur les classes d'algèbres équationnelles et les théorèmes de représentation. Compt. rend. 244, 1862—1863 (1957). [2] Algèbres de Boole et calcul des propositions. Compt. rend. 244, 1999—2002 (1957).

NOVÁK, J.: [1] On the Cartesian product of two compact spaces. Fund. Math. 40, 106—112 (1953).

— and M. NOVOTNÝ: [1] On the convergence in σ-algebras of point-sets. Czechoslovack. Mat. Z. 3, 291—296 (1953).

OGASAWARA, T.: [1] Compact metric Boolean algebras and vector lattices. J. Sci. Hiroshima Univ. Ser. A 11, 125—128 (1942).

OLMSTED, J. M. H.: [1] Lebesgue theory on a Boolean algebra. Trans. Am. math. Soc. 51, 164—193 (1942).

ONICESCU, O.: [1] Notes sur les b-algèbres. An. Univ. ,,C.I. Parhon" Bucureşti. Ser. Şti. Nat. 22, 17—22 (1959); Rev. Math. pures Appl. 4, 345—350 (1959).

OXTOBY, J. C., and S. ULAM: [1] On the equivalence of any set of first category to a set of measure zero. Fund. Math. 31, 201—206 (1938).

PANKAJAM, S.: [1] On symmetric functions of n elements in a Boolean algebra. J. Indian Math. Soc. N. s. 2, 198—210 (1937). [2] On symmetric functions of m symmetric functions in a Boolean algebra. Proc. Indian Acad. Sci. Sect. A 9,

95—102 (1939). [3] Ideal theory in Boolean algebra and its application to deductive systems. Proc. Indian Acad. Sci. Sect. A **14**, 670—684 (1941).
PAUC, CH.: [1] Construction de mesures. Compt. rend. **222**, 123—125 (1946). [2] Compléments à la représentation ensembliste d'une algèbre et d'une σ-algèbre booléennes. Compt. rend. **225**, 219—221 (1947). [3] Darstellungs- und Struktursätze für Boolesche Verbände und σ-Verbände. Arch.Math. **1**, 29—41 (1948). [4] Les théorèmes fort et faible de Vitali et les conditions d'évanescence de halos. Compt. rend. **232**, 1727—1729 (1951).
PEŁCZYŃSKI, A., and Z. SEMADENI: [1] Spaces of continuous functions (III) (Spaces $C(\Omega)$ for Ω without perfect subsets). Studia Math. **18**, 211—222 (1959).
PENNING, CH. J.: [1] Boolean metric spaces, doctoral dissertation. Technische Hogeschool te Delft, 1960.
PEREMANS, W.: [1] Embedding of a distributive lattice into a Boolean algebra. Ind. Math. **19**, 73—81 (1957).
PETTIS, B. J.: [1] On the extension of measures. Ann. Math. **54**, 186—197 (1951).
PIERCE, R. S.: [1] The Boolean algebra of regular open sets. Canad. J. Math. **5**, 95—100 (1953). [2] Representation theorems for certain Boolean algebras. Proc. Am. math. Soc. **10**, 42—50 (1959). [3] Distributivity in Boolean algebras. Pacific J. Math. **7**, 983—992 (1957). [4] A note on complete Boolean algebras. Proc. Am. math. Soc. **9**, 892—896 (1958). [5] A generalization of atomic Boolean algebras. Pacific J. Math. **9**, 175—182 (1959). [6] Distributivity and the normal completion of Boolean algebras. Pacific J. Math. **8**, 133—140 (1958). [7] Some questions about complete Boolean algebras. Proc. Symp. pure Math. **2**, 129—140 (1961). [8] A note on free algebras. Proc. Am. math. Soc. **14**, 845—846 (1963).
POMÉ, R., e G. RUSPA: [1] Un teorema di algebra logica. Atti accad. sci. Torino Classe sci. fis. mat. e nat. **95**, 336—342 (1960/61).
POSPÍŠIL, B.: [1] Von den Verteilungen auf Booleschen Ringen. Math. Ann. **118**, 32—40 (1941). [2] Eine Bemerkung über stetige Verteilungen. Čas. mat. fys. **70**, 68—72 (1941). [3] Wesentliche Primideale in vollständigen Ringen. Fund. Math. **33**, 66—74 (1945). [4] Über die meßbaren Funktionen. Math. Ann. **117**, 327—355 (1939—1941). [5] Remark on bicompact spaces. Ann. Math. **38**, 845—846 (1937). [6] On bicompact spaces. Publ. Faculté Sci. Univ. Masaryk **270**, 11—16 (1939).
POVAROV, G. N.: [1] On functional separability of Boolean functions. Dokl. Akad. Nauk SSSR (NS) **94**, 801—803 (1954). [2] Sur l'invariance des fonctions booleiennes par rapport à des groupes. An. St. Univ. „Al. I. Cuza" Iaşi Sect. I (N. S.) **4**, 39—44 (1958).
RAINWATER, J.: [1] A note on projective resolutions. Proc. Am. math. Soc. **10**, 734—735 (1959).
RASIOWA, H.: [1] Algebraic treatment of the functional calculi of HEYTING and LEWIS. Fund. Math. **38**, 99—126 (1951). [2] A proof of the compactness theorem for arithmetical classes. Fund. Math. **39**, 8—14 (1952). [3] Constructive theories. Bull. Acad. Polon. Sci. Cl. III, **2**, 121—124 (1954). [4] Algebraic models ot axiomatic theories. Fund. Math. **41**, 291—310 (1955). [5] A proof of ε-theorems. Bull. Acad. Pol. Sci. **3**, 299—302 (1955). [6] On the ε-theorems. Fund. Math. **43**, 156—165 (1956); Errata: Fund. Math. **44**, 333 (1957).
— and R. SIKORSKI: [1] A proof of the completeness theorem of GÖDEL. Fund. Math. **37**, 193—200 (1950). [2] A proof of the Skolem-Löwenheim theorem. Fund. Math. **38**, 230—232 (1951). [3] On satisfiability and decidability in non-classical functional calculi. Bull. Acad. Pol. Sci. Cl. III, **1**, 229—231 (1953). [4] Algebraic treatment of the notion of satisfiability. Fund. Math. **40**, 62—95 (1953). [5] On existential theorems in non-classical functional calculi. Fund.

Math. **41**, 21—28 (1954). [6] An application of lattices to logic. Fund. Math. **42**, 83—100 (1955). [7] On the isomorphism of Lindenbaum algebras with fields of sets. Coll. Math. **5**, 143—158 (1958). [8] Formalisierte intuitionistische elementare Theorien. Constructivity in Mathematics (Proc. Colloquium at Amsterdam 1957) Amsterdam 1959, 241—249. [9] The Mathematics of Metamathematics, Monografie Matematyczne, Warszawa 1963.

REICHBACH, J.: [1] Completeness of the functional calculus of first order. Studia Log. **2**, 229—250 (1955).

RIBEIRO, H.: [1] A remark on Boolean algebras with operators. Am. J. Math. **74**, 163—167 (1952). [2] Topological groups and Boolean algebras with operators. Univ. Lisboa Rev. Fac. Ci. A. **4**, 195—200 (1955).

RIDDER, J.: [1] Zur Maß- und Integrationstheorie in Strukturen. Ind. Math. **8**, 64—81 (1946). [2] Einige Anwendungen des Dualitätsprinzips in topologischen Strukturen. Ind. Math. **9**, 341—350 (1947). [3] Eine Bemerkung über Maß in Strukturen. Ind. Math. **9**, 315—319 (1947).

RIEGER, L.: [1] On the lattice theory of Brouwerian propositional logic. Acta Fac. Nat. Univ. Carol. Prague No. 189 (1949). [2] A note on topological representations of distributive lattices. Časopis pest. mat. fys. **74**, 51—61 (1949). [3] On the lattice theory of Brouwerian propositional logic. Spisy vyd. prirod. fak. Univ. Karlovy 189, 1—40 (1949). [4] On countable generalised σ-algebras, with a new proof of Gòdel's completeness theorem. Czech. math. J. **1**, 29—40 (1951). [5] On free $\aleph_\xi$-complete Boolean algebras. Fund. Math. **38**, 35—52 (1951). [6] Some remarks on automorphisms of Boolean algebras. Fund. Math. **38**, 209—216 (1951). [7] On a fundamental theorem of mathematical logic. Časopis Pest. Mat. **80**, 217—231 (1955). [8] On Suslin algebras and their representations. Czech. Math. J. **5**, (80), 99—142 (1955) (Russian). [9] A remark on the s.c. free closure algebras. Czech. Math. J. **7**, 16—20 (1957) (Russian).

RIVKIND, YA. I.: [1] Dense sublattices of normed Boolean algebras. Grodnenskiĭ Gos. Ped. Inst. Uc. Zap. **1**, 59—66 (1955) (Russian). [2] Real functions on Boolean algebras with a measure. Grodnen. Gos. Ped. Inst. Uč. Zap. Ser. Mat. **2**, 89—101 (1957) (Russian).

ROSE, A.: [1] A lattice-theoretic characterization of three-valued logic. J. London Math. Soc. **25**, 255—259 (1950).

RUBIN, J. E.: [1] Remarks about a closure algebra in which closed elements are open. Proc. Am. math. Soc. **7**, 30—34 (1956).

— and D. SCOTT: [1] Some topological theorems equivalent to the Boolean prime ideal theorem. Bull. Am. math. Soc. **60**, 389 (1954).

RUDEANU, S.: [1] Boolean equations and their applications to the study of bridge-circuits, (I). Bull. Math. Soc. Math. Phys. R.P.R. **3**, 447—473 (1959); (II) Communic. Acad. R.P.R. **6**, 611—618 (1961) (Rumanian). [2] On definition of Boolean algebras by means of binary operations. Rev. Math. pur. Appl. **6**, 171—183 (1961) (Rumanian).

RUDIN, W.: [1] Continuous functions on compact sets without perfect subsets. Proc. Am. Math. Soc. **8**, 39—42 (1957). [2] Homogeneity problems in the theory of Čech compactification. Duke Math. J. **23**, 409—419 (1956).

RUPRECHT, E.: [1] Über die Charakterisierung normaler und vollständig regulärer topologischer Boole-Verbände mit Hilfe quasi-stetiger Ortsfunktionen. Math. Nachr. **13**, 289—308 (1955).

SACHS, S.: [1] The lattice of subalgebras of a Boolean algebra. Canad. J. Math. **14**, 451—460.

SAGALOVIČ, YU. L.: [1] On group invariance of Boolean functions. Uspehi Mat. Nauk **14**, 191—195 (1959) (Russian).

SAKS, S.: [1] On some functionals. Trans. Am. math. Soc. 35, 549—550 (1933); addition 965—970.

ŠANIN, N. A.: [1] On products of topological spaces. Trudy Mat. Inst. Steklova 24 (1948) (Russian).

SCOTT, D.: [1] The independence of certain distributive laws in Boolean algebras. Trans. Am. math. Soc. 84, 258—261 (1957). [2] Prime ideal theorems for rings, lattices and Boolean algebras. Bull. Am. math. Soc. 60, 390 (1954). [3] A new characterization of α-representable Boolean algebras. Bull. Am. Soc. 61, 522—523 (1955). [4] Measurable cardinals and constructible sets. Bull. Acad. Pol. Sci. 9, 521—524 (1961).

— and A. TARSKI: [1] The sentential calculus with infinitely long expressions. Coll. Math. 6, 165—170 (1958).

SEGAL, I. E.: [1] Abstract probability spaces and a theorem of Kolmogoroff. Am. J. Math. 76, 721—732.

SEMADENI, Z.: [1] Sur les ensembles clairsemés. Rozprawy Matematyczne 19, 1—39 (1939). [2] Functions with sets of points of discontinuity belonging to a fixed ideal. Fund. Math. 52, 25—39 (1963). [3] Spaces of continuous functions (VI) (Localization of multiplicative linear functionals). Studia Math. 23, 51—84 (1963). [4] Free and direct objects. Bull. Am. math. Soc. 69, 63—66 (1963). [5] Projectivity, injectivity and duality. Rozprawy Matematyczne 35, 1—47 (1963). [6] On weak convergence of measures and σ-complete Boolean algebras. Coll. Math. (to appear).

SHEFFER, H. M.: [1] A set of five independent postulates for Boolean algebras with application to logical constants. Trans. Am. math. Soc. 14, 481—488 (1913).

SHERMAN, S.: [1] On denumerably independent families of Borel fields. Am. J. Math. 72, 612—614 (1950).

SHOLANDER, M.: [1] Postulates for distributive lattices. Canad. J. Math. 3, 28—30 (1951). [2] Postulates for Boolean algebras. Canad. J. Math. 5, 460—464 (1953).

SIERPIŃSKI, W.: [1] Fonctions additives non complètement additives et fonctions non mesurables. Fund. Math. 30, 96—99 (1938). [2] Hypothèse du continu. Warszawa-Lwów 1934. [3] Sur les ensembles presque contenus les uns dans les autres. Fund. Math. 35, 141—150 (1948). [4] Sur les projections des ensembles complémentaires aux ensembles (A). Fund. Math. 11, 117—122 (1928). [5] Sur une suite transfinie d'ensembles de nombres naturals. Fund. Math. 33, 9—11 (1945). [6] Sur la dualité entre la première catégorie et la mesure nulle. Fund. Math. 22, 276—280 (1934).

SIKORSKI, R.: [1] A generalization of theorem of Banach and Cantor-Bernstein. Coll. Math. 1, 140—144 (1948). [2] Sur les corps de Boole topologiques. Compt. rend. 226, 1675—1676 (1948). [3] Sur la convergence des suites d'homomorphies. Compt. rend. 226, 1792—1793 (1948). [4] On the representation of Boolean algebras as fields of sets. Fund. Math. 35, 247—256 (1948). [5] A theorem on extension of homomorphisms. Ann. Soc. Pol. Math. 21, 332—335 (1948). [6] On the inducing of homomorphisms by mappings. Fund. Math. 36, 7—22 (1949). [7] Closure algebras. Fund. Math. 36, 165—206 (1949). [8] A theorem on the structure of homomorphisms. Fund. Math. 36, 245—247 (1949). [9] The integral in a Boolean algebra. Coll. Math. 2, 20—26 (1949). [10] On an unsolved problem from the theory of Boolean algebras. Coll. Math. 2, 27—29 (1949). [11] Independent fields and Cartesian products. Studia Math. 11, 171—184 (1950). [12] Remarks on some topological spaces of high power. Fund. Math. 37, 125—136 (1950). [13] Cartesian products of Boolean algebras. Fund. Math. 37, 25—54 (1950). [14] On an analogy between measures and homomorphisms. Ann. Soc. Pol. Math. 23, 1—20 (1950). [15] On measures in

Cartesian products of Boolean algebras. Coll. Math. **2**, 124–129 (1951). [16] Dimension theory in closure algebras. Fund. Math. **38**, 153–166 (1951). [17] A note to Rieger's paper: On free $\aleph_\xi$-complete Boolean algebras. Fund. Math. **38**, 53–54 (1951). [18] Homomorphisms, mappings and retracts. Coll. Math. **2**, 202–211 (1951). [19] Products of abstract algebras. Fund. Math. **39**, 211–228 (1952). [20] Closure homomorphisms and interior mappings. Fund. Math. **41**, 12–20 (1954). [21] On σ-complete Boolean algebras. Bull. Acad. Pol. Sci. Cl. III **3**, 7–9 (1955). [22] A theorem on non-classical functional calculi. Bull. Acad. Pol. Sci. Cl. III **4**, 649–650 (1956). [23] On the Herbrand theorem. Coll. Math. **6**, 55–59 (1958). [24] Some applications of interior mappings. Fund. Math. **45**, 200–212 (1958). [25] Distributivity and representability. Fund. Math. **48**, 91–103 (1959). [26] Der Heytingsche Prädikatenkalkül und metrische Räume. Constructivity in Mathematics. Proc. Coll. Amsterdam **1957**, 250–253. [27] Representation and distributivity of Boolean algebras. Coll. Math. **8**, 1–13 (1961). [28] A topological characterization of open theories. Bull. Acad. Pol. Sci. **9**, 259–260 (1961). [29] On open theories. Coll Math. **9**, 171–182 (1962). [30] On representations of Lindenbaum algebras. Prace Matematyczne **7**, 97–105 (1962). [31] On dense subsets of Boolean algebras. Coll. Math. **10**, 189–192 (1963). [32] On extensions and products of Boolean algebras. Fund. Math. **53**, 99–116 (1963). [33] A few problems on Boolean algebras. Coll. Math. **11**, 25–28 (1963).
— and T. TRACZYK: [1] On some Boolean algebras. Bull. Acad. Pol. Sci. Cl. III **4**, 489–492 (1956). [2] On free products of α-distributive Boolean algebras. Coll. Math. **11**, 13–16 (1963).
SŁOWIKOWSKI, W., and W. ZAWADOWSKI: [1] A generalization of maximal ideals method of Stone and Gelfand. Fund. Math. **42**, 215–231 (1955).
SMITH, E. C. jr.: [1] A distributivity condition for Boolean algebras. Ann. Math. **64**, 551–561 (1956).
— and A. TARSKI: [1] Higher degrees of distributivity and completeness in Boolean algebras. Trans. Am. math. Soc. **84**, 230–257 (1957).
SPECKER, E.: [1] Endenverbände von Räumen und Gruppen. Math. Ann. **122**, 167–174 (1950). [2] Zur Axiomatik der Mengenlehre. Z. math. Logik u. Grundl. Math. **3**, 173–210 (1957).
STABLER, E. R.: [1] Sets of postulates for Boolean rings. Am. Math. Monthly **48**, 20–28 (1941). [2] Boolean representation theory. Am. Math. Monthly **51**, 129–132 (1944).
STAMM, E.: [1] Beitrag zur Algebra der Logik. Monatsh. Math. Phys. **22**, 137–149 (1911).
STEINHAUS, H.: [1] Sur les distances des points des ensembles de mesure positive. Fund. Math. **1**, 93–104 (1920).
STONE, M. H.: [1] Boolean algebras and their application to topology. Proc. nat. Acad. Sci. (Wash.) **20**, 197–202 (1934). [2] Subsumption of the theory of Boolean algebras under the theory of rings. Proc. nat. Acad. Sci. (Wash.) **21**, 103–105 (1935). [3] Postulates for Boolean algebras and generalized Boolean algebras. Am. J. Math. **57**, 703–732 (1935). [4] Applications of Boolean algebras to topology. Rec. Math. Moscou, N. S. **1**, 765–771 (1936). [5] The theory of representations for Boolean algebras. Trans. Am. math. Soc. **40**, 37–111 (1936). [6] Applications of the theory of Boolean rings to general topology. Trans. Am. math. Soc. **41**, 321–364 (1937). [7] Algebraic characterization of special Boolean rings. Fund. Math. **29**, 223–303 (1937). [8] Note on formal logic. Am. J. Math. **59**, 506–514 (1937). [9] Topological representation of distributive lattices and Brouwerian logics. Čas. mat. fys. **67**, 1–25

(1937). [10] The representation of Boolean algebras. Bull. Am. math Soc 44, 807—816 (1938). [11] On characteristic functions of families of sets. Fund. Math. 33, 27—33 (1945). [12] Free Boolean rings and algebras. An. Acad. Brasil. Ci. 26, 9—17 (1954). [13] Boundedness properties in function-lattices. Canad. J. Math. 1, 176—186 (1949).

SUBRAHMANYAN, N. V.: [1] Structure theory for a generalized Boolean ring. Math. Ann. 141, 297—310 (1960).

SUSSMAN, I.: [1] A generalization of Boolean rings. Math. Ann. 136, 326—338. (1958)

SZASZ, G.: [1] Bevezetés a háleólméletbe (Introduction to lattice theory) Budapest 1959.

TAKEUCHI, K.: [1] The free Boolean σ-algebra with countable generators. Math. J. Tokyo 1, 77—79 (1953).

TARSKI, A.: [1] Une contribution a la théorie de la mesure. Fund. Math. 15, 42—50 (1930). [2] Zur Grundlegung der Booleschen Algebra. Fund. Math. 24, 177—198 (1935). [3] Über additive und multiplikative Mengenkörper und Mengenfunktionen. C. R. Soc. Sci. Lettr. Varsovie Cl. III 30, 151—181 (1937). [4] Der Aussagenkalkül und die Topologie. Fund. Math. 31, 103—134 (1938). [5] Einige Bemerkungen zur Axiomatik der Booleschen Algebra. C. R. Soc. Sci. Lettr. Varsovie 31, 33—35 (1938). [6] Ideale in vollständigen Mengenkörpern. Fund. Math. 32, 45—63 (1939); (II) 33, 51—65 (1945). [7] Über unerreichbare Kardinalzahlen. Fund. Math. 30, 68—89 (1938). [8] Cardinal algebras. New York 1949, [9] A lattice-theoretical fix-point theorem and its applications. Pacific J. Math. 5, 285—309 (1955). [10] Equationally complete rings and relation algebras. Ind. Math. 18, 39—46 (1956). [11] Algebraische Fassung des Maßproblems. Fund. Math. 31, 47—66 (1938). [12] Metamathematical proofs of some representation theorems for Boolean algebras. Bull. Am. math. Soc. 61, 523 (1955). [13] On the calculus of relations. J. Symb. Logic 6, 73—89 (1941). [14] Über das absolute Maß linearer Punktmengen. Fund. Math. 30, 218—234 (1938). [15] Prime ideal theorems for Boolean algebras and the axioms of choice. Bull. Am. math. Soc. 60, 390—391 (1954). [16] Prime ideal theorems for set algebras and ordering principles. Bull. Am. math. Soc. 60, 391 (1954). [17] Prime ideal theorems for set algebras and the axiom of choice. Bull. Am. math. Soc. 60, 391 (1954). [18] Some problems and results relevant to the foundations of set theory. Logic, Methodology and Philosophy of Sciences. Proceedings of the 1960 International Congress, Stanford 1962, 126—136.

TERASAKA, H.: [1] Theorie der topologischen Verbände: Ein Versuch zur Formalisierung der allgemeinen Topologie und der Theorie der reellen Funktionen. Proc. Imp. Acad. Jap. 13, 401—405 (1937). [2] Die Theorie der topologischen Verbände. Reprint of Fund. Math. 33 (1939); Coll. Papers Fac. Sci. Osaka Univ. Ser. A 8 (1940).

THEODORESCU, R.: [1] Remarques sur les homomorphismes aléatoires. An. Univ. „C.I. Parhon" Bucureşti. Ser. Şti. Nat. 22, 55—58 (1959).

TOMITA, M.: [1] Measure theory of complete Boolean algebras. Mem. Fac. Sci. Kyusyu Univ. A. 7, 51—60 (1952).

TRACZYK, T.: [1] On homomorphisms not induced by mappings. Bull. Acad. Pol. Sci. 6, 103—106 (1958). [2] On the approximations of mappings by Baire mappings. Coll. Math. 8, 67—70 (1961). [3] On axioms and some properties of Post algebras. Bull. Acad. Pol. Sci. 10, 509—512 (1962). [4] Axioms and some properties of Post algebras. Coll. Math. 10, 193—209 (1963). [5] Minimal extensions of weakly distributive Boolean algebras. Coll. Math. 11, 17—24 (1963). [6] Some theorems on independence in Post algebras. Bull. Acad. Pol. Sci. 11, 3—8 (1963).

ULAM, S.: [1] Zur Maßtheorie in der allgemeinen Mengenlehre. Fund. Math. **16**, 140—150 (1930). [2] Concerning functions of sets. Fund. Math. **14**, 231—233 (1929).

VAIDYANATHASWAMY, R.: [1] On the group-operations of a Boolean algebra. J. Indian math. Soc. N. S. **2**, 250—254 (1937).

VARSAVSKY, O.: [1] Quantifiers and equivalence relations. Rev. Mat. Guyana **2**, 29—51 (1958).

VILLE, J.: [1] Eléments de l'algèbre de Boole. Publ. Inst. Statist. Univ. Paris **4**, 107—140 (1955).

VINHA NOVAIS, J. A.: [1] Introduction to Boolean algebras. Gaz. Mat. Lisboa **18**, 1—8 (1957).

VINOKUROV, V. G.: [1] Representation of Boolean algebras and spaces with measure. Mat. Sbornik **56**, 375—391 (1962) (Russian).

VLADIMIROV, D. A.: [1] On the countable additivity of a Boolean measure. Vestnik Leningrad. Univ. **16**, 5—15 (1961) (Russian).

VREDENDUIN, P. G. J.: [1] Verbände. Euclides, Groningen **33**, 129—152 (1958) (Dutch).

WALLACE, A. D.: [1] Boolean rings and cohomology. Proc. Am. math. Soc. **4**, 475 (1953).

WARD, A. J.: [1] On relations between certain intrisic topologies in partially ordered sets. Proc. Cambridge Phil. Soc. **51**, 254—261 (1955).

WECKEN, F.: [1] Abstrakte Integrale und fastperiodische Funktionen. Math. Z. **45**, 377—404 (1939).

WHITEMAN, A.: [1] Postulates for Boolean algebra in terms of ternary rejection. Bull. Am. math. Soc. **43**, 293—298 (1937). [2] Postulates for Boolean algebras involving the operations of complete disjunction. Bull. Am. math. Soc. **43**, 293—298 (1937).

WOLK, E. S.: [1] Dedekind completeness and a fixed-point theorem. Canad. J. Math. **9**, 400—405 (1957).

WOOYENAKA, Y.: [1] On Newman algebra. Proc. Japan Acad. **30**, 170—175 (1954); **30**, 562—565 (1954); **31**, 66—69 (1955).

WRIGHT, F. B.: [1] Ideals in a polyadic algebra. Proc. Am. math. Soc. **8**, 544—546 (1957). [2] Some remarks on Boolean duality. Portugal. Math. **16**, 109—117 (1957). [3] Recurrence theorems and operators on Boolean algebras. Proc. London Math. Soc. **11**, 385—401 (1961). [4] Polarity and duality. Pacific J. Math. **10**, 723—730 (1960).

YAQUB, F. M.: [1] Free extensions of Boolean algebras. Pacific J. Math. **13**, 761—771 (1963).

ZARICKIJ, M. A.: [1] Boolean algebras with closure and Boolean algebras with derivation. Dopovidi Akad. Nauk Ukrain RSR **1955**, 3—6 (Ukrain.).

ZEMMER, J. L.: [1] Some remarks on p-rings and their Boolean geometry. Pacific J. Math. **29**, 193—208 (1956).

ZINK, R. E.: [1] On the structure of measure spaces. Acta Math. **107**, 53—71 (1962).

List of symbols

Author index

Subject index

Offsetdruck Julius Beltz, Weinheim/Bergstr

Ergebnisse der Mathematik und ihrer Grenzgebiete

GPSR Compliance

The European Union's (EU) General Product Safety Regulation (GPSR) is a set of rules that requires consumer products to be safe and our obligations to ensure this.

If you have any concerns about our products, you can contact us on ProductSafety@springernature.com

In case Publisher is established outside the EU, the EU authorized representative is:

Springer Nature Customer Service Center GmbH
Europaplatz 3
69115 Heidelberg, Germany

Batch number: 08954017

Printed by Printforce, the Netherlands